VOLUME FOUR HUNDRED AND THIRTY-TWO

Methods in ENZYMOLOGY

Lipidomics and Bioactive Lipids: Mass-Spectrometry–Based Lipid Analysis

METHODS IN ENZYMOLOGY

Editors-in-Chief

JOHN N. ABELSON AND MELVIN I. SIMON

Division of Biology
California Institute of Technology
Pasadena, California

Founding Editors

SIDNEY P. COLOWICK AND NATHAN O. KAPLAN

VOLUME FOUR HUNDRED AND THIRTY-TWO

METHODS IN
ENZYMOLOGY

Lipidomics and
Bioactive Lipids:
Mass-Spectrometry–
Based Lipid Analysis

EDITED BY

H. ALEX BROWN
Departments of Pharmacology and Chemistry
Vanderbilt University School of Medicine
Nashville, Tennessee

ELSEVIER

AMSTERDAM • BOSTON • HEIDELBERG • LONDON
NEW YORK • OXFORD • PARIS • SAN DIEGO
SAN FRANCISCO • SINGAPORE • SYDNEY • TOKYO
Academic Press is an imprint of Elsevier

Academic Press is an imprint of Elsevier
525 B Street, Suite 1900, San Diego, California 92101-4495, USA
84 Theobald's Road, London WC1X 8RR, UK

This book is printed on acid-free paper. ∞

Copyright © 2007, Elsevier Inc. All Rights Reserved.

No part of this publication may be reproduced or transmitted in any form or by any means, electronic or mechanical, including photocopy, recording, or any information storage and retrieval system, without permission in writing from the Publisher.

The appearance of the code at the bottom of the first page of a chapter in this book indicates the Publisher's consent that copies of the chapter may be made for personal or internal use of specific clients. This consent is given on the condition, however, that the copier pay the stated per copy fee through the Copyright Clearance Center, Inc. (www.copyright.com), for copying beyond that permitted by Sections 107 or 108 of the U.S. Copyright Law. This consent does not extend to other kinds of copying, such as copying for general distribution, for advertising or promotional purposes, for creating new collective works, or for resale. Copy fees for pre-2007 chapters are as shown on the title pages. If no fee code appears on the title page, the copy fee is the same as for current chapters.
0076-6879/2007 $35.00

Permissions may be sought directly from Elsevier's Science & Technology Rights Department in Oxford, UK: phone: (+44) 1865 843830, fax: (+44) 1865 853333, E-mail: permissions@elsevier.com. You may also complete your request on-line via the Elsevier homepage (http://elsevier.com), by selecting "Support & Contact" then "Copyright and Permission" and then "Obtaining Permissions."

For information on all Elsevier Academic Press publications
visit our Web site at www.books.elsevier.com

ISBN: 978-0-12-373895-0

PRINTED IN THE UNITED STATES OF AMERICA
07 08 09 10 9 8 7 6 5 4 3 2 1

Working together to grow
libraries in developing countries
www.elsevier.com | www.bookaid.org | www.sabre.org

ELSEVIER   BOOK AID International   Sabre Foundation

Contents

Contributors	*xi*
Preface	*xvii*
Volumes in Series	*xix*

1. Qualitative Analysis and Quantitative Assessment of Changes in Neutral Glycerol Lipid Molecular Species Within Cells   1
Jessica Krank, Robert C. Murphy, Robert M. Barkley, Eva Duchoslav, and Andrew McAnoy

1.	Introduction	2
2.	Reagents	3
3.	Methods	4
4.	Results	7
5.	Conclusions	19
	Acknowledgments	19
	References	19

2. Glycerophospholipid Identification and Quantitation by Electrospray Ionization Mass Spectrometry   21
Pavlina T. Ivanova, Stephen B. Milne, Mark O. Byrne, Yun Xiang, and H. Alex Brown

1.	Introduction	22
2.	Nomenclature	25
3.	Mass Spectrometry	26
4.	General Strategy for Phospholipid Isolation and Mass Spectral Analysis	27
5.	Extraction and Mass Spectral Analysis of Global Glycerophospholipids	28
6.	Polyphosphoinositide Extraction and Mass Spectral Analysis	35
7.	Computational Analysis of Mass Spectral Data	41
	Acknowledgments	54
	References	54

3. **Detection and Quantitation of Eicosanoids via High Performance Liquid Chromatography-Electrospray Ionization-Mass Spectrometry** 59

Raymond Deems, Matthew W. Buczynski, Rebecca Bowers-Gentry, Richard Harkewicz, and Edward A. Dennis

1. Introduction	60
2. Methods	62
3. Results and Discussion	74
Acknowledgments	80
References	81

4. **Structure-Specific, Quantitative Methods for Analysis of Sphingolipids by Liquid Chromatography–Tandem Mass Spectrometry: "Inside-Out" Sphingolipidomics** 83

M. Cameron Sullards, Jeremy C. Allegood, Samuel Kelly, Elaine Wang, Christopher A. Haynes, Hyejung Park, Yanfeng Chen, and Alfred H. Merrill, Jr.

1. Introduction: An Overview of Sphingolipid Structures and Nomenclature	86
2. Analysis of Sphingolipids by Mass Spectrometry	89
3. Analysis of Sphingolipids by "OMIC" Approaches	92
4. Materials and Methods	95
5. Materials	98
6. Extraction	98
7. Identification of the Molecular Species by Tandem Mass Spectrometry	100
8. Quantitation by LC-ESI-MS/MS Using Multireaction Monitoring	101
9. Analysis of (Dihydro)Ceramides, (Dihydro)Sphingomyelins, and (Dihydro)Monohexosyl-Ceramides in Positive Ion Mode	104
10. Other Methods	109
Acknowledgments	111
References	112

5. **Analysis of Ubiquinones, Dolichols, and Dolichol Diphosphate-Oligosaccharides by Liquid Chromatography-Electrospray Ionization-Mass Spectrometry** 117

Teresa A. Garrett, Ziqiang Guan, and Christian R. H. Raetz

1. Introduction	118
2. Materials	121
3. Liquid Chromatography-Mass Spectrometry	122
4. Preparation of Lipid Extracts	122
5. LC-MS Detection and Quantification of Coenzyme Q	123
6. LC-MS Detection and Quantification of Dolichol	130

7. LC-MS and LC-MS/MS Characterization of Dolichol
   Diphosphate-Linked Oligosaccharides 136
   Acknowledgment 140
   References 140

6. Extraction and Analysis of Sterols in Biological Matrices by High Performance Liquid Chromatography Electrospray Ionization Mass Spectrometry 145

Jeffrey G. McDonald, Bonne M. Thompson, Erin C. McCrum, and David W. Russell

1. Introduction 146
2. Supplies and Reagents 148
3. Extraction of Lipids from Cultured Cells and Tissues 149
4. Saponification of Lipid Extracts 151
5. Solid-Phase Extraction 152
6. Analysis by HPLC-ESI-MS 153
7. Quantitation 156
8. Data 157
9. Discussion, Nuances, Caveats, and Pitfalls 162
Acknowledgments 168
References 169

7. The Lipid Maps Initiative in Lipidomics 171

Kara Schmelzer, Eoin Fahy, Shankar Subramaniam, and Edward A. Dennis

1. Introduction 172
2. Building Infrastructure in Lipidomics 173
3. Classification, Nomenclature, and Structural Representation of Lipids 174
4. Mass Spectrometry as a Platform for Lipid Molecular Species 177
5. Future Plans 180
Acknowledgments 182
References 182

8. Basic Analytical Systems for Lipidomics by Mass Spectrometry in Japan 185

Ryo Taguchi, Mashahiro Nishijima, and Takao Shimizu

1. Introduction 186
2. Lipid Bank and Related Databases 187
3. Strategies for Lipid Identification and Quantitative Analysis by Mass Spectrometry 188

4. Several Practical Lipidomics Methods by Mass Spectrometry 190
5. Strategies for Identification of Individual Molecular Species in Glycerolipids and Glycerophospholipids by Lipid Search 192
6. Quantitative Analysis or Profiling of Lipid Molecular Species 197
7. Application of Several Different Methods in Lipidomics 199
8. Future Program for Lipidomics 208
Acknowledgments 209
References 209

9. The European Lipidomics Initiative: Enabling Technologies 213
Gerrit van Meer, Bas R. Leeflang, Gerhard Liebisch, Gerd Schmitz, and Felix M. Goñi

1. Introduction 214
2. Methods of Lipidomics: Quantitative Analysis in Time 215
3. Imaging Lipids 218
4. Methods to Study the Physical Properties of Lipids 219
5. Data Handling and Standardization 224
6. Perspectives 226
Acknowledgments 227
References 227

10. Lipidomic Analysis of Signaling Pathways 233
Michael J. O. Wakelam, Trevor R. Pettitt, and Anthony D. Postle

1. General Considerations 235
2. Lipid Extraction 235
3. HPLC-MS Analysis 237
4. General Phospholipid HPLC 237
5. Ceramide, Diradylglycerol, and Monoradylglycerol Separation 239
6. Phosphoinositide Separation 240
7. Analysis of Phospholipid Synthesis by ESI-MS/MS Using Stable Isotopes 241
8. Labeling Protocols 242
9. Lipid Extraction for ESI-MS/MS 243
10. LC-MS Analysis (*methyl-d_9*)-Choline 243
11. MRM Analysis of (*methyl-d_9*)-Choline Enrichment in Cell and Tissue Phosphorylcholine 244
12. ESI-MS/MS Analysis of Native and Newly Synthesized Phospholipids 244
13. Data Analysis 245
References 246

11. Bioinformatics for Lipidomics 247
Eoin Fahy, Dawn Cotter, Robert Byrnes, Manish Sud, Andrea Maer, Joshua Li, David Nadeau, Yihua Zhau, and Shankar Subramaniam

1. Introduction 248
2. Lipid Structure Databases 249
3. Lipid-Associated Protein/Gene Databases 254
4. Tools for Lipidomics 256
5. Lipid Pathways 268
6. Challenges for Future Lipid Informatics 270
Acknowledgments 272
References 272

12. Mediator Lipidomics: Search Algorithms for Eicosanoids, Resolvins, and Protectins 275
Charles N. Serhan, Yan Lu, Song Hong, and Rong Yang

1. Introduction: Metabolomics and Mediator Lipidomics–Informatics 276
2. Engineered Animals and Human Tissues 278
3. New Chemical Mediator Pathways in Resolution of Inflammation 279
4. Logic Diagram to Identify PUFA-Derived Lipid Mediators: Eicosanoids, Resolvins, and Protectins 284
5. Cognoscitive-Contrast-Angle Algorithm 286
6. Theoretical Database for Novel LM and Search Algorithm 292
7. Discussion 308
8. Conclusions and Next Steps 311
Acknowledgments 313
References 313

13. A Guide to Biochemical Systems Modeling of Sphingolipids for the Biochemist 319
Kellie J. Sims, Fernando Alvarez-Vasquez, Eberhard O. Voit, and Yusuf A. Hannun

1. Introduction 320
2. System Map: Specifying the Model 322
3. Symbolic Equations: From Words and Pictures to Equations 327
4. Numerical Equations: From the Symbolic to a Computational Model 331
5. Model Analysis: Steady State, Stability, and Sensitivity 341
6. Simulation: What Happens If. . .? 345
7. Conclusion 346

8. Epilogue	346
Acknowledgments	348
References	348

14. Quantitation and Standardization of Lipid Internal Standards for Mass Spectroscopy 351
Jeff D. Moore, William V. Caufield, and Walter A. Shaw

1. Introduction	352
2. Lipid Handling Guidelines	352
3. Chemical Characterization of Lipid Stocks	353
4. Preparation of Working Lipid Standards	355
5. Packaging of Lipid Standards	359
6. Quality Control and Stability Testing	361
7. Discussion	364
References	366

| *Author Index* | *369* |
| *Subject Index* | *381* |

Contributors

Jeremy C. Allegood
Schools of Biology, Chemistry, and Biochemistry, and the Parker H. Petit Institute for Bioengineering and Bioscience, Georgia Institute of Technology, Atlanta, Georgia

Fernando Alvarez-Vasquez
Department of Biostatistics, Bioinformatics, and Epidemiology, and Department of Biochemistry and Molecular Biology, Medical University of South Carolina, Charleston, South Carolina

Robert M. Barkley
Department of Pharmacology, University of Colorado at Denver and Health Sciences Center, Aurora, Colorado

Rebecca Bowers-Gentry
Departments of Chemistry, Biochemistry, and Pharmacology, University of California, San Diego, La Jolla, California

H. Alex Brown
Departments of Pharmacology and Chemistry, Vanderbilt University School of Medicine, Nashville, Tennessee

Matthew W. Buczynski
Departments of Chemistry, Biochemistry, and Pharmacology, University of California, San Diego, La Jolla, California

Mark O. Byrne
Departments of Pharmacology and Chemistry, Vanderbilt University and School of Medicine, Nashville, Tennessee

Robert Byrnes
San Diego Supercomputer Center, University of California, San Diego, La Jolla, California

William V. Caufield
Avanti Polar Lipids, Inc., Alabaster, Alabama

Yanfeng Chen
Schools of Biology, Chemistry, and Biochemistry, and the Parker H. Petit Institute for Bioengineering and Bioscience, Georgia Institute of Technology, Atlanta, Georgia

Dawn Cotter
San Diego Supercomputer Center, University of California, San Diego, La Jolla, California

Raymond Deems
Departments of Chemistry, Biochemistry, and Pharmacology, University of California, San Diego, La Jolla, California

Edward A. Dennis
Department of Chemistry, Biochemistry, and Pharmacology, University of California, San Diego, La Jolla, California

Eva Duchoslav
Applied Research Group, MDS Sciex, Concord, Ontario, Canada

Eoin Fahy
San Diego Supercomputer Center, University of California, San Diego, La Jolla, California

Teresa A. Garrett
Department of Biochemistry, Duke University Medical Center, Durham, North Carolina

Felix M. Goñi
Unidad de Biofisica (CSIC-UPV/EHU), and Department of Biochemistry and Molecular Biology, University of the Basque Country, Leioa, Spain

Ziqiang Guan
Department of Biochemistry, Duke University Medical Center, Durham, North Carolina

Yusuf A. Hannun
Department of Biochemistry and Molecular Biology, Medical University of South Carolina, Charleston, South Carolina

Richard Harkewicz
Departments of Chemistry, Biochemistry, and Pharmacology, University of California, San Diego, La Jolla, California

Christopher A. Haynes
Schools of Biology, Chemistry, and Biochemistry, and the Parker H. Petit Institute for Bioengineering and Bioscience, Georgia Institute of Technology, Atlanta, Georgia

Song Hong
Center for Experimental Therapeutics and Reperfusion Injury, Department of Anesthesiology, Perioperative, and Pain Medicine, Brigham and Women's Hospital and Harvard Medical School, Boston, Massachusetts

Pavlina T. Ivanova
Departments of Pharmacology and Chemistry, Vanderbilt University School of Medicine, Nashville, Tennessee

Samuel Kelly
Schools of Biology, Chemistry, and Biochemistry, and the Parker H. Petit Institute for Bioengineering and Bioscience, Georgia Institute of Technology, Atlanta, Georgia

Jessica Krank
Department of Pharmacology, University of Colorado at Denver and Health Sciences Center, Aurora, Colorado

Bas R. Leeflang
Bijvoet Center, Utrecht University, Utrecht, The Netherlands

Joshua Li
San Diego Supercomputer Center, University of California, San Diego, La Jolla, California

Gerhard Liebisch
Institute of Clinical Chemistry and Laboratory Medicine, University Hospital Regensburg, Regensburg, Germany

Yan Lu
Center for Experimental Therapeutics and Reperfusion Injury, Department of Anesthesiology, Perioperative, and Pain Medicine, Brigham and Women's Hospital and Harvard Medical School, Boston, Massachusetts

Andrea Maer
San Diego Supercomputer Center, University of California, San Diego, La Jolla, California

Andrew McAnoy
Department of Pharmacology, University of Colorado at Denver and Health Sciences Center, Aurora, Colorado

Erin C. McCrum
Department of Molecular Genetics, University of Texas Southwestern Medical Center at Dallas, Dallas, Texas

Jeffrey G. McDonald
Department of Molecular Genetics, University of Texas Southwestern Medical Center at Dallas, Dallas, Texas

Alfred H. Merrill, Jr.
Schools of Biology, Chemistry, and Biochemistry, and the Parker H. Petit Institute for Bioengineering and Bioscience, Georgia Institute of Technology, Atlanta, Georgia

Stephen B. Milne
Departments of Pharmacology and Chemistry, Vanderbilt University School of Medicine, Nashville, Tennessee

Jeff D. Moore
Avanti Polar Lipids, Inc., Alabaster, Alabama

Robert C. Murphy
Department of Pharmacology, University of Colorado at Denver and Health Sciences Center, Aurora, Colorado

David Nadeau
San Diego Supercomputer Center, University of California, San Diego, La Jolla, California

Mashahiro Nishijima
National Institute of Health Sciences, Tokyo, Japan

Hyejung Park
Schools of Biology, Chemistry, and Biochemistry, and the Parker H. Petit Institute for Bioengineering and Bioscience, Georgia Institute of Technology, Atlanta, Georgia

Trevor R. Pettitt
Institute for Cancer Studies, Birmingham University, Birmingham, United Kingdom

Anthony D. Postle
School of Chemistry, University of Southampton, Southampton, United Kingdom

Christian R. H. Raetz
Department of Biochemistry, Duke University Medical Center, Durham, North Carolina

David W. Russell
Department of Molecular Genetics, University of Texas Southwestern Medical Center at Dallas, Dallas, Texas

Kara Schmelzer
Department of Chemistry, Biochemistry, and Pharmacology, University of California, San Diego, La Jolla, California

Gerd Schmitz
Institute of Clinical Chemistry and Laboratory Medicine, University Hospital Regensburg, Regensburg, Germany

Charles N. Serhan
Center for Experimental Therapeutics and Reperfusion Injury, Department of Anesthesiology, Perioperative, and Pain Medicine, Brigham and Women's Hospital and Harvard Medical School, Boston, Massachusetts

Walter A. Shaw
Avanti Polar Lipids, Inc., Alabaster, Alabama

Takao Shimizu
Department of Molecular Biology, Graduate School of Medicine, The University of Tokyo, Tokyo, Japan

Kellie J. Sims
Department of Biostatistics, Bioinformatics and Epidemiology, and Department of Biochemistry and Molecular Biology, Medical University of South Carolina, Charleston, South Carolina

Shankar Subramaniam
Department of Chemistry, Biochemistry, and Bioengineering, Graduate Program in Bioinformatics and Systems Biology, and San Diego Supercomputer Center, University of California, San Diego, La Jolla, California

Manish Sud
San Diego Supercomputer Center, University of California, San Diego, La Jolla, California

M. Cameron Sullards
Schools of Biology, Chemistry, and Biochemistry, and the Parker H. Petit Institute for Bioengineering and Bioscience, Georgia Institute of Technology, Atlanta, Georgia

Ryo Taguchi
Department of Metabolome, Graduate School of Medicine, The University of Tokyo, and Core Research for Evolutional Science and Technology (CREST), Japan Science and Technology Agency, Tokyo, Japan

Bonne M. Thompson
Department of Molecular Genetics, University of Texas Southwestern Medical Center at Dallas, Dallas, Texas

Gerrit van Meer
Bijvoet Center and Institute of Biomembranes, Utrecht University, Utrecht, The Netherlands

Eberhard O. Voit
The Wallace H. Coulter Department of Biomedical Engineering, Georgia Institute of Technology and Emory University, Atlanta, Georgia

Michael J. O. Wakelam
The Babraham Institute, Babraham Research Campus, Cambridge, and Institute for Cancer Studies, Birmingham University, Birmingham, United Kingdom

Elaine Wang
Schools of Biology, Chemistry, and Biochemistry, and the Parker H. Petit Institute for Bioengineering and Bioscience, Georgia Institute of Technology, Atlanta, Georgia

Yun Xiang
Departments of Pharmacology and Chemistry, Vanderbilt University School of Medicine, Nashville, Tennessee

Rong Yang
Center for Experimental Therapeutics and Reperfusion Injury, Department of Anesthesiology, Perioperative, and Pain Medicine, Brigham and Women's Hospital and Harvard Medical School, Boston, Massachusetts

Yihua Zhau
San Diego Supercomputer Center, University of California, San Diego, La Jolla, California

Preface

Lipid metabolism and cellular signaling are highly integrated processes that regulate cell growth, proliferation, and survival. Lipids have essential roles in cellular functions, including determinants of membrane structure, serving as docking sites for cytosolic proteins and allosteric modulators. Abnormalities in lipid composition have established roles in human diseases, including diabetes, coronary disease, obesity, neurodegenerative diseases, and cancer. In the post-genomic era, we look at epigenetic factors and metabolomic biomarkers to better understand the molecular mechanisms of complex cellular processes and realize the benefits of personalized medicine.

Recent advances in lipid profiling and quantitative analysis provide an opportunity to define new roles of lipids in complex biological functions. Lipidomics was developed to be a systems biology approach to better understand contextual changes in lipid composition within an organelle, cell, or tissue as a result of challenge, stress, or metabolism. It provides an approach for determining precursor–product relationships as well as ordering the temporal and spatial events that constitute vital processes. This volume of *Methods in Enzymology* is one of a three-volume set on *Lipidomics and Bioactive Lipids* designed to provide state-of-the-art techniques in profiling and quantification of lipids using mass spectrometry and other analytical techniques used to determine the roles of lipids in cell function and disease. The first volume (432), *Mass-Spectrometry–Based Lipid Analysis,* provides current techniques to profile lipids using qualitative and quantitative approaches. The cell liposome is composed of thousands of molecular species of lipids; thus, generating a detailed description of the membrane composition presents both analytical and bioinformatic challenges. This volume includes the methodologies developed by the National Institute of General Medicine large-scale collaborative initiative, LIPID MAPS (www.lipidmaps.org), as well as an overview of international lipidomics projects. The second volume (433), *Specialized Analytical Methods and Lipids in Disease,* presents applications of lipid analysis to understanding disease processes, in addition to describing more specialized analytical approaches. The third volume (434), *Lipids and Cell Signaling,* is a series of chapters focused on lipid-signaling molecules and enzymes.

The goal of these volumes is to provide a guide to techniques used in profiling and quantification of cellular lipids with an emphasis on lipid signaling pathways. Many of the leaders in the emerging field of lipidomics have contributed to these volumes, and I am grateful for their comments in shaping the content. I hope that this guide will satisfy the needs of students who are interested in lipid structure and function as well as experienced researchers. It must be noted that many of the solvents, reagents, and instrumentation described in these chapters have the potential to be harmful to health. Readers should consult material safety data sheets, follow instrument instructions, and be properly trained in laboratory procedures before attempting any of the methods described.

H. ALEX BROWN

Methods in Enzymology

Volume I. Preparation and Assay of Enzymes
Edited by Sidney P. Colowick and Nathan O. Kaplan

Volume II. Preparation and Assay of Enzymes
Edited by Sidney P. Colowick and Nathan O. Kaplan

Volume III. Preparation and Assay of Substrates
Edited by Sidney P. Colowick and Nathan O. Kaplan

Volume IV. Special Techniques for the Enzymologist
Edited by Sidney P. Colowick and Nathan O. Kaplan

Volume V. Preparation and Assay of Enzymes
Edited by Sidney P. Colowick and Nathan O. Kaplan

Volume VI. Preparation and Assay of Enzymes *(Continued)*
Preparation and Assay of Substrates
Special Techniques
Edited by Sidney P. Colowick and Nathan O. Kaplan

Volume VII. Cumulative Subject Index
Edited by Sidney P. Colowick and Nathan O. Kaplan

Volume VIII. Complex Carbohydrates
Edited by Elizabeth F. Neufeld and Victor Ginsburg

Volume IX. Carbohydrate Metabolism
Edited by Willis A. Wood

Volume X. Oxidation and Phosphorylation
Edited by Ronald W. Estabrook and Maynard E. Pullman

Volume XI. Enzyme Structure
Edited by C. H. W. Hirs

Volume XII. Nucleic Acids (Parts A and B)
Edited by Lawrence Grossman and Kivie Moldave

Volume XIII. Citric Acid Cycle
Edited by J. M. Lowenstein

Volume XIV. Lipids
Edited by J. M. Lowenstein

Volume XV. Steroids and Terpenoids
Edited by Raymond B. Clayton

VOLUME XVI. Fast Reactions
Edited by KENNETH KUSTIN

VOLUME XVII. Metabolism of Amino Acids and Amines
(Parts A and B)
Edited by HERBERT TABOR AND CELIA WHITE TABOR

VOLUME XVIII. Vitamins and Coenzymes (Parts A, B, and C)
Edited by DONALD B. MCCORMICK AND LEMUEL D. WRIGHT

VOLUME XIX. Proteolytic Enzymes
Edited by GERTRUDE E. PERLMANN AND LASZLO LORAND

VOLUME XX. Nucleic Acids and Protein Synthesis (Part C)
Edited by KIVIE MOLDAVE AND LAWRENCE GROSSMAN

VOLUME XXI. Nucleic Acids (Part D)
Edited by LAWRENCE GROSSMAN AND KIVIE MOLDAVE

VOLUME XXII. Enzyme Purification and Related Techniques
Edited by WILLIAM B. JAKOBY

VOLUME XXIII. Photosynthesis (Part A)
Edited by ANTHONY SAN PIETRO

VOLUME XXIV. Photosynthesis and Nitrogen Fixation (Part B)
Edited by ANTHONY SAN PIETRO

VOLUME XXV. Enzyme Structure (Part B)
Edited by C. H. W. HIRS AND SERGE N. TIMASHEFF

VOLUME XXVI. Enzyme Structure (Part C)
Edited by C. H. W. HIRS AND SERGE N. TIMASHEFF

VOLUME XXVII. Enzyme Structure (Part D)
Edited by C. H. W. HIRS AND SERGE N. TIMASHEFF

VOLUME XXVIII. Complex Carbohydrates (Part B)
Edited by VICTOR GINSBURG

VOLUME XXIX. Nucleic Acids and Protein Synthesis (Part E)
Edited by LAWRENCE GROSSMAN AND KIVIE MOLDAVE

VOLUME XXX. Nucleic Acids and Protein Synthesis (Part F)
Edited by KIVIE MOLDAVE AND LAWRENCE GROSSMAN

VOLUME XXXI. Biomembranes (Part A)
Edited by SIDNEY FLEISCHER AND LESTER PACKER

VOLUME XXXII. Biomembranes (Part B)
Edited by SIDNEY FLEISCHER AND LESTER PACKER

VOLUME XXXIII. Cumulative Subject Index Volumes I–XXX
Edited by MARTHA G. DENNIS AND EDWARD A. DENNIS

VOLUME XXXIV. Affinity Techniques (Enzyme Purification: Part B)
Edited by WILLIAM B. JAKOBY AND MEIR WILCHEK

VOLUME XXXV. Lipids (Part B)
Edited by JOHN M. LOWENSTEIN

VOLUME XXXVI. Hormone Action (Part A: Steroid Hormones)
Edited by BERT W. O'MALLEY AND JOEL G. HARDMAN

VOLUME XXXVII. Hormone Action (Part B: Peptide Hormones)
Edited by BERT W. O'MALLEY AND JOEL G. HARDMAN

VOLUME XXXVIII. Hormone Action (Part C: Cyclic Nucleotides)
Edited by JOEL G. HARDMAN AND BERT W. O'MALLEY

VOLUME XXXIX. Hormone Action (Part D: Isolated Cells, Tissues, and Organ Systems)
Edited by JOEL G. HARDMAN AND BERT W. O'MALLEY

VOLUME XL. Hormone Action (Part E: Nuclear Structure and Function)
Edited by BERT W. O'MALLEY AND JOEL G. HARDMAN

VOLUME XLI. Carbohydrate Metabolism (Part B)
Edited by W. A. WOOD

VOLUME XLII. Carbohydrate Metabolism (Part C)
Edited by W. A. WOOD

VOLUME XLIII. Antibiotics
Edited by JOHN H. HASH

VOLUME XLIV. Immobilized Enzymes
Edited by KLAUS MOSBACH

VOLUME XLV. Proteolytic Enzymes (Part B)
Edited by LASZLO LORAND

VOLUME XLVI. Affinity Labeling
Edited by WILLIAM B. JAKOBY AND MEIR WILCHEK

VOLUME XLVII. Enzyme Structure (Part E)
Edited by C. H. W. HIRS AND SERGE N. TIMASHEFF

VOLUME XLVIII. Enzyme Structure (Part F)
Edited by C. H. W. HIRS AND SERGE N. TIMASHEFF

VOLUME XLIX. Enzyme Structure (Part G)
Edited by C. H. W. HIRS AND SERGE N. TIMASHEFF

VOLUME L. Complex Carbohydrates (Part C)
Edited by VICTOR GINSBURG

VOLUME LI. Purine and Pyrimidine Nucleotide Metabolism
Edited by PATRICIA A. HOFFEE AND MARY ELLEN JONES

VOLUME LII. Biomembranes (Part C: Biological Oxidations)
Edited by SIDNEY FLEISCHER AND LESTER PACKER

VOLUME LIII. Biomembranes (Part D: Biological Oxidations)
Edited by SIDNEY FLEISCHER AND LESTER PACKER

VOLUME LIV. Biomembranes (Part E: Biological Oxidations)
Edited by SIDNEY FLEISCHER AND LESTER PACKER

VOLUME LV. Biomembranes (Part F: Bioenergetics)
Edited by SIDNEY FLEISCHER AND LESTER PACKER

VOLUME LVI. Biomembranes (Part G: Bioenergetics)
Edited by SIDNEY FLEISCHER AND LESTER PACKER

VOLUME LVII. Bioluminescence and Chemiluminescence
Edited by MARLENE A. DELUCA

VOLUME LVIII. Cell Culture
Edited by WILLIAM B. JAKOBY AND IRA PASTAN

VOLUME LIX. Nucleic Acids and Protein Synthesis (Part G)
Edited by KIVIE MOLDAVE AND LAWRENCE GROSSMAN

VOLUME LX. Nucleic Acids and Protein Synthesis (Part H)
Edited by KIVIE MOLDAVE AND LAWRENCE GROSSMAN

VOLUME 61. Enzyme Structure (Part H)
Edited by C. H. W. HIRS AND SERGE N. TIMASHEFF

VOLUME 62. Vitamins and Coenzymes (Part D)
Edited by DONALD B. MCCORMICK AND LEMUEL D. WRIGHT

VOLUME 63. Enzyme Kinetics and Mechanism (Part A: Initial Rate and Inhibitor Methods)
Edited by DANIEL L. PURICH

VOLUME 64. Enzyme Kinetics and Mechanism
(Part B: Isotopic Probes and Complex Enzyme Systems)
Edited by DANIEL L. PURICH

VOLUME 65. Nucleic Acids (Part I)
Edited by LAWRENCE GROSSMAN AND KIVIE MOLDAVE

VOLUME 66. Vitamins and Coenzymes (Part E)
Edited by DONALD B. MCCORMICK AND LEMUEL D. WRIGHT

VOLUME 67. Vitamins and Coenzymes (Part F)
Edited by DONALD B. MCCORMICK AND LEMUEL D. WRIGHT

VOLUME 68. Recombinant DNA
Edited by RAY WU

VOLUME 69. Photosynthesis and Nitrogen Fixation (Part C)
Edited by ANTHONY SAN PIETRO

VOLUME 70. Immunochemical Techniques (Part A)
Edited by HELEN VAN VUNAKIS AND JOHN J. LANGONE

VOLUME 71. Lipids (Part C)
Edited by JOHN M. LOWENSTEIN

VOLUME 72. Lipids (Part D)
Edited by JOHN M. LOWENSTEIN

VOLUME 73. Immunochemical Techniques (Part B)
Edited by JOHN J. LANGONE AND HELEN VAN VUNAKIS

VOLUME 74. Immunochemical Techniques (Part C)
Edited by JOHN J. LANGONE AND HELEN VAN VUNAKIS

VOLUME 75. Cumulative Subject Index Volumes XXXI, XXXII, XXXIV–LX
Edited by EDWARD A. DENNIS AND MARTHA G. DENNIS

VOLUME 76. Hemoglobins
Edited by ERALDO ANTONINI, LUIGI ROSSI-BERNARDI, AND EMILIA CHIANCONE

VOLUME 77. Detoxication and Drug Metabolism
Edited by WILLIAM B. JAKOBY

VOLUME 78. Interferons (Part A)
Edited by SIDNEY PESTKA

VOLUME 79. Interferons (Part B)
Edited by SIDNEY PESTKA

VOLUME 80. Proteolytic Enzymes (Part C)
Edited by LASZLO LORAND

VOLUME 81. Biomembranes (Part H: Visual Pigments and Purple Membranes, I)
Edited by LESTER PACKER

VOLUME 82. Structural and Contractile Proteins (Part A: Extracellular Matrix)
Edited by LEON W. CUNNINGHAM AND DIXIE W. FREDERIKSEN

VOLUME 83. Complex Carbohydrates (Part D)
Edited by VICTOR GINSBURG

VOLUME 84. Immunochemical Techniques (Part D: Selected Immunoassays)
Edited by JOHN J. LANGONE AND HELEN VAN VUNAKIS

VOLUME 85. Structural and Contractile Proteins (Part B: The Contractile Apparatus and the Cytoskeleton)
Edited by DIXIE W. FREDERIKSEN AND LEON W. CUNNINGHAM

VOLUME 86. Prostaglandins and Arachidonate Metabolites
Edited by WILLIAM E. M. LANDS AND WILLIAM L. SMITH

VOLUME 87. Enzyme Kinetics and Mechanism (Part C: Intermediates, Stereo-chemistry, and Rate Studies)
Edited by DANIEL L. PURICH

VOLUME 88. Biomembranes (Part I: Visual Pigments and Purple Membranes, II)
Edited by LESTER PACKER

VOLUME 89. Carbohydrate Metabolism (Part D)
Edited by WILLIS A. WOOD

VOLUME 90. Carbohydrate Metabolism (Part E)
Edited by WILLIS A. WOOD

VOLUME 91. Enzyme Structure (Part I)
Edited by C. H. W. HIRS AND SERGE N. TIMASHEFF

VOLUME 92. Immunochemical Techniques (Part E: Monoclonal Antibodies and General Immunoassay Methods)
Edited by JOHN J. LANGONE AND HELEN VAN VUNAKIS

VOLUME 93. Immunochemical Techniques (Part F: Conventional Antibodies, Fc Receptors, and Cytotoxicity)
Edited by JOHN J. LANGONE AND HELEN VAN VUNAKIS

VOLUME 94. Polyamines
Edited by HERBERT TABOR AND CELIA WHITE TABOR

VOLUME 95. Cumulative Subject Index Volumes 61–74, 76–80
Edited by EDWARD A. DENNIS AND MARTHA G. DENNIS

VOLUME 96. Biomembranes [Part J: Membrane Biogenesis: Assembly and Targeting (General Methods; Eukaryotes)]
Edited by SIDNEY FLEISCHER AND BECCA FLEISCHER

VOLUME 97. Biomembranes [Part K: Membrane Biogenesis: Assembly and Targeting (Prokaryotes, Mitochondria, and Chloroplasts)]
Edited by SIDNEY FLEISCHER AND BECCA FLEISCHER

VOLUME 98. Biomembranes (Part L: Membrane Biogenesis: Processing and Recycling)
Edited by SIDNEY FLEISCHER AND BECCA FLEISCHER

VOLUME 99. Hormone Action (Part F: Protein Kinases)
Edited by JACKIE D. CORBIN AND JOEL G. HARDMAN

VOLUME 100. Recombinant DNA (Part B)
Edited by RAY WU, LAWRENCE GROSSMAN, AND KIVIE MOLDAVE

VOLUME 101. Recombinant DNA (Part C)
Edited by RAY WU, LAWRENCE GROSSMAN, AND KIVIE MOLDAVE

VOLUME 102. Hormone Action (Part G: Calmodulin and Calcium-Binding Proteins)
Edited by ANTHONY R. MEANS AND BERT W. O'MALLEY

VOLUME 103. Hormone Action (Part H: Neuroendocrine Peptides)
Edited by P. MICHAEL CONN

VOLUME 104. Enzyme Purification and Related Techniques (Part C)
Edited by WILLIAM B. JAKOBY

VOLUME 105. Oxygen Radicals in Biological Systems
Edited by LESTER PACKER

VOLUME 106. Posttranslational Modifications (Part A)
Edited by FINN WOLD AND KIVIE MOLDAVE

VOLUME 107. Posttranslational Modifications (Part B)
Edited by FINN WOLD AND KIVIE MOLDAVE

VOLUME 108. Immunochemical Techniques (Part G: Separation and Characterization of Lymphoid Cells)
Edited by GIOVANNI DI SABATO, JOHN J. LANGONE, AND HELEN VAN VUNAKIS

VOLUME 109. Hormone Action (Part I: Peptide Hormones)
Edited by LUTZ BIRNBAUMER AND BERT W. O'MALLEY

VOLUME 110. Steroids and Isoprenoids (Part A)
Edited by JOHN H. LAW AND HANS C. RILLING

VOLUME 111. Steroids and Isoprenoids (Part B)
Edited by JOHN H. LAW AND HANS C. RILLING

VOLUME 112. Drug and Enzyme Targeting (Part A)
Edited by KENNETH J. WIDDER AND RALPH GREEN

VOLUME 113. Glutamate, Glutamine, Glutathione, and Related Compounds
Edited by ALTON MEISTER

VOLUME 114. Diffraction Methods for Biological Macromolecules (Part A)
Edited by HAROLD W. WYCKOFF, C. H. W. HIRS, AND SERGE N. TIMASHEFF

VOLUME 115. Diffraction Methods for Biological Macromolecules (Part B)
Edited by HAROLD W. WYCKOFF, C. H. W. HIRS, AND SERGE N. TIMASHEFF

VOLUME 116. Immunochemical Techniques
(Part H: Effectors and Mediators of Lymphoid Cell Functions)
Edited by GIOVANNI DI SABATO, JOHN J. LANGONE, AND HELEN VAN VUNAKIS

VOLUME 117. Enzyme Structure (Part J)
Edited by C. H. W. HIRS AND SERGE N. TIMASHEFF

VOLUME 118. Plant Molecular Biology
Edited by ARTHUR WEISSBACH AND HERBERT WEISSBACH

VOLUME 119. Interferons (Part C)
Edited by SIDNEY PESTKA

VOLUME 120. Cumulative Subject Index Volumes 81–94, 96–101

VOLUME 121. Immunochemical Techniques (Part I: Hybridoma Technology and Monoclonal Antibodies)
Edited by JOHN J. LANGONE AND HELEN VAN VUNAKIS

VOLUME 122. Vitamins and Coenzymes (Part G)
Edited by FRANK CHYTIL AND DONALD B. MCCORMICK

VOLUME 123. Vitamins and Coenzymes (Part H)
Edited by FRANK CHYTIL AND DONALD B. MCCORMICK

VOLUME 124. Hormone Action (Part J: Neuroendocrine Peptides)
Edited by P. MICHAEL CONN

VOLUME 125. Biomembranes (Part M: Transport in Bacteria, Mitochondria, and Chloroplasts: General Approaches and Transport Systems)
Edited by SIDNEY FLEISCHER AND BECCA FLEISCHER

VOLUME 126. Biomembranes (Part N: Transport in Bacteria, Mitochondria, and Chloroplasts: Protonmotive Force)
Edited by SIDNEY FLEISCHER AND BECCA FLEISCHER

VOLUME 127. Biomembranes (Part O: Protons and Water: Structure and Translocation)
Edited by LESTER PACKER

VOLUME 128. Plasma Lipoproteins (Part A: Preparation, Structure, and Molecular Biology)
Edited by JERE P. SEGREST AND JOHN J. ALBERS

VOLUME 129. Plasma Lipoproteins (Part B: Characterization, Cell Biology, and Metabolism)
Edited by JOHN J. ALBERS AND JERE P. SEGREST

VOLUME 130. Enzyme Structure (Part K)
Edited by C. H. W. HIRS AND SERGE N. TIMASHEFF

VOLUME 131. Enzyme Structure (Part L)
Edited by C. H. W. HIRS AND SERGE N. TIMASHEFF

VOLUME 132. Immunochemical Techniques (Part J: Phagocytosis and Cell-Mediated Cytotoxicity)
Edited by GIOVANNI DI SABATO AND JOHANNES EVERSE

VOLUME 133. Bioluminescence and Chemiluminescence (Part B)
Edited by MARLENE DELUCA AND WILLIAM D. MCELROY

VOLUME 134. Structural and Contractile Proteins (Part C: The Contractile Apparatus and the Cytoskeleton)
Edited by RICHARD B. VALLEE

VOLUME 135. Immobilized Enzymes and Cells (Part B)
Edited by KLAUS MOSBACH

VOLUME 136. Immobilized Enzymes and Cells (Part C)
Edited by KLAUS MOSBACH

VOLUME 137. Immobilized Enzymes and Cells (Part D)
Edited by KLAUS MOSBACH

VOLUME 138. Complex Carbohydrates (Part E)
Edited by VICTOR GINSBURG

VOLUME 139. Cellular Regulators (Part A: Calcium- and Calmodulin-Binding Proteins)
Edited by ANTHONY R. MEANS AND P. MICHAEL CONN

VOLUME 140. Cumulative Subject Index Volumes 102–119, 121–134

VOLUME 141. Cellular Regulators (Part B: Calcium and Lipids)
Edited by P. MICHAEL CONN AND ANTHONY R. MEANS

VOLUME 142. Metabolism of Aromatic Amino Acids and Amines
Edited by SEYMOUR KAUFMAN

VOLUME 143. Sulfur and Sulfur Amino Acids
Edited by WILLIAM B. JAKOBY AND OWEN GRIFFITH

VOLUME 144. Structural and Contractile Proteins (Part D: Extracellular Matrix)
Edited by LEON W. CUNNINGHAM

VOLUME 145. Structural and Contractile Proteins (Part E: Extracellular Matrix)
Edited by LEON W. CUNNINGHAM

VOLUME 146. Peptide Growth Factors (Part A)
Edited by DAVID BARNES AND DAVID A. SIRBASKU

VOLUME 147. Peptide Growth Factors (Part B)
Edited by DAVID BARNES AND DAVID A. SIRBASKU

VOLUME 148. Plant Cell Membranes
Edited by LESTER PACKER AND ROLAND DOUCE

VOLUME 149. Drug and Enzyme Targeting (Part B)
Edited by RALPH GREEN AND KENNETH J. WIDDER

VOLUME 150. Immunochemical Techniques (Part K: *In Vitro* Models of B and T Cell Functions and Lymphoid Cell Receptors)
Edited by GIOVANNI DI SABATO

VOLUME 151. Molecular Genetics of Mammalian Cells
Edited by MICHAEL M. GOTTESMAN

VOLUME 152. Guide to Molecular Cloning Techniques
Edited by SHELBY L. BERGER AND ALAN R. KIMMEL

VOLUME 153. Recombinant DNA (Part D)
Edited by RAY WU AND LAWRENCE GROSSMAN

VOLUME 154. Recombinant DNA (Part E)
Edited by RAY WU AND LAWRENCE GROSSMAN

VOLUME 155. Recombinant DNA (Part F)
Edited by RAY WU

VOLUME 156. Biomembranes (Part P: ATP-Driven Pumps and Related Transport: The Na, K-Pump)
Edited by SIDNEY FLEISCHER AND BECCA FLEISCHER

VOLUME 157. Biomembranes (Part Q: ATP-Driven Pumps and Related Transport: Calcium, Proton, and Potassium Pumps)
Edited by SIDNEY FLEISCHER AND BECCA FLEISCHER

VOLUME 158. Metalloproteins (Part A)
Edited by JAMES F. RIORDAN AND BERT L. VALLEE

VOLUME 159. Initiation and Termination of Cyclic Nucleotide Action
Edited by JACKIE D. CORBIN AND ROGER A. JOHNSON

VOLUME 160. Biomass (Part A: Cellulose and Hemicellulose)
Edited by WILLIS A. WOOD AND SCOTT T. KELLOGG

VOLUME 161. Biomass (Part B: Lignin, Pectin, and Chitin)
Edited by WILLIS A. WOOD AND SCOTT T. KELLOGG

VOLUME 162. Immunochemical Techniques (Part L: Chemotaxis and Inflammation)
Edited by GIOVANNI DI SABATO

VOLUME 163. Immunochemical Techniques (Part M: Chemotaxis and Inflammation)
Edited by GIOVANNI DI SABATO

VOLUME 164. Ribosomes
Edited by HARRY F. NOLLER, JR., AND KIVIE MOLDAVE

VOLUME 165. Microbial Toxins: Tools for Enzymology
Edited by SIDNEY HARSHMAN

VOLUME 166. Branched-Chain Amino Acids
Edited by ROBERT HARRIS AND JOHN R. SOKATCH

VOLUME 167. Cyanobacteria
Edited by LESTER PACKER AND ALEXANDER N. GLAZER

VOLUME 168. Hormone Action (Part K: Neuroendocrine Peptides)
Edited by P. MICHAEL CONN

VOLUME 169. Platelets: Receptors, Adhesion, Secretion (Part A)
Edited by JACEK HAWIGER

VOLUME 170. Nucleosomes
Edited by PAUL M. WASSARMAN AND ROGER D. KORNBERG

VOLUME 171. Biomembranes (Part R: Transport Theory: Cells and Model Membranes)
Edited by SIDNEY FLEISCHER AND BECCA FLEISCHER

VOLUME 172. Biomembranes (Part S: Transport: Membrane Isolation and Characterization)
Edited by SIDNEY FLEISCHER AND BECCA FLEISCHER

VOLUME 173. Biomembranes [Part T: Cellular and Subcellular Transport: Eukaryotic (Nonepithelial) Cells]
Edited by SIDNEY FLEISCHER AND BECCA FLEISCHER

VOLUME 174. Biomembranes [Part U: Cellular and Subcellular Transport: Eukaryotic (Nonepithelial) Cells]
Edited by SIDNEY FLEISCHER AND BECCA FLEISCHER

VOLUME 175. Cumulative Subject Index Volumes 135–139, 141–167

VOLUME 176. Nuclear Magnetic Resonance (Part A: Spectral Techniques and Dynamics)
Edited by NORMAN J. OPPENHEIMER AND THOMAS L. JAMES

VOLUME 177. Nuclear Magnetic Resonance (Part B: Structure and Mechanism)
Edited by NORMAN J. OPPENHEIMER AND THOMAS L. JAMES

VOLUME 178. Antibodies, Antigens, and Molecular Mimicry
Edited by JOHN J. LANGONE

VOLUME 179. Complex Carbohydrates (Part F)
Edited by VICTOR GINSBURG

VOLUME 180. RNA Processing (Part A: General Methods)
Edited by JAMES E. DAHLBERG AND JOHN N. ABELSON

VOLUME 181. RNA Processing (Part B: Specific Methods)
Edited by JAMES E. DAHLBERG AND JOHN N. ABELSON

VOLUME 182. Guide to Protein Purification
Edited by MURRAY P. DEUTSCHER

VOLUME 183. Molecular Evolution: Computer Analysis of Protein and Nucleic Acid Sequences
Edited by RUSSELL F. DOOLITTLE

VOLUME 184. Avidin-Biotin Technology
Edited by MEIR WILCHEK AND EDWARD A. BAYER

VOLUME 185. Gene Expression Technology
Edited by DAVID V. GOEDDEL

VOLUME 186. Oxygen Radicals in Biological Systems (Part B: Oxygen Radicals and Antioxidants)
Edited by LESTER PACKER AND ALEXANDER N. GLAZER

VOLUME 187. Arachidonate Related Lipid Mediators
Edited by ROBERT C. MURPHY AND FRANK A. FITZPATRICK

VOLUME 188. Hydrocarbons and Methylotrophy
Edited by MARY E. LIDSTROM

VOLUME 189. Retinoids (Part A: Molecular and Metabolic Aspects)
Edited by LESTER PACKER

VOLUME 190. Retinoids (Part B: Cell Differentiation and Clinical Applications)
Edited by LESTER PACKER

VOLUME 191. Biomembranes (Part V: Cellular and Subcellular Transport: Epithelial Cells)
Edited by SIDNEY FLEISCHER AND BECCA FLEISCHER

VOLUME 192. Biomembranes (Part W: Cellular and Subcellular Transport: Epithelial Cells)
Edited by SIDNEY FLEISCHER AND BECCA FLEISCHER

VOLUME 193. Mass Spectrometry
Edited by JAMES A. MCCLOSKEY

VOLUME 194. Guide to Yeast Genetics and Molecular Biology
Edited by CHRISTINE GUTHRIE AND GERALD R. FINK

VOLUME 195. Adenylyl Cyclase, G Proteins, and Guanylyl Cyclase
Edited by ROGER A. JOHNSON AND JACKIE D. CORBIN

VOLUME 196. Molecular Motors and the Cytoskeleton
Edited by RICHARD B. VALLEE

VOLUME 197. Phospholipases
Edited by EDWARD A. DENNIS

VOLUME 198. Peptide Growth Factors (Part C)
Edited by DAVID BARNES, J. P. MATHER, AND GORDON H. SATO

VOLUME 199. Cumulative Subject Index Volumes 168–174, 176–194

VOLUME 200. Protein Phosphorylation (Part A: Protein Kinases: Assays, Purification, Antibodies, Functional Analysis, Cloning, and Expression)
Edited by TONY HUNTER AND BARTHOLOMEW M. SEFTON

VOLUME 201. Protein Phosphorylation (Part B: Analysis of Protein Phosphorylation, Protein Kinase Inhibitors, and Protein Phosphatases)
Edited by TONY HUNTER AND BARTHOLOMEW M. SEFTON

VOLUME 202. Molecular Design and Modeling: Concepts and Applications (Part A: Proteins, Peptides, and Enzymes)
Edited by JOHN J. LANGONE

VOLUME 203. Molecular Design and Modeling: Concepts and Applications (Part B: Antibodies and Antigens, Nucleic Acids, Polysaccharides, and Drugs)
Edited by JOHN J. LANGONE

VOLUME 204. Bacterial Genetic Systems
Edited by JEFFREY H. MILLER

VOLUME 205. Metallobiochemistry (Part B: Metallothionein and Related Molecules)
Edited by JAMES F. RIORDAN AND BERT L. VALLEE

VOLUME 206. Cytochrome P450
Edited by MICHAEL R. WATERMAN AND ERIC F. JOHNSON

VOLUME 207. Ion Channels
Edited by BERNARDO RUDY AND LINDA E. IVERSON

VOLUME 208. Protein–DNA Interactions
Edited by ROBERT T. SAUER

VOLUME 209. Phospholipid Biosynthesis
Edited by EDWARD A. DENNIS AND DENNIS E. VANCE

VOLUME 210. Numerical Computer Methods
Edited by LUDWIG BRAND AND MICHAEL L. JOHNSON

VOLUME 211. DNA Structures (Part A: Synthesis and Physical Analysis of DNA)
Edited by DAVID M. J. LILLEY AND JAMES E. DAHLBERG

VOLUME 212. DNA Structures (Part B: Chemical and Electrophoretic Analysis of DNA)
Edited by DAVID M. J. LILLEY AND JAMES E. DAHLBERG

VOLUME 213. Carotenoids (Part A: Chemistry, Separation, Quantitation, and Antioxidation)
Edited by LESTER PACKER

VOLUME 214. Carotenoids (Part B: Metabolism, Genetics, and Biosynthesis)
Edited by LESTER PACKER

VOLUME 215. Platelets: Receptors, Adhesion, Secretion (Part B)
Edited by JACEK J. HAWIGER

VOLUME 216. Recombinant DNA (Part G)
Edited by RAY WU

VOLUME 217. Recombinant DNA (Part H)
Edited by RAY WU

VOLUME 218. Recombinant DNA (Part I)
Edited by RAY WU

VOLUME 219. Reconstitution of Intracellular Transport
Edited by JAMES E. ROTHMAN

VOLUME 220. Membrane Fusion Techniques (Part A)
Edited by NEJAT DÜZGÜNEŞ

VOLUME 221. Membrane Fusion Techniques (Part B)
Edited by NEJAT DÜZGÜNEŞ

VOLUME 222. Proteolytic Enzymes in Coagulation, Fibrinolysis, and Complement Activation (Part A: Mammalian Blood Coagulation Factors and Inhibitors)
Edited by LASZLO LORAND AND KENNETH G. MANN

VOLUME 223. Proteolytic Enzymes in Coagulation, Fibrinolysis, and Complement Activation (Part B: Complement Activation, Fibrinolysis, and Nonmammalian Blood Coagulation Factors)
Edited by LASZLO LORAND AND KENNETH G. MANN

VOLUME 224. Molecular Evolution: Producing the Biochemical Data
Edited by ELIZABETH ANNE ZIMMER, THOMAS J. WHITE, REBECCA L. CANN, AND ALLAN C. WILSON

VOLUME 225. Guide to Techniques in Mouse Development
Edited by PAUL M. WASSARMAN AND MELVIN L. DEPAMPHILIS

VOLUME 226. Metallobiochemistry (Part C: Spectroscopic and Physical Methods for Probing Metal Ion Environments in Metalloenzymes and Metalloproteins)
Edited by JAMES F. RIORDAN AND BERT L. VALLEE

VOLUME 227. Metallobiochemistry (Part D: Physical and Spectroscopic Methods for Probing Metal Ion Environments in Metalloproteins)
Edited by JAMES F. RIORDAN AND BERT L. VALLEE

VOLUME 228. Aqueous Two-Phase Systems
Edited by HARRY WALTER AND GÖTE JOHANSSON

VOLUME 229. Cumulative Subject Index Volumes 195–198, 200–227

VOLUME 230. Guide to Techniques in Glycobiology
Edited by WILLIAM J. LENNARZ AND GERALD W. HART

VOLUME 231. Hemoglobins (Part B: Biochemical and Analytical Methods)
Edited by JOHANNES EVERSE, KIM D. VANDEGRIFF, AND ROBERT M. WINSLOW

VOLUME 232. Hemoglobins (Part C: Biophysical Methods)
Edited by JOHANNES EVERSE, KIM D. VANDEGRIFF, AND ROBERT M. WINSLOW

VOLUME 233. Oxygen Radicals in Biological Systems (Part C)
Edited by LESTER PACKER

VOLUME 234. Oxygen Radicals in Biological Systems (Part D)
Edited by LESTER PACKER

VOLUME 235. Bacterial Pathogenesis (Part A: Identification and Regulation of Virulence Factors)
Edited by VIRGINIA L. CLARK AND PATRIK M. BAVOIL

VOLUME 236. Bacterial Pathogenesis (Part B: Integration of Pathogenic Bacteria with Host Cells)
Edited by VIRGINIA L. CLARK AND PATRIK M. BAVOIL

VOLUME 237. Heterotrimeric G Proteins
Edited by RAVI IYENGAR

VOLUME 238. Heterotrimeric G-Protein Effectors
Edited by RAVI IYENGAR

VOLUME 239. Nuclear Magnetic Resonance (Part C)
Edited by THOMAS L. JAMES AND NORMAN J. OPPENHEIMER

VOLUME 240. Numerical Computer Methods (Part B)
Edited by MICHAEL L. JOHNSON AND LUDWIG BRAND

VOLUME 241. Retroviral Proteases
Edited by LAWRENCE C. KUO AND JULES A. SHAFER

VOLUME 242. Neoglycoconjugates (Part A)
Edited by Y. C. LEE AND REIKO T. LEE

VOLUME 243. Inorganic Microbial Sulfur Metabolism
Edited by HARRY D. PECK, JR., AND JEAN LEGALL

VOLUME 244. Proteolytic Enzymes: Serine and Cysteine Peptidases
Edited by ALAN J. BARRETT

VOLUME 245. Extracellular Matrix Components
Edited by E. RUOSLAHTI AND E. ENGVALL

VOLUME 246. Biochemical Spectroscopy
Edited by KENNETH SAUER

VOLUME 247. Neoglycoconjugates (Part B: Biomedical Applications)
Edited by Y. C. LEE AND REIKO T. LEE

VOLUME 248. Proteolytic Enzymes: Aspartic and Metallo Peptidases
Edited by ALAN J. BARRETT

VOLUME 249. Enzyme Kinetics and Mechanism (Part D: Developments in Enzyme Dynamics)
Edited by DANIEL L. PURICH

VOLUME 250. Lipid Modifications of Proteins
Edited by PATRICK J. CASEY AND JANICE E. BUSS

VOLUME 251. Biothiols (Part A: Monothiols and Dithiols, Protein Thiols, and Thiyl Radicals)
Edited by LESTER PACKER

VOLUME 252. Biothiols (Part B: Glutathione and Thioredoxin; Thiols in Signal Transduction and Gene Regulation)
Edited by LESTER PACKER

VOLUME 253. Adhesion of Microbial Pathogens
Edited by RON J. DOYLE AND ITZHAK OFEK

VOLUME 254. Oncogene Techniques
Edited by PETER K. VOGT AND INDER M. VERMA

VOLUME 255. Small GTPases and Their Regulators (Part A: Ras Family)
Edited by W. E. BALCH, CHANNING J. DER, AND ALAN HALL

VOLUME 256. Small GTPases and Their Regulators (Part B: Rho Family)
Edited by W. E. BALCH, CHANNING J. DER, AND ALAN HALL

VOLUME 257. Small GTPases and Their Regulators (Part C: Proteins Involved in Transport)
Edited by W. E. BALCH, CHANNING J. DER, AND ALAN HALL

VOLUME 258. Redox-Active Amino Acids in Biology
Edited by JUDITH P. KLINMAN

VOLUME 259. Energetics of Biological Macromolecules
Edited by MICHAEL L. JOHNSON AND GARY K. ACKERS

VOLUME 260. Mitochondrial Biogenesis and Genetics (Part A)
Edited by GIUSEPPE M. ATTARDI AND ANNE CHOMYN

VOLUME 261. Nuclear Magnetic Resonance and Nucleic Acids
Edited by THOMAS L. JAMES

VOLUME 262. DNA Replication
Edited by JUDITH L. CAMPBELL

VOLUME 263. Plasma Lipoproteins (Part C: Quantitation)
Edited by WILLIAM A. BRADLEY, SANDRA H. GIANTURCO, AND JERE P. SEGREST

VOLUME 264. Mitochondrial Biogenesis and Genetics (Part B)
Edited by GIUSEPPE M. ATTARDI AND ANNE CHOMYN

VOLUME 265. Cumulative Subject Index Volumes 228, 230–262

VOLUME 266. Computer Methods for Macromolecular Sequence Analysis
Edited by RUSSELL F. DOOLITTLE

VOLUME 267. Combinatorial Chemistry
Edited by JOHN N. ABELSON

VOLUME 268. Nitric Oxide (Part A: Sources and Detection of NO; NO Synthase)
Edited by LESTER PACKER

VOLUME 269. Nitric Oxide (Part B: Physiological and Pathological Processes)
Edited by LESTER PACKER

VOLUME 270. High Resolution Separation and Analysis of Biological Macromolecules (Part A: Fundamentals)
Edited by BARRY L. KARGER AND WILLIAM S. HANCOCK

VOLUME 271. High Resolution Separation and Analysis of Biological Macromolecules (Part B: Applications)
Edited by BARRY L. KARGER AND WILLIAM S. HANCOCK

VOLUME 272. Cytochrome P450 (Part B)
Edited by ERIC F. JOHNSON AND MICHAEL R. WATERMAN

VOLUME 273. RNA Polymerase and Associated Factors (Part A)
Edited by SANKAR ADHYA

VOLUME 274. RNA Polymerase and Associated Factors (Part B)
Edited by SANKAR ADHYA

VOLUME 275. Viral Polymerases and Related Proteins
Edited by LAWRENCE C. KUO, DAVID B. OLSEN, AND STEVEN S. CARROLL

VOLUME 276. Macromolecular Crystallography (Part A)
Edited by CHARLES W. CARTER, JR., AND ROBERT M. SWEET

VOLUME 277. Macromolecular Crystallography (Part B)
Edited by CHARLES W. CARTER, JR., AND ROBERT M. SWEET

VOLUME 278. Fluorescence Spectroscopy
Edited by LUDWIG BRAND AND MICHAEL L. JOHNSON

VOLUME 279. Vitamins and Coenzymes (Part I)
Edited by DONALD B. MCCORMICK, JOHN W. SUTTIE, AND CONRAD WAGNER

VOLUME 280. Vitamins and Coenzymes (Part J)
Edited by DONALD B. MCCORMICK, JOHN W. SUTTIE, AND CONRAD WAGNER

VOLUME 281. Vitamins and Coenzymes (Part K)
Edited by DONALD B. MCCORMICK, JOHN W. SUTTIE, AND CONRAD WAGNER

VOLUME 282. Vitamins and Coenzymes (Part L)
Edited by DONALD B. MCCORMICK, JOHN W. SUTTIE, AND CONRAD WAGNER

VOLUME 283. Cell Cycle Control
Edited by WILLIAM G. DUNPHY

VOLUME 284. Lipases (Part A: Biotechnology)
Edited by BYRON RUBIN AND EDWARD A. DENNIS

VOLUME 285. Cumulative Subject Index Volumes 263, 264, 266–284, 286–289

VOLUME 286. Lipases (Part B: Enzyme Characterization and Utilization)
Edited by BYRON RUBIN AND EDWARD A. DENNIS

VOLUME 287. Chemokines
Edited by RICHARD HORUK

VOLUME 288. Chemokine Receptors
Edited by RICHARD HORUK

VOLUME 289. Solid Phase Peptide Synthesis
Edited by GREGG B. FIELDS

VOLUME 290. Molecular Chaperones
Edited by GEORGE H. LORIMER AND THOMAS BALDWIN

VOLUME 291. Caged Compounds
Edited by GERARD MARRIOTT

VOLUME 292. ABC Transporters: Biochemical, Cellular, and Molecular Aspects
Edited by SURESH V. AMBUDKAR AND MICHAEL M. GOTTESMAN

VOLUME 293. Ion Channels (Part B)
Edited by P. MICHAEL CONN

VOLUME 294. Ion Channels (Part C)
Edited by P. MICHAEL CONN

VOLUME 295. Energetics of Biological Macromolecules (Part B)
Edited by GARY K. ACKERS AND MICHAEL L. JOHNSON

VOLUME 296. Neurotransmitter Transporters
Edited by SUSAN G. AMARA

VOLUME 297. Photosynthesis: Molecular Biology of Energy Capture
Edited by LEE MCINTOSH

VOLUME 298. Molecular Motors and the Cytoskeleton (Part B)
Edited by RICHARD B. VALLEE

VOLUME 299. Oxidants and Antioxidants (Part A)
Edited by LESTER PACKER

VOLUME 300. Oxidants and Antioxidants (Part B)
Edited by LESTER PACKER

VOLUME 301. Nitric Oxide: Biological and Antioxidant Activities (Part C)
Edited by LESTER PACKER

VOLUME 302. Green Fluorescent Protein
Edited by P. MICHAEL CONN

VOLUME 303. cDNA Preparation and Display
Edited by SHERMAN M. WEISSMAN

VOLUME 304. Chromatin
Edited by PAUL M. WASSARMAN AND ALAN P. WOLFFE

VOLUME 305. Bioluminescence and Chemiluminescence (Part C)
Edited by THOMAS O. BALDWIN AND MIRIAM M. ZIEGLER

VOLUME 306. Expression of Recombinant Genes in
Eukaryotic Systems
Edited by JOSEPH C. GLORIOSO AND MARTIN C. SCHMIDT

VOLUME 307. Confocal Microscopy
Edited by P. MICHAEL CONN

VOLUME 308. Enzyme Kinetics and Mechanism (Part E: Energetics of
Enzyme Catalysis)
Edited by DANIEL L. PURICH AND VERN L. SCHRAMM

VOLUME 309. Amyloid, Prions, and Other Protein Aggregates
Edited by RONALD WETZEL

VOLUME 310. Biofilms
Edited by RON J. DOYLE

VOLUME 311. Sphingolipid Metabolism and Cell Signaling (Part A)
Edited by ALFRED H. MERRILL, JR., AND YUSUF A. HANNUN

VOLUME 312. Sphingolipid Metabolism and Cell Signaling (Part B)
Edited by ALFRED H. MERRILL, JR., AND YUSUF A. HANNUN

VOLUME 313. Antisense Technology (Part A: General Methods, Methods of
Delivery, and RNA Studies)
Edited by M. IAN PHILLIPS

VOLUME 314. Antisense Technology (Part B: Applications)
Edited by M. IAN PHILLIPS

VOLUME 315. Vertebrate Phototransduction and the Visual Cycle (Part A)
Edited by KRZYSZTOF PALCZEWSKI

VOLUME 316. Vertebrate Phototransduction and the Visual Cycle (Part B)
Edited by KRZYSZTOF PALCZEWSKI

VOLUME 317. RNA-Ligand Interactions (Part A: Structural Biology Methods)
Edited by DANIEL W. CELANDER AND JOHN N. ABELSON

VOLUME 318. RNA-Ligand Interactions (Part B: Molecular Biology Methods)
Edited by DANIEL W. CELANDER AND JOHN N. ABELSON

VOLUME 319. Singlet Oxygen, UV-A, and Ozone
Edited by LESTER PACKER AND HELMUT SIES

VOLUME 320. Cumulative Subject Index Volumes 290–319

VOLUME 321. Numerical Computer Methods (Part C)
Edited by MICHAEL L. JOHNSON AND LUDWIG BRAND

VOLUME 322. Apoptosis
Edited by JOHN C. REED

VOLUME 323. Energetics of Biological Macromolecules (Part C)
Edited by MICHAEL L. JOHNSON AND GARY K. ACKERS

VOLUME 324. Branched-Chain Amino Acids (Part B)
Edited by ROBERT A. HARRIS AND JOHN R. SOKATCH

VOLUME 325. Regulators and Effectors of Small GTPases (Part D: Rho Family)
Edited by W. E. BALCH, CHANNING J. DER, AND ALAN HALL

VOLUME 326. Applications of Chimeric Genes and Hybrid Proteins (Part A: Gene Expression and Protein Purification)
Edited by JEREMY THORNER, SCOTT D. EMR, AND JOHN N. ABELSON

VOLUME 327. Applications of Chimeric Genes and Hybrid Proteins (Part B: Cell Biology and Physiology)
Edited by JEREMY THORNER, SCOTT D. EMR, AND JOHN N. ABELSON

VOLUME 328. Applications of Chimeric Genes and Hybrid Proteins (Part C: Protein–Protein Interactions and Genomics)
Edited by JEREMY THORNER, SCOTT D. EMR, AND JOHN N. ABELSON

VOLUME 329. Regulators and Effectors of Small GTPases (Part E: GTPases Involved in Vesicular Traffic)
Edited by W. E. BALCH, CHANNING J. DER, AND ALAN HALL

VOLUME 330. Hyperthermophilic Enzymes (Part A)
Edited by MICHAEL W. W. ADAMS AND ROBERT M. KELLY

VOLUME 331. Hyperthermophilic Enzymes (Part B)
Edited by MICHAEL W. W. ADAMS AND ROBERT M. KELLY

VOLUME 332. Regulators and Effectors of Small GTPases (Part F: Ras Family I)
Edited by W. E. BALCH, CHANNING J. DER, AND ALAN HALL

VOLUME 333. Regulators and Effectors of Small GTPases (Part G: Ras Family II)
Edited by W. E. BALCH, CHANNING J. DER, AND ALAN HALL

VOLUME 334. Hyperthermophilic Enzymes (Part C)
Edited by MICHAEL W. W. ADAMS AND ROBERT M. KELLY

VOLUME 335. Flavonoids and Other Polyphenols
Edited by LESTER PACKER

VOLUME 336. Microbial Growth in Biofilms (Part A: Developmental and Molecular Biological Aspects)
Edited by RON J. DOYLE

VOLUME 337. Microbial Growth in Biofilms (Part B: Special Environments and Physicochemical Aspects)
Edited by RON J. DOYLE

VOLUME 338. Nuclear Magnetic Resonance of Biological Macromolecules (Part A)
Edited by THOMAS L. JAMES, VOLKER DÖTSCH, AND ULI SCHMITZ

VOLUME 339. Nuclear Magnetic Resonance of Biological Macromolecules (Part B)
Edited by THOMAS L. JAMES, VOLKER DÖTSCH, AND ULI SCHMITZ

VOLUME 340. Drug–Nucleic Acid Interactions
Edited by JONATHAN B. CHAIRES AND MICHAEL J. WARING

VOLUME 341. Ribonucleases (Part A)
Edited by ALLEN W. NICHOLSON

VOLUME 342. Ribonucleases (Part B)
Edited by ALLEN W. NICHOLSON

VOLUME 343. G Protein Pathways (Part A: Receptors)
Edited by RAVI IYENGAR AND JOHN D. HILDEBRANDT

VOLUME 344. G Protein Pathways (Part B: G Proteins and Their Regulators)
Edited by RAVI IYENGAR AND JOHN D. HILDEBRANDT

VOLUME 345. G Protein Pathways (Part C: Effector Mechanisms)
Edited by RAVI IYENGAR AND JOHN D. HILDEBRANDT

VOLUME 346. Gene Therapy Methods
Edited by M. IAN PHILLIPS

VOLUME 347. Protein Sensors and Reactive Oxygen Species (Part A: Selenoproteins and Thioredoxin)
Edited by HELMUT SIES AND LESTER PACKER

VOLUME 348. Protein Sensors and Reactive Oxygen Species (Part B: Thiol Enzymes and Proteins)
Edited by HELMUT SIES AND LESTER PACKER

VOLUME 349. Superoxide Dismutase
Edited by LESTER PACKER

VOLUME 350. Guide to Yeast Genetics and Molecular and Cell Biology (Part B)
Edited by CHRISTINE GUTHRIE AND GERALD R. FINK

VOLUME 351. Guide to Yeast Genetics and Molecular and Cell Biology (Part C)
Edited by CHRISTINE GUTHRIE AND GERALD R. FINK

VOLUME 352. Redox Cell Biology and Genetics (Part A)
Edited by CHANDAN K. SEN AND LESTER PACKER

VOLUME 353. Redox Cell Biology and Genetics (Part B)
Edited by CHANDAN K. SEN AND LESTER PACKER

VOLUME 354. Enzyme Kinetics and Mechanisms (Part F: Detection and Characterization of Enzyme Reaction Intermediates)
Edited by DANIEL L. PURICH

VOLUME 355. Cumulative Subject Index Volumes 321–354

VOLUME 356. Laser Capture Microscopy and Microdissection
Edited by P. MICHAEL CONN

VOLUME 357. Cytochrome P450, Part C
Edited by ERIC F. JOHNSON AND MICHAEL R. WATERMAN

VOLUME 358. Bacterial Pathogenesis (Part C: Identification, Regulation, and Function of Virulence Factors)
Edited by VIRGINIA L. CLARK AND PATRIK M. BAVOIL

VOLUME 359. Nitric Oxide (Part D)
Edited by ENRIQUE CADENAS AND LESTER PACKER

VOLUME 360. Biophotonics (Part A)
Edited by GERARD MARRIOTT AND IAN PARKER

VOLUME 361. Biophotonics (Part B)
Edited by GERARD MARRIOTT AND IAN PARKER

VOLUME 362. Recognition of Carbohydrates in Biological Systems (Part A)
Edited by YUAN C. LEE AND REIKO T. LEE

VOLUME 363. Recognition of Carbohydrates in Biological Systems (Part B)
Edited by YUAN C. LEE AND REIKO T. LEE

VOLUME 364. Nuclear Receptors
Edited by DAVID W. RUSSELL AND DAVID J. MANGELSDORF

VOLUME 365. Differentiation of Embryonic Stem Cells
Edited by PAUL M. WASSAUMAN AND GORDON M. KELLER

VOLUME 366. Protein Phosphatases
Edited by SUSANNE KLUMPP AND JOSEF KRIEGLSTEIN

VOLUME 367. Liposomes (Part A)
Edited by NEJAT DÜZGÜNEŞ

VOLUME 368. Macromolecular Crystallography (Part C)
Edited by CHARLES W. CARTER, JR., AND ROBERT M. SWEET

VOLUME 369. Combinational Chemistry (Part B)
Edited by GUILLERMO A. MORALES AND BARRY A. BUNIN

VOLUME 370. RNA Polymerases and Associated Factors (Part C)
Edited by SANKAR L. ADHYA AND SUSAN GARGES

VOLUME 371. RNA Polymerases and Associated Factors (Part D)
Edited by SANKAR L. ADHYA AND SUSAN GARGES

VOLUME 372. Liposomes (Part B)
Edited by NEJAT DÜZGÜNEŞ

VOLUME 373. Liposomes (Part C)
Edited by NEJAT DÜZGÜNEŞ

VOLUME 374. Macromolecular Crystallography (Part D)
Edited by CHARLES W. CARTER, JR., AND ROBERT W. SWEET

VOLUME 375. Chromatin and Chromatin Remodeling Enzymes (Part A)
Edited by C. DAVID ALLIS AND CARL WU

VOLUME 376. Chromatin and Chromatin Remodeling Enzymes (Part B)
Edited by C. DAVID ALLIS AND CARL WU

VOLUME 377. Chromatin and Chromatin Remodeling Enzymes (Part C)
Edited by C. DAVID ALLIS AND CARL WU

VOLUME 378. Quinones and Quinone Enzymes (Part A)
Edited by HELMUT SIES AND LESTER PACKER

VOLUME 379. Energetics of Biological Macromolecules (Part D)
Edited by JO M. HOLT, MICHAEL L. JOHNSON, AND GARY K. ACKERS

VOLUME 380. Energetics of Biological Macromolecules (Part E)
Edited by JO M. HOLT, MICHAEL L. JOHNSON, AND GARY K. ACKERS

VOLUME 381. Oxygen Sensing
Edited by CHANDAN K. SEN AND GREGG L. SEMENZA

VOLUME 382. Quinones and Quinone Enzymes (Part B)
Edited by HELMUT SIES AND LESTER PACKER

VOLUME 383. Numerical Computer Methods (Part D)
Edited by LUDWIG BRAND AND MICHAEL L. JOHNSON

VOLUME 384. Numerical Computer Methods (Part E)
Edited by LUDWIG BRAND AND MICHAEL L. JOHNSON

VOLUME 385. Imaging in Biological Research (Part A)
Edited by P. MICHAEL CONN

VOLUME 386. Imaging in Biological Research (Part B)
Edited by P. MICHAEL CONN

VOLUME 387. Liposomes (Part D)
Edited by NEJAT DÜZGÜNEŞ

VOLUME 388. Protein Engineering
Edited by DAN E. ROBERTSON AND JOSEPH P. NOEL

VOLUME 389. Regulators of G-Protein Signaling (Part A)
Edited by DAVID P. SIDEROVSKI

VOLUME 390. Regulators of G-Protein Signaling (Part B)
Edited by DAVID P. SIDEROVSKI

VOLUME 391. Liposomes (Part E)
Edited by NEJAT DÜZGÜNEŞ

VOLUME 392. RNA Interference
Edited by ENGELKE ROSSI

VOLUME 393. Circadian Rhythms
Edited by MICHAEL W. YOUNG

VOLUME 394. Nuclear Magnetic Resonance of Biological
Macromolecules (Part C)
Edited by THOMAS L. JAMES

VOLUME 395. Producing the Biochemical Data (Part B)
Edited by ELIZABETH A. ZIMMER AND ERIC H. ROALSON

VOLUME 396. Nitric Oxide (Part E)
Edited by LESTER PACKER AND ENRIQUE CADENAS

VOLUME 397. Environmental Microbiology
Edited by JARED R. LEADBETTER

VOLUME 398. Ubiquitin and Protein Degradation (Part A)
Edited by RAYMOND J. DESHAIES

VOLUME 399. Ubiquitin and Protein Degradation (Part B)
Edited by RAYMOND J. DESHAIES

VOLUME 400. Phase II Conjugation Enzymes and Transport Systems
Edited by HELMUT SIES AND LESTER PACKER

VOLUME 401. Glutathione Transferases and Gamma Glutamyl Transpeptidases
Edited by HELMUT SIES AND LESTER PACKER

VOLUME 402. Biological Mass Spectrometry
Edited by A. L. BURLINGAME

VOLUME 403. GTPases Regulating Membrane Targeting and Fusion
Edited by WILLIAM E. BALCH, CHANNING J. DER, AND ALAN HALL

VOLUME 404. GTPases Regulating Membrane Dynamics
Edited by WILLIAM E. BALCH, CHANNING J. DER, AND ALAN HALL

VOLUME 405. Mass Spectrometry: Modified Proteins and Glycoconjugates
Edited by A. L. BURLINGAME

VOLUME 406. Regulators and Effectors of Small GTPases: Rho Family
Edited by WILLIAM E. BALCH, CHANNING J. DER, AND ALAN HALL

VOLUME 407. Regulators and Effectors of Small GTPases: Ras Family
Edited by WILLIAM E. BALCH, CHANNING J. DER, AND ALAN HALL

VOLUME 408. DNA Repair (Part A)
Edited by JUDITH L. CAMPBELL AND PAUL MODRICH

VOLUME 409. DNA Repair (Part B)
Edited by JUDITH L. CAMPBELL AND PAUL MODRICH

VOLUME 410. DNA Microarrays (Part A: Array Platforms and Web-Bench Protocols)
Edited by ALAN KIMMEL AND BRIAN OLIVER

VOLUME 411. DNA Microarrays (Part B: Databases and Statistics)
Edited by ALAN KIMMEL AND BRIAN OLIVER

VOLUME 412. Amyloid, Prions, and Other Protein Aggregates (Part B)
Edited by INDU KHETERPAL AND RONALD WETZEL

VOLUME 413. Amyloid, Prions, and Other Protein Aggregates (Part C)
Edited by INDU KHETERPAL AND RONALD WETZEL

VOLUME 414. Measuring Biological Responses with Automated Microscopy
Edited by JAMES INGLESE

VOLUME 415. Glycobiology
Edited by MINORU FUKUDA

VOLUME 416. Glycomics
Edited by MINORU FUKUDA

VOLUME 417. Functional Glycomics
Edited by MINORU FUKUDA

VOLUME 418. Embryonic Stem Cells
Edited by IRINA KLIMANSKAYA AND ROBERT LANZA

VOLUME 419. Adult Stem Cells
Edited by IRINA KLIMANSKAYA AND ROBERT LANZA

VOLUME 420. Stem Cell Tools and Other Experimental Protocols
Edited by IRINA KLIMANSKAYA AND ROBERT LANZA

VOLUME 421. Advanced Bacterial Genetics: Use of Transposons and Phage for Genomic Engineering
Edited by KELLY T. HUGHES

VOLUME 422. Two-Component Signaling Systems, (Part A)
Edited by MELVIN I. SIMON, BRIAN R. CRANE, AND ALEXANDRINE CRANE

VOLUME 423. Two-Component Signaling Systems, (Part B)
Edited by MELVIN I. SIMON, BRIAN R. CRANE, AND ALEXANDRINE CRANE

VOLUME 424. RNA Editing
Edited by JONATHA M. GOTT

VOLUME 425. RNA Modification
Edited by JONATHA M. GOTT

VOLUME 426. Integrins
Edited by DAVID A. CHERESH

VOLUME 427. MicroRNA Methods
Edited by JOHN J. ROSSI

VOLUME 428. Osmosensing and Osmosignaling
Edited by HELMUT SIES AND DIETER HÄUSSINGER

VOLUME 429. Translation Initiation: Extract Systems and Molecular Genetics
Edited by JON LORSCH

VOLUME 430. Translation Initiation: Reconstituted Systems and Biophysical Methods
Edited by JON LORSCH

VOLUME 431. Translation Initiation: Cell Biology, High-Throughput Methods, and Chemical-Based Approaches
Edited by JON LORSCH

VOLUME 432. Lipidomics and Bioactive Lipids: Mass-Spectrometry–Based Lipid Analysis
Edited by H. ALEX BROWN

VOLUME 433. Lipidomics and Bioactive Lipids: Specialized Analytical Methods and Lipids in Disease (in preparation)
Edited by H. ALEX BROWN

VOLUME 434. Lipidomics and Bioactive Lipids: Lipids and Cell Signaling (in preparation)
Edited by H. ALEX BROWN

VOLUME 435. Oxygen Biology and Hypoxia (in preparation)
Edited by HELMUT SIES and BERNHARD BRÜNE

VOLUME 436. Globins and Other Nitric Oxide-Reactive Proteins, (Part A) (in preparation)
Edited by ROBERT K. POOLE

VOLUME 437. Globins and Other Nitric Oxide-Reactive Proteins, (Part B) (in preparation)
Edited by ROBERT K. POOLE

VOLUME 438. Small GTPases in Diseases, Part A (in preparation)
Edited by WILLIAM E. BALCH, CHANNING J. DER, AND ALAN HALL

CHAPTER ONE

QUALITATIVE ANALYSIS AND QUANTITATIVE ASSESSMENT OF CHANGES IN NEUTRAL GLYCEROL LIPID MOLECULAR SPECIES WITHIN CELLS

Jessica Krank,* Robert C. Murphy,* Robert M. Barkley,* Eva Duchoslav,[†] and Andrew McAnoy*

Contents

1. Introduction	2
2. Reagents	3
2.1. Cell culture	3
2.2. Standards	3
2.3. Extraction and purification	3
3. Methods	4
3.1. Cell culture	4
4. Results	7
4.1. Qualitative analysis	7
4.2. Quantitative analysis	11
5. Conclusions	19
Acknowledgments	19
References	19

Abstract

Triacylglycerols (TAGs) and diacylglycerols (DAGs) are present in cells as a complex mixture of molecular species that differ in the nature of the fatty acyl groups esterified to the glycerol backbone. In some cases, the molecular weights of these species are identical, confounding assignments of identity and quantity by molecular weight. Electrospray ionization results in the formation of $[M+NH_4]^+$ ions that can be collisionally activated to yield an abundant product ion corresponding to the loss of ammonia plus one of the fatty acyl groups as a free carboxylic acid. A method was developed using tandem mass

---

* Department of Pharmacology, University of Colorado at Denver and Health Sciences Center, Aurora, Colorado
[†] Applied Research Group, MDS Sciex, Concord, Ontario, Canada

spectrometry (MS) and neutral loss scanning to analyze the complex mixture of TAGs and DAGs present in cells and to quantitatively determine changes in TAGs and DAGs molecular species containing identical fatty acyl groups in an experimental series. Eighteen different deuterium-labeled internal standards were synthesized to serve to normalize the ion signal for each neutral loss scan. An example of the application of this method was in the quantitative analysis of TAG and DAG molecular species present in RAW 264.7 cells treated with a Toll-4 receptor ligand, Kdo$_2$-lipid A, in a time course study.

1. Introduction

Diacylglycerols and triacylglycerols represent major classes of glyceryl lipids that are present in all mammalian cells. Class analysis of these compounds has largely centered around total DAG or TAG content based upon a spectrophotometric-linked enzymatic assay; however, it is possible to determine individual molecular species of both of these classes of glyceryl lipids by MS. The major challenge in analysis of DAGs and TAGs is the total number of molecular species present within a cell, which results in multiple components appearing at the same molecular weight or truly isobaric compounds. Thus, the measurement of only the molecular ion species by MS, either the ammonium adduct ion or the alkali attachment ion (Han and Gross, 2001; McAnoy et al., 2005), would be insufficient to uniquely identifying each component. Qualitative analysis can be carried out even when faced with complex mixtures of these neutral lipids if one employs tandem MS in three steps (MS^3) (McAnoy et al., 2005). However, quantitative analysis is confounded by the isobaric nature of multiple species, and several strategies have been developed through which changes in molecular species can be assessed. The method described here employs a series of experiments using deuterium-labeled internal standards where the deuterium atoms are placed in the glycerol backbone, and the abundance of each glyceryl lipid is compared to the signal for the internal standard. Alternative methods are available to assess TAG and DAG molecular species based upon chromatographic separation of individual molecular species (Phillips et al., 1984) and analysis by gas chromatographs-mass spectrometers (GC-MS) (Myher et al., 1988) or liquid chromatographs-mass spectrometers (LC-MS) using electrospray ionization (ESI) or atmospheric pressure chemical ionization (Byrdwell et al., 1996; Hsu and Turk, 1999). The application of the method presented here will be exemplified by glyceryl lipid analysis in the RAW 264.7 cell line. These macrophage-like cells were originally derived from tumors induced in male BALB/c mice by the Abelson murine leukemia virus and respond to the Toll-4 receptor agonist Kdo$_2$-lipid A (Raetz et al., 2006).

2. Reagents

2.1. Cell culture

RAW 264.7 cells were obtained from American Type Culture Collection (Manassas, VA). Tissue culture reagents included high-glucose Dulbecco's Modified Eagle's Medium (DMEM) (Cellgro, Herndon, VA), fetal calf serum (Hyclone, Logan, UT), Dulbecco's Phosphate-Buffered Saline (D-PBS; Cellgro, Herndon, VA), and penicillin/streptomycin (Cellgro, Herndon, VA). The cells were carried in culture in either 150-cm^2 tissue culture flasks with filter caps or 100-mm^2 tissue culture dishes (Fisher Scientific, Fair Lawn, NJ).

2.2. Standards

All deuterium-labeled glyceryl lipids were [1,1,2,3,3]-d_5 in the glycerol backbone. The following lipids were obtained from Avanti Polar Lipids (Alabaster, AL): Kdo_2-lipid A; d_5-DAG mixture containing d_5-14:0/14:0 DAG, d_5-15:0/15:0 DAG, d_5-16:0/16:0 DAG, d_5-17:0/17:0 DAG, d_5-19:0/19:0 DAG, d_5-20:0/20:0 DAG, d_5-20:2/20:2 DAG, d_5-20:4/20:4 DAG, and d_5-20:5/20:5 DAG; d_5-TAG mixture containing d_5-14:0/16:1/14:0 TAG, d_5-15:0/18:1/15:0 TAG, d_5-16:0/18:0/16:0 TAG, d_5-17:0/17:1/17:0 TAG, d_5-19:0/12:0/19:0 TAG, d_5-20:0/20:1/20:0 TAG, d_5-20:2/18:3/20:2 TAG, d_5-20:4/18:2/20:4 TAG, and d_5-20:5/22:6/20:5 TAG.

2.3. Extraction and purification

High performance liquid chromatography (HPLC)–grade ammonium acetate (NH_4OAc), chloroform ($CHCl_3$), ethyl acetate, hexanes, isooctane, isopropanol, methanol, and methylene chloride (CH_2Cl_2) were obtained from Fisher Scientific (Fair Lawn, NJ). Optima-grade toluene and water also were obtained from Fisher Scientific (Fair Lawn, NJ). Discovery DSC-Si silica, and Discovery DSC-NH_2 amino propyl solid phase extraction cartridges, and the vacuum manifold were obtained from Supelco (Bellefonte, PA). Other equipment included a benchtop vortexer (Vortex Genie 2, VWR Scientific, West Chester, PA), Fluoropore 0.2-μm FG membrane filters (Millipore, Billerica, MA), 96-well polypropylene plates (Eppendorf, Westbury, NY), and Quant-iT DNA analysis kit (Molecular Probes, Eugene, OR).

3. Methods

3.1. Cell culture

RAW 264.7 cells were grown in T-150 flasks containing 30 ml of complete growth medium consisting of 500 ml of high-glucose DMEM supplemented with 10% FBS and 1% penicillin/streptomycin. The cells were maintained in a humidified incubator at 37° with a 5% CO_2 atmosphere. They were split such that 2×10^6 cells were seeded in a new flask each time the cells reached 80% confluency. Briefly, when the cells reached this confluency (approximately every 2 to 3 days), the old growth medium was removed, and 10 ml of fresh medium, which had been previously warmed to 37°, was added. The cells were then scraped from the flask using a large disposable cell scraper. A 10-μl aliquot was counted using a hemocytometer. The volume of media determined to contain 2×10^6 cells was then added to a new T-150 flask, and the volume was adjusted to 30 ml with complete growth medium. The flasks were then placed back into the incubator.

For quantitative analysis, three additional T-150 flasks were seeded and allowed to grow until they were approximately 80% confluent. Once confluent, the medium was removed, 10 ml of fresh medium was added to each flask, and the cells were scraped as before. The scraped cells from the three flasks were then combined, and a 10 μl aliquot was removed and counted. Cells (5×10^6) were then added to each of the 15 100-mm^2 tissue culture dishes. The volume of media in the dishes was adjusted to 5 ml with complete growth media. The dishes were then placed in the incubator and allowed to grow for 30 hr.

3.1.1. Kdo$_2$-lipid A preparation

A working solution of Kdo$_2$-lipid A, the active component of bacterial endotoxin (Raetz, 1990), was prepared at a concentration of 100 μg/ml in DPBS. This solution was sonicated for 5 min before each experiment to achieve a uniformly opalescent suspension.

3.1.2. Stimulation and harvesting

RAW 264.7 cells were treated with Kdo$_2$-lipid A to examine the changes in the neutral glyceryl lipids upon activation of the Toll-4 receptor. Cells were stimulated by the addition of 5 μl of the Kdo$_2$-lipid A working solution for a final concentration of 100 ng/ml. They were then incubated for 0.5, 1, 2, 4, 8, 12, and 24 hr. The same volume of DPBS was added to the corresponding control plates. At each time point, a 0.5 ml aliquot of the growth media was removed from each dish and reserved for tumor necrosis factor alfa (TNF-α) analysis. The dishes containing the control and Kdo$_2$-lipid A–treated cells were washed twice with ice-cold DPBS and then scraped into fresh DPBS at a concentration of approximately 16×10^6 cells per ml of buffer. A 50-μl aliquot was reserved for DNA analysis using a fluorescence assay (Ahn et al., 1996).

3.1.3. Quantitation of DNA

The total amount of DNA in each time course sample was determined using the Quant-iT kit. A standard curve was constructed using the included standards and was linear from 0 to 100 ng. A portion (10 μl) of each reserved sample aliquot and each standard was added to the black 96-well plates. The working solution was prepared as a 1:200 dilution of reagent into the included buffer. An aliquot (190 μl) of this solution was added to each well. Samples were then read on a fluorometer with an excitation wavelength maxima of 510 nm and an emission wavelength maxima of 527 nm. The fluorescence of the samples was compared to a calibration curve generated from the fluorescence of the standards to determine the amount of DNA in each sample.

3.1.4. Extraction of total lipids

Total lipids were extracted from the cell suspension (Bligh and Dyer, 1959). Ice-cold methanol (2.5 ml) was added to each 1 ml of DPBS containing the scraped cell suspension. A volume containing 600 pmol of each of the 18 d_5-labeled DAG and TAG internal standards in toluene/methanol (1:1) was added to this suspension for samples being analyzed quantitatively. A monophasic solution was formed by the addition of 1.25 ml of CH_2Cl_2 per ml of cell suspension. This solution was vortexed for 30 s using a benchtop vortexer. After the addition of water (1.0 ml) and CH_2Cl_2 (1.25 ml), the sample was vortexed for an additional 30 s. The phases were then separated by centrifugation at 1000 rpm for 5 min. The lower organic phase was removed to a clean tube using a glass transfer pipette. The upper aqueous phase was reextracted with an additional 2 ml of CH_2Cl_2. This solution was vortexed and centrifuged as before. The lower organic layer was then combined with that obtained in the first extraction. The total lipid extract was taken to dryness under a gentle stream of nitrogen.

3.1.5. Solid-phase extraction of lipid classes

Total lipids from samples obtained for qualitative analysis were fractionated using Discovery-NH_2 solid-phase extraction cartridges following a previously published method (Kaluzny et al., 1985). Cartridges containing 500 mg of solid phase were conditioned with 9 ml of hexane. The total lipid extract from 100×10^6 RAW 264.7 cells was loaded in 200 μl of $CHCl_3$. Glyceryl lipids were eluted with 6 ml of $CHCl_3$/IPA (2:1). This fraction was dried under a gentle stream of nitrogen. The samples were dissolved in 1 ml of $CHCl_3$. A 100-μl aliquot of this sample was diluted into 900 μl of $CHCl_3$/MeOH (1:1) to which 20 μl of 0.1 M of NH_4OAc was added for a final concentration of 5 mM. These samples were used for qualitative studies.

Glyceryl lipids for quantitative analysis were obtained from the total lipid extract of approximately 8 × 10^6 RAW 264.7 cells by solid-phase extraction using silica cartridges following a previously published method (Ingalls et al., 1993). Cartridges of 100 mg of material were conditioned with 4 ml of isooctane/ethyl acetate (80:1). The samples were redissolved in 1 ml of isooctane/ethyl acetate (75:25), sonicated for 30 s, and loaded onto the cartridges using a vacuum manifold with the flow-through collected in a clean tube. The neutral lipid fraction was eluted into the same tube with 4 ml of isooctane/ethyl acetate (75:25). This fraction contained the TAGs and DAGs as well as cholesterol and cholesteryl esters, while nonesterified fatty acids, monoacylglycerols, and phospholipids were retained on the cartridge. The samples were then taken to dryness under a gentle stream of nitrogen. The samples were dissolved in 200 μl of toluene/methanol (1:1) with 1 mM NH_4OAc for MS analysis using an API 4000 Q TRAP (Applied Biosystems, Foster City, CA).

3.1.6. Mass spectrometry

Qualitative analyses were performed on an LTQ, linear ion trap mass spectrometer (Thermo-Finnigan, San Jose, CA). Samples were introduced into the electrospray source using a drawn microcapillary at a flow rate of 1 μl/min. The mass spectrometer was operated in positive ion mode with a spray voltage of 2.4 kV, a capillary temperature of 250°, a capillary voltage of 29.0 V, a 1.5-u ion isolation window, and a 100-ms maximum inject time. Scans (typically 50) were averaged for the MS spectra, and approximately 100 scans were averaged for the MS^2 and MS^3 spectra.

Neutral loss survey experiments were performed on an API 4000 Q TRAP equipped with a NanoMate nanoelectrospray ionization source (Advion Biosciences, Ithaca, NY). The NanoMate was operated in positive ion mode with a spray voltage of 1.35 kV, vented headspace, and pressure of 0.30 psi, which resulted in a flow rate of approximately 250 nl/min. The mass spectrometer was operated with a step size of 0.10 u, a curtain gas setting of 10, a collision gas setting of medium, a declustering potential of 120, an entrance potential of 10, collision energy of 37, and collision cell exit potential of 6.

The mass-to-charge (m/z) range from m/z 500 to 1200 was scanned over 6 s, giving a total of 30 scans for each neutral loss period. These scans were averaged before submission to the Lipid Profiler software (Ejsing et al., 2006) for integration and subsequent analysis. This Applied Biosystems program was modified specifically to process this data format.

4. Results

4.1. Qualitative analysis

Mass spectrometric analysis of the neutral lipid extract from RAW cells yielded a very complex spectrum. From m/z 700 to 1000, peaks corresponded to ammoniated TAG species at nearly every even mass (Fig. 1.1). Each of these m/z values could be assigned a total number of carbon atoms and double bonds based on the mass; however, this designation did not indicate individual molecular species. Multiple isobaric species could exist, differing in the exact individual fatty acyl groups present, positional distribution of fatty acyl groups, double bond location, and geometry. These species could also be straight chained or branched fatty acyl groups. In order to improve the characterization of molecular species to individual fatty acyl groups present (total fatty acyl carbons and total number of double bonds in each fatty acyl group), MS^3 was employed.

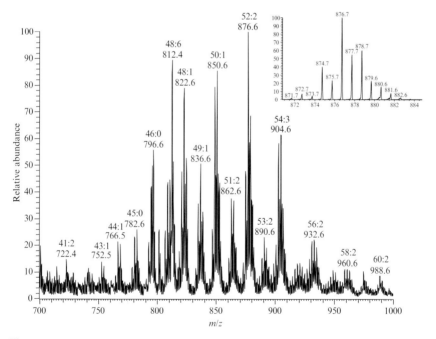

Figure 1.1 Full-scan mass spectrum of the neutral lipid extract from RAW 264.7 cells showing several envelopes of molecular ions corresponding to triacylglycerols. The envelope corresponding to the series of triacylglycerols containing 52 carbon atoms is shown at high resolution (inset).

For example, the signal at m/z 876 (Fig. 1.1) would correspond to many isobaric combinations of fatty acyl chains with a total of 52 carbons and 2 double bonds or a total of 53 carbons and 9 double bonds within the three esterified fatty acyl groups. To identify all of the isobaric species present, this strategy used both MS^2 and MS^3 experiments. For the MS^2 case, the TAG $[M+NH_4]^+$ parent ion (m/z 876) was selected to undergo collision-induced dissociation (CID) (Fig. 1.2). In general, when $[M+NH_4]^+$ ions are activated, these ions undergo the neutral loss of a fatty acyl group and ammonia, yielding a corresponding DAG fragment ion (Byrdwell and Neff, 2002; Cheng et al., 1998). The CID of m/z 876 yielded several peaks (m/z 549, 575, 577, 603, 605, 619, 631, and 647), each of which corresponded to the loss of a unique fatty acid from one of the isobaric TAG species present at this m/z ratio. The most abundant product ions corresponded to the neutral losses of 18:1 (m/z 577), 16:0 (m/z 603), and 18:0 (m/z 575). The difference in the mass of the $[M+NH_4]^+$ precursor ion and the product ion was used to identify the fatty acyl group lost from the total number of carbon atoms and double bonds. The neutral loss masses for several common fatty acyl groups are shown in Table 1.1; however, this neutral loss does not provide

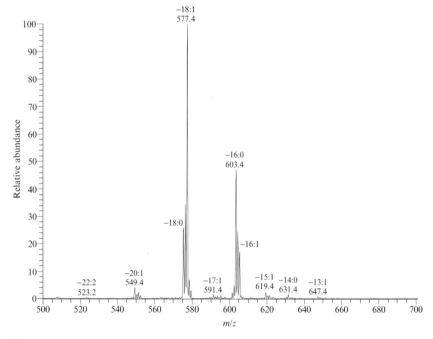

Figure 1.2 Product ion spectrum from the collision-induced dissociation of the triacylglycerol ion at m/z 876 from the full scan at a collision energy of 30 V. The ion at m/z 876 undergoes loss of several fatty acyl groups as indicated, with the loss of 16:0 (palmitic acid) and 18:1 (oleic acid) being the most abundant.

information to assign double bond or alkyl branching in the lost fatty acyl group, nor does it indicate to which glycerol carbon atom it was esterified.

From the MS^2 spectrum, a product ion can be selected for further fragmentation in the linear ion trap. Collision-induced dissociation of the DAG product ions yielded information about both of the remaining fatty

Table 1.1 Neutral loss mass corresponding to common fatty acyl groups esterified to triacylglycerol and diacylglycerol molecular species after collisional activation and observed by tandem mass spectrometry (MS/MS)

Fatty acyl substituent[a]	Neutral loss (u)[b] $RCOOH+NH_3$
14:0[a]	245
15:0	259
16:1	271
16:0	273
17:0	287
18:4	293
18:3	295
18:2	297
18:1	299
18:0	301
19:1	313
19:0	315
20:5	319
20:4	321
20:3	323
20:2	325
20:1	327
20:0	329
21:0	343
22:6	345
22:5	347
22:4	349
22:3	351
22:2	353
22:1	355
22:0	357
24:1	383
24:0	385

[a] Designation of total carbon atoms in the fatty acyl group: total number of double bonds.
[b] Neutral loss in daltons (u) as $RCOOH+NH_3$, which corresponds to the mass of the fatty acyl group as a free carboxylic acid plus the mass of ammonia.

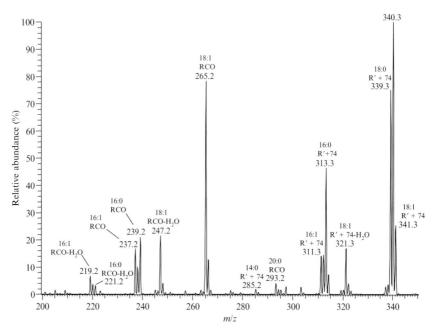

Figure 1.3 MS/MS/MS spectrum of the diacyl product ion at m/z 577 resulting from the collision-induced dissociation of the ion at m/z 876 in the mass spectrum. The ion at m/z 577 was selected in the linear ion trap to undergo collision-induced dissociation to determine the nature of the final fatty acyl species present in the isobaric molecules. Both acylium ions ([RCO]$^+$) and ions retaining the glycerol backbone ([R′+74]$^+$) are present in this spectrum.

acyl substituents. The mass of the MS^3 product ions related directly to the final esterified acyl chain, and the difference in mass between the MS^2 precursor and MS^3 product ion gave information about the fatty acyl chain that was lost as a neutral species. As shown for the initial MS^2 product ion at m/z 577 (Fig. 1.3), collision-induced dissociation led to a number of MS^3 product ions corresponding to the remaining fatty acyl chains as acylium ions ([RCO]$^+$) and as fatty acids esterified to the glycerol backbone ([R′+74]$^+$) with their associated losses of water. For example, the presence of an 18:1 fatty acyl component in the MS^2 product ion m/z 577 was revealed by the abundant MS^3 product ion at m/z 265 that corresponds to $C_{17}H_{33}C=O^+$, while the presence of 16:0 was revealed by the MS^3 product ion at m/z 239 ($C_{15}H_{31}C=O^+$) and 313 ([$C_{15}H_{31}COO-CH_2CH_2CHO$] + H$^+$). The mechanism for the formation of these ions had been described previously from labeling studies in which the hydrogen atoms on the glycerol backbone were replaced with deuterium atoms (McAnoy et al., 2005).

From the information obtained from the MS^2 and MS^3 spectra, a table containing all of the isobaric species identified for a specific m/z value was generated. As shown in Table 1.2, for the three most abundant fragment ions observed in the MS^2 spectrum of the ion at m/z 876 (m/z 577, 603, and 575), the DAG ions, subjected to MS^3, yielded product ions with unique structural information. From a single [M+NH$_4$]$^+$ ion, the identity of approximately 100 distinct molecular species (not counting stereoisomers and double-bond isomers) could be determined. Some species such as 16:0/18:1/18:1 or 16:0/18:0/18:2 would be predicted by the total carbons and double bonds present in the species and the relatively common nature of the fatty acids; however, the presence of these molecular species is directly supported by product ions from several MS^3 spectra. Other species were seen that would not be expected without mass spectral evidence. These included 17:1/18:1/17:0, which was observed in the MS^3 spectra of m/z 577 (loss of 18:1) and 591 (loss of 17:1), and 16:1/14:0/22:1, which was observed in the MS^3 spectra of m/z 605 (loss of 16:1) and 631 (loss of 14:0). Each of the molecular species in Table 1.2 was confirmed by ions in multiple spectra. Some species were repeated in Table 1.2 because their presence was suggested by ions in more than one MS^3 spectrum; however, species corresponding to 53:9 could not be found.

4.2. Quantitative analysis

A quantitative method for accurate determination of changes in molecular species components in the complex mixture of neutral lipids containing a number of isobaric species was developed utilizing the characteristic fragmentation pathways. Deuterium-labeled internal standards, whose masses fell into regions of the spectrum that did not contain ions corresponding to ammoniated DAG and TAG molecular species, afforded a means for normalization of neutral loss data for 18 different fatty acyl groups.

Neutral loss experiments were performed to follow changes in Kdo$_2$-lipid A–treated cells in terms of DAG and TAG molecular species that contained at least one of the 18 different fatty acyl groups present in the deuterium-labeled internal standards. Each neutral loss corresponding to one of the 18 fatty acyl groups in the deuterated internal standards (Table 1.3) was monitored for a period of 3 min, yielding a characteristic total ion current (Fig. 1.4). For each period in the experiment, a unique neutral loss spectrum was obtained, which revealed all of the [M+NH$_4$]$^+$ ions in the complex mixture that contained a single fatty acyl group as well as some indication of the abundance of this [M+NH$_4$]$^+$ ion relative to a deuterium-labeled internal standard (Fig. 1.5). For example, the presence of 18:1 esterified to DAG molecular species gave abundant ions at m/z 612, 638, and 668. The ion at m/z 612 would correspond to an 18:1/16:0 DAG, m/z 638 an 18:1/18:1 DAG, and m/z 668 an 18:1/20:0 DAG molecular species (Fig. 1.5A). The abundance of these molecular species compared to the internal standard at m/z 827.8 was monitored in a series of experiments

Table 1.2 Triacylglycerol molecular species identified from the MS/MS/MS (MS³) analysis of the three most abundant diacylglycerol product ions observed in the tandem mass spectrometry (MS/MS) (MS²) spectrum of the [M+NH₄]⁺ triacylglycerol observed at m/z 876 in the full mass spectrum of RAW 264.7 cells

MS² product ion[a]	Triacylglycerol[b]	MS³ ion type	MS² product ion[a]	Triacylglycerol[b]	MS³ ion type	MS² product ion[a]	Triacylglycerol[b]	MS³ ion type
577 (−18:1)	18:1/18:1/16:0	RCO	603 (−16:0)	16:0/18:1/18:1	RCO	575 (−18:0)	18:0/18:1/16:1	R′+74
	18:1/14:1/20:0	R′+74−H₂O		16:0/14:1/22:1	R′+74−H₂O		18:0/18:1/16:1	RCO
	18:1/18:1/16:0	R′+74		16:0/18:1/18:1	R′+74		18:0/14:1/20:1	R′+74−H₂O
	18:1/18:0/16:1	R′+74		16:0/18:1/18:1	RCO−H₂O		18:0/16:1/18:1	R′+74
	18:1/18:1/16:0	RCO−H₂O		16:0/18:1/18:1	R′+74−H₂O		18:0/16:1/18:1	RCO
	18:1/16:0/18:1	RCO		16:0/20:1/16:1	R′+74		18:0/12:1/22:1	R′+74−H₂O
	18:1/12:0/22:1	R′+74−H₂O		16:0/16:0/20:2	R′+74		18:0/16:0/18:2	R′+74
	18:1/22:1/12:0	RCO		16:0/18:2/18:0	R′+74−H₂O		18:0/16:1/18:1	RCO−H₂O
	18:1/17:3/18:5	R′+74		16:0/18:0/18:2	R′+74		18:0/18:1/16:1	RCO−H₂O
	18:1/18:1/16:0	R′+74−H₂O		16:0/18:2/18:0	R′+74		18:0/18:1/16:1	R′+74−H₂O
	18:1/16:1/18:0	RCO		16:0/19:0/17:2	R′+74		18:0/22:1/12:1	RCO
	18:1/12:1/22:0	R′+74−H₂O		16:0/17:2/19:0	R′+74		18:0/18:2/16:0	R′+74
	18:1/16:1/18:0	R′+74		16:0/18:0/18:2	R′+74−H₂O		18:0/19:0/15:2	R′+74−H₂O
	18:1/16:1/18:0			16:0/20:2/16:0			18:0/20:1/14:1	RCO

Composition	Product	Composition	Product	Composition	Product
18:1/20:1/14:0	RCO-H₂O, RCO				
18:1/16:1/18:0	R'+74-H₂O	16:0/18:2/18:0	R'+74-H₂O, RCO	18:0/16:1/18:1	R'+74-H₂O
18:1/19:0/15:1	R'+74-H₂O	16:0/19:0/17:2	RCO-H₂O	18:0/18:2/16:0	R'+74-H₂O
18:1/22:1/12:0	RCO-H₂O	16:0/18:2/18:0	RCO-H₂O	18:0/22:2/12:0	RCO
18:1/16:0/18:1	RCO-H₂O	16:0/20:1/16:1	RCO	18:0/18:2/16:0	RCO-H₂O
18:1/15:1/19:0	R'+74	16:0/16:1/20:1	R'+74-H₂O	18:0/19:0/15:2	RCO
18:1/14:0/20:1	R'+74	16:0/20:2/16:0	R'+74		
18:1/18:0/16:1	RCO	16:0/21:0/15:2	R'+74-H₂O	18:0/14:2/20:0	RCO-H₂O
18:1/14:0/20:1	R'+74-H₂O	16:0/20:2/16:0	RCO	18:0/14:0/20:2	R'+74-H₂O
18:1/20:0/14:1	RCO	16:0/21:0/15:2	RCO-H₂O	18:0/22:1/12:1	RCO-H₂O
18:1/16:0/18:1	R'+74-H₂O	16:0/16:2/20:0	R'+74-H₂O		
18:1/22:0/12:1	RCO	16:0/16:1/20:1	RCO		
18:1/18:0/16:1	R'+74-H₂O	16:0/16:1/20:1	R'+74		
		16:0/20:2/16:0	RCO-H₂O		
18:1/23:0/11:1	RCO-H₂O	16:0/14:0/22:2	R'+74		
18:1/15:1/19:0	RCO-H₂O	16:0/15:1/21:1	RCO-H₂O		
18:1/20:1/14:0		16:0/20:1/16:1			

(continued)

Table 1.2 (continued)

MS^2 product ion[a]	Triacylglycerol[b]	MS^3 ion type	MS^2 product ion[a]	Triacylglycerol[b]	MS^3 ion type	MS^2 product ion[a]	Triacylglycerol[b]	MS^3 ion type
					RCO-H_2O			
	18:1/19:0/15:1	RCO-H_2O		16:0/16:1/20:1	RCO-H_2O			
	18:1/14:1/20:0	RCO		16:0/15:1/21:1	RCO			
	18:1/15:1/19:0	RCO		16:0/20:1/16:1	R'+74-H_2O			
	18:1/12:0/22:1	R'+74		16:0/21:1/15:1	RCO-H_2O			
	18:1/13:1/21:0	RCO		16:0/19:1/17:1	RCO			
	18:1/17:1/17:0	RCO		16:0/15:1/21:1	R'+74-H_2O			
	18:1/15:1/19:0	R'+74-H_2O		16:0/13:0/23:2	R'+74			
	18:1/14:0/20:1	RCO-H_2O		16:0/17:1/19:1	RCO			
	18:1/18:0/16:1	RCO-H_2O						

[a] The neutral loss from m/z 876 to the MS^2 product ion revealed one of the fatty acyl groups in this $[M+NH_4]^+$ species in parentheses under the mass of the MS^2 product ion.

[b] Each TAG column has the first entry in the TAG molecular species abbreviation, by default, as the species that was lost as a neutral in MS^2. The second entry in the abbreviation is the acyl group indicated by the presence of the specified ion type in the MS^3 spectrum (column 3) and the third entry of the abbreviation was calculated from the mass difference of the two.

Analysis of Glyceryl Lipids

Table 1.3 [1,1,2,3,3-d5] Glycerol-labeled glyceryl lipids used as internal standards for neutral loss mass spectrometric analysis

Triacylglycerol fatty acyl composition	m/z [M+NH$_4$]$^+$	Diacylglycerol fatty acyl composition	m/z [M+NH$_4$]$^+$
19:0/12:0/19:0[a]	857.83	19:0/19:0	675.66
14:0/16:1/14:0	771.73	20:0/20:0	703.70
15:0/18:1/15:0	771.73	14:0/14:0	535.51
16:0/18:0/16:0	857.83	15:0/15:0	563.54
17:0/17:1/17:0	869.83	16:0/16:0	591.57
20:4/18:2/20:4	949.80	17:0/17:0	619.57
20:0/20:1/20:0	995.97	20:2/20:2	695.60
20:2/18:3/20:2	955.84	20:4/20:4	687.54
20:5/22:6/20:5	993.77	20:5/20:5	683.51

[a] The acyl order as indicated: sn-1/sn-2/sn-3.

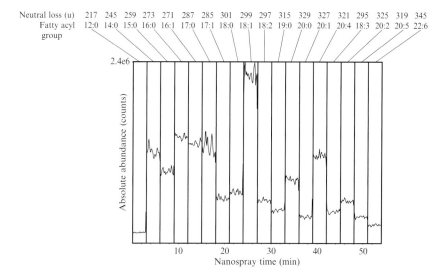

Figure 1.4 Total ion current from a neutral loss survey in the time course of glyceryl lipid changes upon activation of the Toll-4 receptor in RAW 264.7 cells. Each column represents a 3-min period in which the neutral loss of a different fatty acyl moiety from the diacylglycerol and triacylglycerol species was monitored.

covering the course of Kdo$_2$-lipid A stimulation, revealing changes in each DAG species taking place. However, it would not be possible to state relative abundance of 18:1/18:1 DAG and 18:1/16:0 DAG to each other because of the difference in neutral loss behavior due to structural features such as esterification position on the glycerol backbone (Li and Evans, 2006).

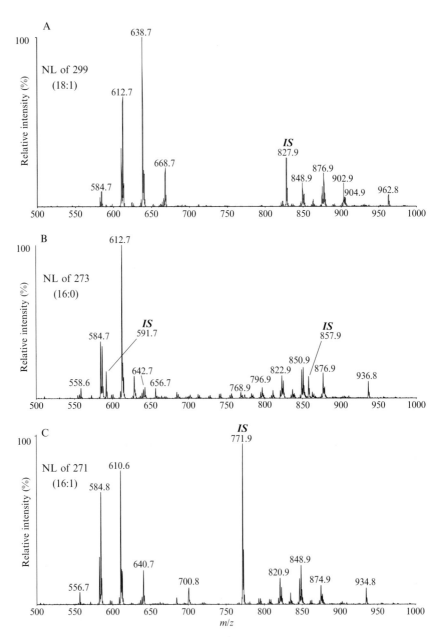

Figure 1.5 Spectra showing the [M+NH$_4$]$^+$ ions detected by the neutral loss of an 18:1, 16:0 or 16:1 fatty acyl group upon collisional activation. In this case, the diacylglycerol species can be uniquely identified, but the triacylglycerol molecular species can only be partially elucidated.

Analysis of Glyceryl Lipids 17

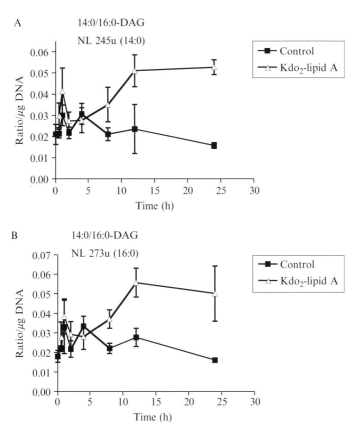

Figure 1.6 (A) Quantitative analysis showing the increase in abundance of the 14:0/16:0 diacylglycerol over time in Kdo_2-lipid A–treated cells compared to control measured from the neutral loss of 14:0. (B) Quantitative analysis showing the increase in abundance of the 14:0/16:0 diacylglycerol over time in Kdo_2-lipid A–treated cells compared to control measured from the neutral loss of 16:0.

Each neutral loss experiment revealed common and new molecular species, such as species containing 16:0 (Fig. 1.5B) and 16:1 (Fig. 1.5C). Thus, these spectra gave unique identification of DAG species, but only partial characterization of TAG species.

The Lipid Profiler software package was used to integrate the ion abundance for each m/z, to correct abundances due to ^{13}C isotopes, and to identify DAG and TAG ions and the internal standards in the averaged spectra from each period in the neutral loss survey. The corrected DAG and TAG abundances were normalized to the abundance of the internal standards.

Changes in the abundance of specific species compared to the internal standard over time in the control and Kdo_2-lipid A–treated cells were thereby assessed in a quantitative way (Fig. 1.6). This quantitative approach

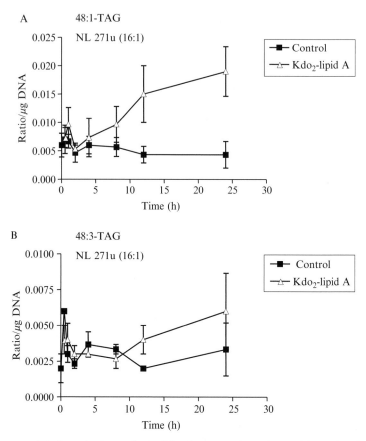

Figure 1.7 (A) Quantitative analysis of the changes in the 48:1 triacylglycerol molecular species which contains a 16:1 fatty acyl group. The abundance of this species changes in Kdo_2-lipid A–treated cells compared to controls. (B) Quantitative analysis of the changes in the 48:3 triacylglycerol molecular species that contains a 16:1 fatty acyl group. The abundance of this species in Kdo_2-lipid A–treated cells does not change compared to control.

revealed time-dependent changes in the 14:0/16:0 DAG molecular species in Kdo_2-lipid A–treated RAW cells, as measured by the neutral loss of 14:0 (Fig. 1.6A) or 16:0 (Fig. 1.6B). The abundance of 14:0/16:0 DAG in Kdo_2-lipid A–treated cells was greater than that in control samples, with very similar time dependence, regardless of which neutral loss was monitored.

The neutral loss of 16:1 was common to several TAG molecular species; however, in some cases, a significant increase was observed between the ratios of abundances for molecular species in treated and control samples at 24 hr, as is the case for the 48:1 TAG molecular species that contain 16:1 (Fig. 1.7A). For the 48:3 TAG containing 16:1 (Fig. 1.7B), though, the relative abundance was constant for treated cells compared to controls at 24 h.

5. CONCLUSIONS

The analysis of glyceryl lipids present within mammalian cells is very challenging. However, the power of combining MS^2 and MS^3 can be used to uniquely identify TAGs and DAGs and to quantitate molecular species with a specific total of fatty acyl carbon atoms and double bonds and containing a specific fatty acyl group from a very complex biological sample in the neutral lipid extract of RAW 264.7 macrophages. This approach can be used to precisely assess changes within populations of molecular species that cannot be determined by measurement of molecular ion species such as $[M+H]^+$, $[M+Li]^+$, or $[M+NH_4]^+$ abundances alone. However, these stable isotope-controlled MS^2 and MS^3 experiments do not provide molar concentration data and, in the case of TAG molecules, do not analyze the complex mixture to the extent of providing molecular species information, stereochemistry, or even positional analysis.

ACKNOWLEDGMENTS

This work was supported in part by the LIPID MAPS Large Scale Collaborative Grant from the National Institutes of Health (GM069338).

REFERENCES

Ahn, S. J., Costa, J., and Emanuel, J. R. (1996). PicoGreen quantitation of DNA: Effective evaluation of samples pre- or post-PCR. *Nucleic Acids Res.* **24,** 2623–2625.

Bligh, E. G., and Dyer, W. J. (1959). A rapid method of total lipid extraction and purification. *Can. J. Biochem. Physiol.* **37,** 911–917.

Byrdwell, W. C., Emken, E. A., Neff, W. E., and Adlof, R. O. (1996). Quantitative analysis of triglycerides using atmospheric pressure chemical ionization-mass spectrometry. *Lipids* **31,** 919–935.

Byrdwell, W. C., and Neff, W. E. (2002). Dual parallel electrospray ionization and atmospheric pressure chemical ionization mass spectrometry (MS), MS/MS, and MS/MS/MS for the analysis of triacylglycerols and triacylglycerol oxidation products. *Rapid Commun. Mass Spectrom.* **16,** 300–319.

Cheng, C., Gross, M. L., and Pittenauer, E. (1998). Complete structural elucidation of triacylglycerols by tandem sector mass spectrometry. *Anal. Chem.* **70,** 4417–4426.

Ejsing, C. S., Duchoslav, E., Sampaio, J., Simons, K., Bonner, R., Thiele, C., Ekroos, K., and Shevchenko, A. (2006). Automated identification and quantification of glycerophospholipid molecular species by multiple precursor ion scanning. *Anal. Chem.* **78,** 6202–6214.

Han, X., and Gross, R. W. (2001). Quantitative analysis and molecular species fingerprinting of triacylglyceride molecular species directly from lipid extracts of biological samples by electrospray ionization tandem mass spectrometry. *Anal. Biochem.* **295,** 88–100.

Hsu, F. F., and Turk, J. (1999). Structural characterization of triacylglycerols as lithiated adducts by electrospray ionization mass spectrometry using low-energy collisionally induced dissociation on a triple stage quadrupole instrument. *J. Am. Soc. Mass Spectrom.* **10,** 587–599.

Ingalls, S. T., Kriaris, M. S., Xu, Y., DeWulf, D. W., Tserng, K. Y., and Hoppel, C. L. (1993). Method for isolation of non-esterified fatty acids and several other classes of plasma lipids by column chromatography on silica gel. *J. Chromatogr.* **619,** 9–19.

Kaluzny, M. A., Duncan, L. A., Merritt, M. V., and Epps, D. E. (1985). Rapid separation and purification of lipid classes in high yield and purity using bonded phase columns. *J. Lipid Res.* **26,** 135–140.

Li, X., Collins, E. J., and Evans, J. J. (2006). Examining the collision-induced decomposition spectra of ammoniated triglycerides as a function of fatty acid chain length and degree of unsaturation. II. The PXP/YPY series. *Rapid Commun. Mass Spectrom.* **20,** 171–177.

McAnoy, A. M., Wu, C. C., and Murphy, R. C. (2005). Direct qualitative analysis of triacylglycerols by electrospray mass spectrometry using a linear ion trap. *J. Am. Soc. Mass Spectrom.* **16,** 1498–1509.

Myher, J. J., Kuksis, A., Marai, L., and Sandra, P. (1988). Identification of the more complex triacylglycerols in bovine milk fat by gas chromatography-mass spectrometry using polar capillary columns. *J. Chromatogr.* **452,** 93–118.

Phillips, F. C., Erdahl, W. L., Schmidt, J. A., and Privett, O. S. (1984). Quantitative analysis of triglyceride species of vegetable oils by high performance liquid chromatography via a flame ionization detector. *Lipids* **19,** 880–887.

Raetz, C. R. H. (1990). Biochemistry of endotoxins. *Ann. Rev. Biochem.* **59,** 129–170.

Raetz, C. R., Garrett, T. A., Reynolds, C. M., Shaw, W. A., Moore, J. D., Smith, D. C., Jr, Ribeiro, A. A., Murphy, R. C., Ulevitch, R. J., Fearns, C., Reichart, D., Glass, C. K., *et al.* (2006). Kdo_2-Lipid A of *Escherichia coli*, a defined endotoxin that activates macrophages via TLR-4. *J. Lipid Res.* **47,** 1097–1111.

CHAPTER TWO

Glycerophospholipid Identification and Quantitation by Electrospray Ionization Mass Spectrometry

Pavlina T. Ivanova, Stephen B. Milne, Mark O. Byrne, Yun Xiang, and H. Alex Brown

Contents

1. Introduction	22
2. Nomenclature	25
3. Mass Spectrometry	26
4. General Strategy for Phospholipid Isolation and Mass Spectral Analysis	27
5. Extraction and Mass Spectral Analysis of Global Glycerophospholipids	28
5.1. Phospholipid extraction from cultured cells	29
5.2. Phospholipid extraction from tissue	30
5.3. Direct infusion mass spectrometry of phospholipid extracts	30
5.4. LC-MS analysis (quantitation) of phospholipid extracts	33
6. Polyphosphoinositide Extraction and Mass Spectral Analysis	35
6.1. Extraction of polyphosphoinositides from cultured cells	37
6.2. Extraction of polyphosphoinositides from tissue	37
6.3. Direct-infusion mass spectral analysis of polyphosphoinositides	38
6.4. Deacylation of GPInsPn lipids	39
6.5. LC-MS analysis of deacylated GPInsPn compounds	39
7. Computational Analysis of Mass Spectral Data	41
7.1. Direct infusion (intra-source separation)	43
7.2. LC-MS data analysis	47
Acknowledgments	54
References	54

Departments of Pharmacology and Chemistry, Vanderbilt University School of Medicine, Nashville, Tennessee

Abstract

Glycerophospholipids are the structural building blocks of the cellular membrane. In addition to creating a protective barrier around the cell, lipids are precursors of intracellular signaling molecules that modulate membrane trafficking and are involved in transmembrane signal transduction. Phospholipids are also increasingly recognized as important participants in the regulation and control of cellular function and disease. Analysis and characterization of lipid species by mass spectrometry (MS) have evolved and advanced with improvements in instrumentation and technology. Key advances, including the development of "soft" ionization techniques for MS such as electrospray ionization (ESI), matrix-assisted laser desorption/ionization (MALDI), and tandem mass spectrometry (MS/MS), have facilitated the analysis of complex lipid mixtures by overcoming the earlier limitations. ESI-MS has become the technique of choice for the analysis of multi-component mixtures of lipids from biological samples due to its exceptional sensitivity and capacity for high throughput.

This chapter covers qualitative and quantitative MS methods used for the elucidation of glycerophospholipid identity and quantity in cell or tissue extracts. Sections are included on the extraction, MS analysis, and data analysis of glycerophospholipids and polyphosphoinositides.

1. INTRODUCTION

Lipids are defined as a wide variety of biological molecules demonstrating different structure and function. It is commonly accepted that lipids are molecules highly soluble in organic solvents; however, many classes (such as the highly polar polyphosphoinositides and molecules with large hydrophilic domains like lipopolysaccharide) exist that are not soluble in these solvents (Dowhan, 1997; Dowhan and Bogdanov, 2002). Their primary role has been thought of as a central component of a semipermeable membrane. Multiple functions and adaptability of cellular membranes require a wide spectrum of lipids to support the necessary environment. The physical and chemical properties of the membranes directly affect the cellular processes, making the role of the lipids dynamic rather than just a simple inert barrier. Additionally, phospholipids are linked to important physiological processes such as bioenergetics, signal transduction across the cell membrane, and cellular recognition. Upon activation of a number of enzymes (i.e., phospholipases, kinases, and phosphatases), membrane phospholipids generate signaling lipids. Their formation is generally initialized by a ligand binding to a cell surface receptor leading to activation of phospholipases or kinases. The synthesized lipid second messengers act on specific proteins to manipulate cell function. Signaling lipids are quickly metabolized to limit the response or to initiate another signaling cascade.

The major lipid constituents of most membranes are phosphate-containing glycerol-based lipids generally referred to as glycerophospholipids. Phospholipids are grouped into classes depending on the "headgroup" identity. The major classes (Fig. 2.1) of phospholipids found in a mammalian cell membrane include glycerophosphatidic acid (GPA), glycerophosphocholine (GPCho), glycerophosphoethanolamine (GPEtn), glycerophosphoinositol (GPIns), glycerophosphoglycerol (GPGro), and glycerophosphoserine (GPSer). A multitude of molecular species exist within each class containing different combinations of fatty acids in the sn-1 and sn-2 position of the glycerol backbone. In addition, fatty acids at sn-1 position can be substituted for ether or vinyl ether moiety (plasmanyl and plasmenyl glycerophospholipids, respectively) in some of the classes.

Comprehensive analysis of phospholipid molecular species has been challenging owing to their large number and diversity, complicating their separation and identification. A liquid–liquid extraction of cellular material results in heterogeneous mixtures of phospholipids, which can be further separated by selective use of organic solvents. Class separation generally involves thin-layer chromatography (TLC), high-performance liquid chromatography (HPLC), and gas chromatography (GC), any of which utilizes known standards and sometimes requires prior derivatization. With the advent of gas chromatography and gas chromatography-MS (GC-MS), class separation by TLC followed by hydrolysis and derivatization made possible the identification of individual fatty acid species. The conventional method is not only time consuming but also requires large amounts of lipids. The analysis of intact polar phospholipids became possible only after development of fast atom bombardment mass spectrometry (FAB-MS). More recently, the introduction of "soft" ESI-MS has greatly simplified the procedure for lipid analysis. This ionization process results in decreased molecular ion decomposition, better reproducibility, and lower detection limits compared to FAB-MS. The great potential of ESI-MS for characterization of phospholipids and other polar lipids has been described in other publications (Brügger *et al.*, 1997; Fridriksson *et al.*, 1999; Han and Gross, 1994). We have expanded the use of direct infusion ESI-MS for identification of over 600 phospholipid species from an unprocessed total lipid extract. Changes in the cellular concentration of diverse lipids can be determined by analysis of the mass spectra via statistical algorithms. The generated lipid arrays represent a qualitative map of molecular species changes after challenge with a biological stimulant (Ivanova *et al.*, 2004). The quantitative analysis of the myriad of phospholipid species belonging to different phospholipid classes involves a chromatographic separation by class to avoid the possible mass overlap occurring during direct infusion analysis. Thus, a class separation by liquid chromatography (LC) followed by MS detection for species identification is very important in the analysis of glycerophospholipids extracted from a complex biological matrix (DeLong

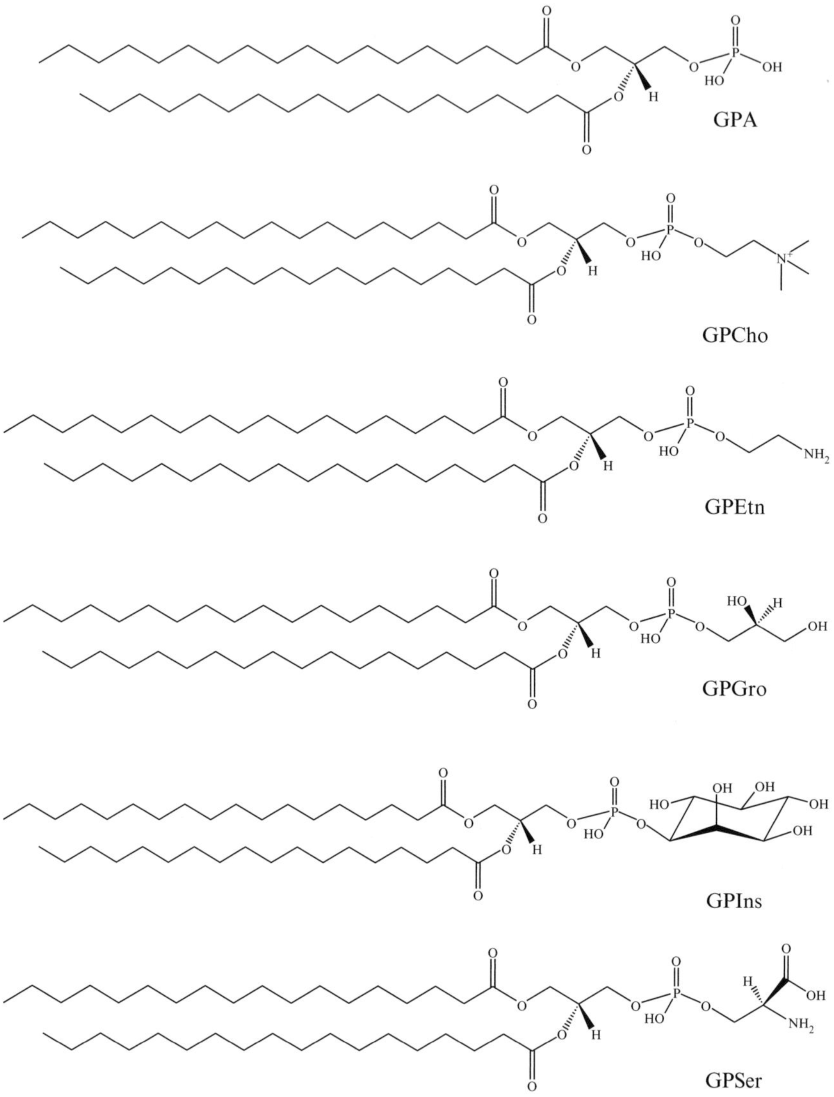

Figure 2.1 Major glycerophospholipid classes. Phospholipids consist of a glycerol backbone, two acyl moieties, and a phosphodiester headgroup. The variety in length and degree of unsaturation of the acyl chains creates a wide spectrum of species within each class. Ethers or vinyl ethers can substitute the *sn*-1 position and generate plasmanyl or plasmenyl species in some classes.

et al., 2001; Lesnefsky *et al.*, 2000; Taguchi *et al.*, 2000; Wang *et al.*, 2004). The importance of phospholipids in cell signaling and organ physiology as well as their association with many diseases that implicate lipid metabolic pathways and enzyme disruption requires novel methodological approaches and development of new technology for their analysis. Analyzing the complex lipid extracts from biological sources and obtaining a lipid profile will give information on the changes in lipid species composition as a result of the biological condition (stimulation by receptor activation; progress of a disease or comparison between normal and transformed cells).

In this chapter, we present a description of the lipidomics technology and its application for identification and quantitation of the major glycerophospholipid classes. Consistent with the spirit of *Methods in Enzymology* volumes, this chapter was designed for readers who may not have extensive experience with MS.

2. Nomenclature

Classification of lipids has been difficult to define due to their diverse structures. Although they are derived from similar biological precursors and have similar physical and chemical properties, lipid classes diverge based on the chemical entities they possess. A categorization of a large spectrum of biologically relevant lipids has been proposed recently by Fahy *et al.* (2005). The simplest class of lipids is the fatty acids. Oxidation of long-chain fatty acids is one of the main sources of energy for mammals. Another group of biologically active lipids used as fuels and in many metabolic processes are the sterols (steroid hormones, vitamins, bile acids). Three other groups are closely related chemically and include neutral lipids (glycerolipids), glycerophospholipids, and sphingolipids. Glycerophospholipids are defined by the presence of a phosphate group esterified to one of the glycerol hydroxyl groups. Naming of the glycerophospholipids involves a stereo-specific numbering (sn) (Hirschmann, 1960; IUPAC-IUB CBN, 1967), and the glycerol backbone is typically acylated or alkylated at the sn-1 and/or sn-2 position (Fig. 2.2). These substituents are generally expressed with the term "radyl." The double bond geometry of the radyl moieties is described with E/Z designation (instead of *trans/cis*), and the lack of one radyl group is denoted as "lyso" glycerophospholipid.

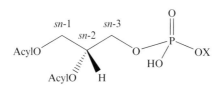

Figure 2.2 Stereo-specific number system for glycerophospholipids (X is the headgroup).

3. MASS SPECTROMETRY

Mass spectrometry is a unique analytical technique, relying on the separation of ions according to both their electrical charge and their total atomic mass by magnetic and electrical fields. A mass spectrometer typically consists of four important components: inlet (for sample introduction), ion source, mass analyzer, and ion detector. The analyte solution is sprayed through a very thin capillary, and the application of a strong electric field generates highly charged droplets. They pass through a heated inert gas in the mass analyzer and separate according to their mass-to-charge ratio (m/z), at which point they are recorded by a detection system. The very high vacuum in the instrument (10^{-5} to 10^{-7} Torr) increases the free path of the ions and ensures mostly monomolecular reactions, which can disclose structural information of the analyzed compound. MS data are largely dependent on the sample introduction and method of ionization, while the type of mass analyzer does not usually alter the observed chemical reactions.

Despite the fact that lipid analysis was one of the first applications of the mass spectrograph after the founding studies by Sir J. J. Thompson and F. W. Aston, the routine analysis of lipids was hindered by their tendency to undergo fragmentation during the process (Fenwick, 1983). The development of FAB-MS allowed the analysis of natural phospholipids; however, along with the generated abundant molecular (i.e., unfragmented) ions, fragment ions were observed, providing some structural information and revealing the great potential of MS/MS (i.e., analysis of molecular and fragment ions generated thereof) for the structural characterization of lipids (Murphy and Harrison, 1994). Thus, the highly attractive MS analysis of phospholipids was rarely used until the introduction of "soft" ionization methods such as MALDI and ESI. "Soft" ionization does not cause extensive fragmentation, meaning that comprehensive detection of an entire range of phospholipids within a complex mixture can be correlated to experimental conditions or disease states. Various ESI-MS methods have been developed for analysis of different classes, subclasses, and individual lipid species from biological extracts. Comprehensive reviews of the methods and their application have recently been published (Murphy *et al.*, 2001; Pulfer and Murphy, 2003; Watson, 2006; Wenk, 2005). The major advantages of ESI-MS are high accuracy, sensitivity, reproducibility, and the applicability of the technique to complex phospholipid solutions without prior derivatization.

ESI-MS was initially developed by Fenn *et al.* (1989) for analysis of biomolecules. It depends on the formation of gaseous ions from polar, thermally labile, and mostly nonvolatile molecules, and thus is completely suitable for phospholipids. The principle of ionization involves a sample in solution that is passed through a very thin capillary or needle at a slow rate (1–300 μl/min). The application of a strong electric field generates significant charge at the end of the capillary that produces a fine spray of highly charged droplets. These droplets then pass through a heated inert gas for desolvation prior to MS analysis of individual ionic species.

Electrospray ion sources can be combined with various mass analyzers such as ion trap (IT), quadrupole, time of flight (TOF), and Fourier transform ion cyclotron (FT-ICR). They all differ in their mass accuracy (the error in the exact mass determination compared to theoretical value) and resolution (the value of mass m divided by the mass difference Δm between two ion profiles with a small mass difference). Today, the most widely used mass spectrometers with ESI sources are triple quadrupole (TQ), quadrupole ion trap (QIT), IT, and quadrupole TOF (Q-TOF) instruments.

The "soft" ESI causes little or no fragmentation, and virtually all phospholipid species are detected as molecular ion species. The identification and structural information about a certain peak is acquired by fragmentation (MS/MS). The ion of interest is subjected to collision-induced dissociation by interaction with a collision gas. In tandem mass instruments, the first mass analyzer (Q1) is used for the selection of the ion of interest, which is then fragmented in the collision cell (Q2), and the second mass analyzer (Q3) is used to separate the fragment ions on the basis of their m/z values, thus creating a product or "daughter" spectrum that provides structural information.

4. General Strategy for Phospholipid Isolation and Mass Spectral Analysis

The analysis and structural identification of glycerophospholipids from cell extracts or tissue biopsies is outlined in Fig. 2.3. Global (GPA, GPCho, GPEtn, GPGro, GPIns, and GPSer) glycerophospholipids are extracted by a modified Bligh and Dyer extraction. The samples are subjected to direct infusion MS, where qualitative changes in lipid species are documented by the construction of lipidomic arrays (Forrester *et al.*, 2004), as well as "ratiomics analysis"—comparison of peak intensities of various species in the same class (Rouzer *et al.*, 2006). Direct infusion is also used in MS/MS for species identification and structural confirmation (acyl chain and headgroup information). Another part of the obtained glycerophospholipid extract is put through LC-ESI-MS analysis, which results in a class separation and quantification of the individual glycerophospholipid species.

Polyphosphoinositides are extracted by a different procedure tailored to separate them from the other phospholipids. The species identification and acyl chain composition of the $GPInsP_n$ is achieved again by direct infusion and MS/MS, while the phosphate headgroup stereochemistry and quantification

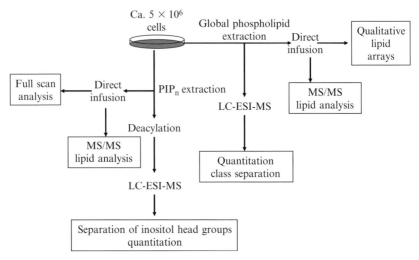

Figure 2.3 General strategy for the extraction and analysis of glycerophospholipids from cell cultures or tissue. Glycerophospholipids or polyphosphoinositides are extracted according to respective procedures and subjected to direct infusion and tandem mass spectrometry analysis for species identification and qualitative lipid arrays construction for lipid changes evaluation between different conditions. Part or all of the same extract is analyzed by liquid chromatography-electrospray ionization-mass spectrometry glycerophospholipid analysis, which results in class separation and quantification. The quantitative analysis of polyphosphoinositides involves deacylation and liquid chromatography-electrospray ionization-mass spectrometry analysis of the glycerophosphoinositides for headgroup regioisomer identification.

can be evaluated by deacylation and LC-ESI-MS analysis of the resulting glycerophosphoinositides. Both LC-ESI-MS analyses utilize chemically defined standards and HPLC grade or higher solvents.

5. EXTRACTION AND MASS SPECTRAL ANALYSIS OF GLOBAL GLYCEROPHOSPHOLIPIDS

Phospholipid analysis begins with their isolation from cell culture or tissues. The use of organic solvents for extraction facilitates the removal of nonlipid components. By use of different organic solvents, the extractions can be modified and tailored for a specific class of lipids (e.g., phospholipids, sphingolipids, lysophospholipids, and polyphosphoinositides) (Christie, 2003). Special precautions should be taken to ensure the deactivation of enzymes and the completeness of lipid recovery. The nonlipid contaminants can be eliminated by washing the lipid-containing layer after phase separation. It is very important to minimize the risk of oxidation of the polyunsaturated fatty acids or lipid hydrolysis during the process of isolation.

Therefore, the extraction of lipids is always rapidly undertaken at low temperature (4°) as soon as possible after removal of the tissue from the living organism or from a cell culture, after the reactions have stopped.

Two structural features of phospholipids are the nonpolar hydrocarbon chains of the fatty acids and the polar phosphate-containing headgroups. The combination of polar and nonpolar groups within the molecule affects the solubility in organic solvent and, thus, their extraction. Hence, the use of a single organic solvent is not suitable for all species. The ability of phospholipids to swell in water and the likely interaction between lipids and proteins indicate that water is an important participant in the extraction process.

Historically, the most widely used extraction solvents were ethanol, as in diethyl ether (3:1) at 55 to 60° for several hours according to the method described by Bloor (1928). However, the solvent led to lipid peroxidation in animal samples and increased the enzymatic reactions on phospholipids during plant extraction. Considering the disadvantages of this solvent system, ethanol-diethyl ether mixtures have been substituted by more efficient solvents. The chemistry of the phospholipid molecules requires the presence of a more polar solvent, such as alcohol, and ultimately a nonpolar solvent, such as chloroform, for a complete extraction. The mixture of chloroform and methanol in various ratios is the most efficient extraction mixture for phospholipids. This approach was developed in the late 1950s by Folch *et al.* (1951) and uses chloroform:methanol in ratio 2:1 and large volumes of water for washing out the nonlipid components. Although the extraction procedure was very efficient, rapid, and conducted at room temperature (or lower), the formation of emulsions was a major drawback. Despite the modification Folch *et al.* made later by including salt in the media, this procedure has been replaced by the most widely adapted method for lipid extraction still in use today. The method of Bligh and Dyer (1959) was originally designed for the extraction of lipids from fish muscle. This method was advantageous for tissues containing a high percentage of water. The method is a variation of Folch's extraction, and calculates the amount of water present in the sample so that the final composition of chloroform: methanol:water is 1:2:0.8, creating a single extraction phase. The whole extraction is very rapid and most efficient. After addition of equal volumes of water and chloroform, the lipids are recovered in the chloroform-rich lower phase, which is separated and rinsed with water/methanol.

5.1. Phospholipid extraction from cultured cells

As an example of our lipidomics approach for phospholipid analysis by ESI-MS, we use RAW 264.7 cells. Phospholipids were extracted using a modified Bligh and Dyer procedure. The method is suitable for extraction from cell culture plates (100 mm) after aspirating the medium and washing the adhered cells twice with 5 ml of ice-cold 1X phosphate-buffered saline (PBS). Approximately 1×10^7 cells are then scraped using 800 µl of cold 0.1 N HCl:CH_3OH (1:1) and transferred into cold 1.5-ml microfuge tubes (# L292351, Laboratory Product Sales, Rochester, NY). Other tubes can be used, but should be checked with pure solvent prior to usage to ensure that impurities (plastic stabilizers, etc.) are not extracted from the plastic. After addition of 400 µl of cold $CHCl_3$, the extraction proceeds with vortexing (1 min) and centrifugation (5 min, 4°, 18,000×g). The lower organic phase is then isolated and solvent evaporated (Labconco Centrivap Concentrator, Kansas City, MO). The resulting lipid film is rapidly reconstituted in 80 µl CH_3OH:$CHCl_3$ (9:1). Prior to analysis, 1 µl of NH_4OH (18M) is added to each sample to ensure protonation of lipid species.

This method is also appropriate for extraction of phospholipids from previously isolated cell pellets while observing the restrictions for working fast and on ice at all times. After washing the adhered cells (from a 100-mm plate) twice with 5-ml, ice-cold 1X PBS, cells are scraped in 1 ml of 1X PBS, centrifuged (600×g, 4°, 5 min), and, after aspirating off the PBS, quickly frozen in liquid nitrogen in the event of transportation or extraction. The extraction of cell pellets follows the same procedure as described above by vortexing the pellet with 800 µl of cold 0.1 N HCl:CH_3OH (1:1) and adding 400 µl of $CHCl_3$ for phospholipid extraction.

5.2. Phospholipid extraction from tissue

Samples from tissue biopsies (20 to 50 mg) can also be extracted by this method. In this instance, the samples are quickly frozen by immersion in liquid nitrogen (stored at −80°, if not extracted immediately). The frozen samples are then placed in a tight-fit glass homogenizer (Kimble/Kontes Glass Co, Vineland, NJ), 800 µl of cold 0.1 N HCl:CH_3OH (1:1) is added, and the sample is homogenized for about 1 min while working on ice. The suspension is transferred to a cold microfuge tube, 400 µl of ice-cold $CHCl_3$ is added, and the extraction proceeds as previously described. Care should be taken in using individual (or cleaned in between) homogenizers to prevent sample cross-contamination.

5.3. Direct infusion mass spectrometry of phospholipid extracts

The characterization of phospholipids from an unprocessed total lipid extract (i.e., extract whose compounds have not been derivatized) by ESI-MS is based on the ability of each lipid class to acquire positive or negative charges when in solution during ESI. Thus, under suitable conditions of sample preparation, all molecular species that exist among cellular lipid classes can be detected in a single run of a total lipid extract. A single molecular ion with an m/z characteristic for the monoisotopic molecular weight is present for each molecular species. Collision-induced dissociation of the peaks of interest yields fragmentation patterns, which are used to unambiguously identify the lipid(s) present at a particular m/z value. A decade of dedicated research has been accomplished by a number of groups showing the potential of ESI-MS of lipids as a direct method to evaluate alterations in the cellular lipome (Ekroos et al., 2002; Han and Gross, 1994, 2003; Kerwin et al., 1994; Milne et al., 2006; Murphy et al., 2001; Ramanadham et al., 1998; Schwudke et al., 2006).

Typically, cellular lipid profiles under two experimental conditions (basal and stimulated, or wild type and mutant) can be monitored over a given time course, and the samples analyzed in both positive and negative ionization mode.

Three major lipid classes can be detected in positive ESI mode: GPCho, GPSer, and GPEtn (Ivanova et al., 2001; Milne et al., 2003). Fragmentation of the choline-containing GPCho results in a characteristic m/z 184 phosphocholine headgroup peak and a $[M+H-59]^+$ peak corresponding to the neutral loss of $(CH_3)_3N$. Relatively small peaks corresponding to the loss of one of the fatty acyl substituents as a ketene $[M-R_2CH=C=O]^+$ (or so-called lyso GPCho) and lyso GPCho-H_2O are also detected. In addition to the diacyl GPCho compounds, a large number of plasmanyl and plasmenyl phosphocholines can also be identified. All together, over 100 phosphatidylcholine lipids have been identified in RAW 264.7 cell extracts. GPSer fragmentation in positive mode produces ions resulting from the neutral loss of the polar headgroup phosphoserine ($[M+H-185]$). Fragmentation of phosphatidylethanolamines and lyso phosphatidylethanolamines in positive mode normally yields one peak, a $[M+H-141]^+$ ion from the neutral loss of the phosphoethanolamine headgroup. Again, plasmanyl and plasmenyl lipids were a large portion of the over 130 GPEtn species identified to date.

Six major lipid classes can be detected in negative ESI mode: GPIns, GPSer, GPGro, GPA, GPEtn, and chloride adducts of GPCho (note that chloride is from decomposition of $CHCl_3$). Additionally, the lyso variants for six of these phospholipids can also be detected in this mode. Negative mode fragmentation of these species yielded a wealth of structural information (Table 2.1). In each case, head group fragmentation, lyso-lipid formation, and fatty acid fragments aided in the lipid identification process (Hsu and Turk, 2000a,b). Phosphatidylinositol fragmentation can generate a wide variety of product ("daughter") ions (Hsu and Turk, 2000c). Four types of lyso-phosphatidic acid and lyso phosphatidylinositols and five characteristic head group fragments can routinely be used in identifying the observed GPIns and lyso GPIns species. In a similar fashion, GPSer and lyso GPSer compounds can be identified from their phosphatidic ($[M-H-Y]^-$) and lyso phosphatidic acid ($[M-H-Y-RCH=C=O]^-$) fragments (Han and Gross, 1995; Larsen et al., 2001; Smith et al., 1995). The product ion spectra of

Table 2.1 Negative product ("daughter") ions routinely observed from fragmentation of glycerophospholipids

	GPA	GPCho (Cl)	GPEtn	GPGro	GPIns	GPSer
[M−H]⁻	X		X	X	X	X
[M−H−RCO=C=O]⁻	X		X	X	X	
[M−H−RCOOH]⁻	X		X	X	X	
[M−H−RCOOH]⁻					X	X
[M−H−headgroup]⁻			X	X	X	X
[M−H−headgroup−RCH=C=O]⁻			X		X	X
[M−H−headgroup−RCOOH]⁻			X		X	
[M−H−RCHCO]⁻			X	X	X	
[M−H−RCHCO−2H₂O]⁻					X	
[RCOO]⁻	X	X	X	X	X	X
Head-group-specific ion	135		140, 196	171	241, 223	
[GP−H₂O−H]⁻ (153)	X	X	X	X	X	X
[H₂PO₄]⁻	X	X	X	X	X	X
[PO₃]⁻	X	X	X	X	X	X
[M+35]⁻ [M+3]⁻ (M+Cl)		X				
[M+CL−CH3]⁻ (M−50) (M−52)		X				

GPGro also contain an ion formed by the loss of sn-2 substituent R$_2$COOH from its phosphatidic acid ion [M-74] (Hsu and Turk, 2001). Fully annotated MS/MS spectra for examples from all six major glycerophospholipid classes are available on the LIPID MAPS public website http://www.lipidmaps.org/data/standards/standards.php?lipidclass=LMGP.

Mass spectral analysis was performed on a Finnigan TSQ Quantum triple quadrupole mass spectrometer (ThermoFinnigan, San Jose, CA) equipped with a Harvard Apparatus syringe pump (Harvard Apparatus, Holliston, MA) and an electrospray source. Samples were analyzed at an infusion rate of 10 µl/min in both positive and negative ionization modes over the range of m/z 400 to 1200. Instrument parameters were optimized with 1, 2-dioctanoyl-sn-Glycero-3-phosphoethanolamine (16:0 GPEtn) prior to analysis. Examples of full-scan spectra in negative and positive instrument modes are shown in Fig. 2.4. Data were collected with the Xcalibur software package (ThermoFinnigan) and analyzed by a software program developed in our research group (see "Computational Analysis of Mass Spectral Data" section for a detailed description of data analysis).

5.4. LC-MS analysis (quantitation) of phospholipid extracts

Quantification of glycerophospholipids is achieved by the use of a LC-MS technique employing synthetic (non-naturally occurring) diacyl (Table 2.2) and lysophospholipid standards (Table 2.3). Typically, 200 ng of each odd-carbon standard is added per 10^7 cells. The extraction of lipids from cell culture or tissue biopsies is the same as described previously for the direct infusion MS analysis (see "Extraction of Glycerophospholipids" section). After solvent evaporation, the resulting lipid film is dissolved in 100 µl of Isopropanol (IPA):Hexane:100 mM NH$_4$CO$_2$H$_{(aq)}$ 58:40:2 (mobile phase A). For our examples, we utilized an Applied Biosystems/MDS SCIEX 4000 Q TRAP hybrid triple quadrupole/linear ion trap mass spectrometer (Applied Biosystems, Foster City, CA). Coupled to this instrument were a Shimadzu (Shimadzu Scientific Instruments, Inc., Columbia, MD) HPLC system consisting of a SCL 10 AVP controller, two LC 10 ADVP pumps and a CTC HTC PAL autosampler (Leap Technologies, Carrboro, NC). All samples were separated on a Phenomenex (Phenomenex, Torrance, CA) Luna Silica column (2 × 250 mm, 5-µ particle size) using a 20-µl sample injection. Lipids were separated using a binary gradient program consisting of IPA:Hexane:100 mM NH$_4$CO$_2$H$_{(aq)}$ 58:40:2 (mobile phase A) and IPA:Hexane:100 mM NH$_4$CO$_2$H$_{(aq)}$ 50:40:10 (mobile phase B). The following LC gradient was used: 0 to 5 min, B = 50%; 5 to 30 min, B = 50 to 100%; 30 to 40 min, B = 100%; 40 to 41 min, B = 100 to 50%; and 41 to 50 min, B = 50%. The mobile phase was delivered at a flow rate of 0.3 ml/min. The MS spectra were acquired in negative instrument mode using a turbo spray source operated at 450° with an ion voltage of −3500 V, and nitrogen as

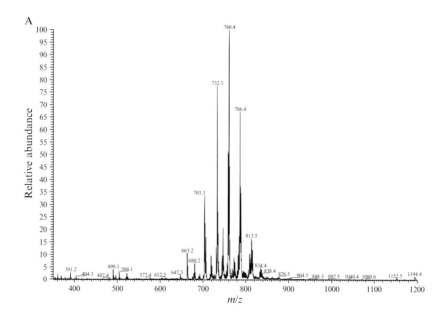

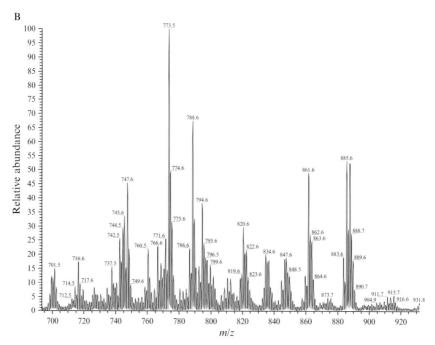

Figure 2.4 Examples of glycerophospholipid direct infusion spectra. (A) Positive instrument mode spectra showing GPCho, GPEtn, and GPSer lipids. (B) Negative instrument mode showing GPA, GPCho (Cl adducts), GPEtn, GPGro, GPIns, and GPSer lipid species.

Phospholipid Mass Spectrometric Analysis   35

Table 2.2 Odd-carbon standards for liquid chromatography-mass spectometry analysis of phospholipids: Twenty-four odd-carbon diacyl standards

	12:0/ 13:0	17:0/14:1 (9Z)	17:0/20:4 (5Z,8Z,11Z,14Z)	21:0/22:6 (4Z,7Z,10Z,13Z,16Z,19Z)
GPA	25:0	31:1	37:4	43:6
GPCho	25:0	31:1	37:4	43:6
GPEtn	25:0	31:1	37:4	43:6
GPGro	25:0	31:1	37:4	43:6
GPIns	25:0	31:1	37:4	43:6
GPSer	25:0	31:1	37:4	43:6

Table 2.3 Odd-carbon standards for liquid chromatography-mass spectrometry analysis of phospholipids: Four lysolipid standards

13:0/0:0	17:1 (9Z)/0:0
13:0 LysoGPA	17:1 LysoGPA
13:0 LysoGPCho	17:1 LysoGPCho

curtain and nebulizer gas. The curtain gas (CUR) was 30 l/hr, and ion source gas 1 and 2 were both 50 l/hr. The declustering potential (DP) was −110 V, and the collision energy (CE) was −5 V. Scan type: EMS, unit resolution for Q1; scan rate: 1000 amu/s; scan range from m/z 350 to 1200, with the ion trap set for dynamic fill time. As an example of this technique, extracted ion chromatograms (XICs) of the four GPGro odd-carbon standards in a RAW 264.7 cell background are shown in Fig. 2.5. The creation of standard curves and quantitation of data will be described in the "LC-MS Data Analysis" section at the end of this chapter.

6. POLYPHOSPHOINOSITIDE EXTRACTION AND MASS SPECTRAL ANALYSIS

Polyphosphoinositides are low-abundance phospholipids in eukaryotic cell membranes that are involved in regulation of distinct cellular processes. In order to minimize the interference of other phospholipids during analysis, a selective two-step extraction is employed. First, the cell material is extracted with neutral solvents and the resulting pellet is extracted with acidified solvents for quantitative recovery of polyphosphoinositides. The majority of noninositol phospholipids are extracted with neutral solvents, and no GPInsP_2 or GPInsP_3 species are detected in this extract. The acyl composition of

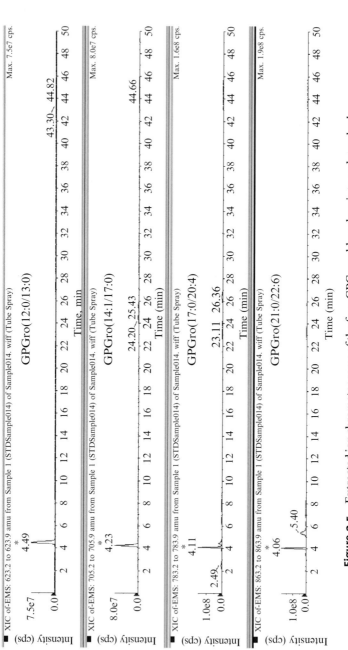

Figure 2-5 Extracted ion chromatograms of the four GPGro odd-carbon internal standards.

polyphoshoinositides is determined by direct infusion and fragmentation MS. The quantification and analysis of headgroup regiosomers is accomplished by LC-MS analysis after deacylation.

6.1. Extraction of polyphosphoinositides from cultured cells

Cells for polyphosphoinositide analysis must be between 60 to 70% confluency at the time of harvesting, unless the experiment specifically requires otherwise. Systematic effects of confluence have been previously observed in our laboratory. Cells from 100 to 150 mm culture plates are washed with 2 ml of ice-cold 1X PBS after aspirating the culture medium. Then, 1.5 ml of 1X PBS is added, and cells are scraped and transferred into a cold 1.5-ml microfuge tube. Cell pellets are collected after centrifugation at $4000 \times g$ for 5 min at $4°$. The procedures are performed on ice, and all solutions and sample tubes are kept on ice at all times. Each pellet is given 400 μl of ice-cold 1:1 $CHCl_3$:CH_3OH and vortexed for 1 min, or until thoroughly mixed. Samples are centrifuged at $7500 \times g$ for 5 min at $4°$, supernatant decanted, and discarded. To the remainder of the cell pellet, 200 μl of 2:1 $CHCl_3$:CH_3OH containing 0.25% 12 N HCl are added. Samples are vortexed for 5 min and then pulse spun. The supernatant is then given 40 μl of 1 N HCl and vortexed for 15 s. The resulting two phases are separated by centrifugation (pulse at $18,000 \times g$). The solvent from the collected lower layer is evaporated in a vacuum centrifuge (Labconco CentriVap Concentrator, Kansas City, MO). The resulting lipid film is rapidly redissolved in 55 μl of 1:1:0.3 $CHCl_3$:CH_3OH:H_2O. Before analysis, 5 μl of 300 mM piperidine (Lytle *et al.*, 2000) is added as an ionization enhancer, and the sample is vortexed and pulse spun.

6.2. Extraction of polyphosphoinositides from tissue

Biopsy samples (20 to 50 mg) are quickly frozen by immersion in liquid nitrogen. Samples can be stored at $-80°$, if not immediately extracted. The frozen samples are homogenized using 500 μl of CH_3OH in a Dounce tissue grinder (Kimble/Kontes Co., Vineland, NJ) for about 1 min working on ice. The suspension is then transferred to a cold 1.5-ml microfuge tube, and 500 μl of cold $CHCl_3$ are added. After vortexing for 1 min at $4°$, samples are centrifuged at $7500 \times g$ for 5 min at $4°$. Supernatant is decanted and discarded. The retained pellet is then given 200 μl of 2:1 $CHCl_3$:CH_3OH containing 0.25% 12 N HCl. Samples are vortexed for 5 min and then pulse spun. Supernatant is transferred to a new cold microfuge tube and, after addition of 40 μl of 1 N HCl, the samples are vortexed for 15 s. The resulting two phases are separated by centrifugation (pulse at $18,000 \times g$). The solvent from the collected lower layer is evaporated in a vacuum centrifuge. The resulting lipid film is rapidly redissolved in 55 μl of 1:1:0.3 $CHCl_3$:CH_3OH:H_2O. Before analysis, 5 μl of

300-mM piperidine (Lytle et al., 2000) is added as an ionization aid, and the sample is vortexed and pulse spun.

6.3. Direct-infusion mass spectral analysis of polyphosphoinositides

Mass spectral analysis was performed on an Applied Biosystems/MDS SCIEX 4000 Q TRAP hybrid, triple-quadrupole, linear ion trap MS (Applied Biosystems, Foster City, CA). The instrument was equipped with a Harvard Apparatus syringe pump and an ESI source. Samples were analyzed at an infusion rate of 10 μl/min in negative ionization mode over the range of m/z 400 to 1200. GPIns$P$ and GPInsP_2 were analyzed by full scan analysis and normalized to 16:0 GPInsP_2 internal standard (Avanti Polar Lipids, Alabaster, AL). Utilizing this method, approximately 30 GPIns$P$ and GPInsP_2 species can be detected. A typical RAW 264.7 full-scan spectra highlighting the GPIns$P$ and GPInsP_2 spectral region is shown in Fig. 2.6. Due to their very low abundance and detection only after stimulation, GPInsP_3 species cannot be identified by full or precursor ion scan (m/z 481). Instead, identification of the individual phosphatidylinositol phosphates present in the total lipid extracts was accomplished by ESI-MS/MS with a collision energy of 50 eV (Milne et al., 2005). Peaks corresponding to known GPInsP_3 compounds were fragmented and manually inspected for the presence of the identification peaks (examples of GPIns$P$, GPInsP_2, and

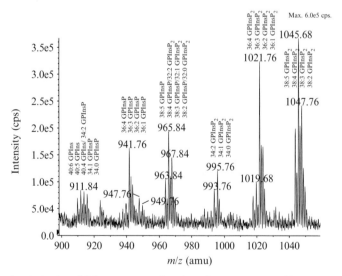

Figure 2.6 GPIns$P$ and GPInsP_2 spectral region of a RAW 264.7 cell extract. Approximately 30 chemically distinct GPIns$P$ and GPInsP_2 species can be identified by tandem mass spectrometry fragmentation in a typical spectrum.

GPInsP_3 fragmentation patterns can be found on the LIPID MAPS public website: http://www.lipidmaps.org/data/standards/standards.php?lipidclass=LMGP). A confirmed identification was achieved when key fragmentation peaks were larger than three times the signal-to-noise ratio (S/N). The limit of detection using this method was found to be less than 9 pmol/ml for 38:4 GPInsP_3 (Avanti Polar Lipids, Alabaster, AL). Data were collected with the Analyst software package (Applied Biosystems).

6.4. Deacylation of GPInsPn lipids

The quantitative analysis of headgroup isomeric GPInsP_2 and GPInsP_3 is achieved by LC-MS analysis of the deacylated polyphosphoinositides and the use of cytidine-5′-monophosphate (CMP; Sigma-Aldrich, St. Louis, MO) as an internal standard. The lower layer of the last step of the extraction protocol that yields separation of phosphoinositides from the rest of the phospholipids is subjected to deacylation as described by others (Clarke and Dawson, 1981; Cunningham et al., 1990). The organic phase (lower layer) is transferred into a glass screw-cap vial and solvent-evaporated under nitrogen. Afterward, 500 µl of freshly prepared methylamine reagent (1-butanol, methanol, 40% aq. methylamine, 1:4:4, v/v) was added to the glass vial and heated at 53° in a heating block (water bath) for 1 hr. After cooling to room temperature, the content is transferred into a microfuge tube and solvent-evaporated in a vacuum centrifuge (Labconco CentriVap Concentrator, Kansas City, MO). The resulting film is resuspended in 500 µl of distilled H$_2$O, vortexed briefly, and centrifugated for 2 min at 18,000×g. Supernatant is transferred to a new microfuge tube. GroPInsP_n are extracted with 500 µl of 1-butanol/petroleum ether/ethyl formate (20:4:1,v/v) to remove the undeacylated lipids and free fatty acids. After mixing (by vortex) for 30 s and centrifugation (18,000×g, 2 min), the lower (aqueous) layer is collected. The extraction is repeated, and the upper phase is discarded. Combined aqueous phases are dried in a vacuum centrifuge (or lyophilized). The samples are resuspended in 36 µl of methanol:water (1:1), mixed with 40 µl of 100-µM CMP and 4 µl of 300-mM piperidine, and analyzed by LC-MS.

The phosphoinositides standards were deacylated likewise with methylamine and dissolved in methanol:water (1:1) containing 10% 300-mM piperidine to produce the stock solution with concentration of 200 µM. The stock solution was further serially diluted to obtain working solution over the range 20 to 200 µM.

6.5. LC-MS analysis of deacylated GPInsPn compounds

Phosphoinositides are precursors in intracellular signaling cascades as well as ligands for membrane-associated proteins that are involved in trafficking and cytoskeletal dynamics (Downes et al., 2005; Michell et al., 2006). Since the mid-1980s, the understanding of inositol-containing glycerophospholipids has elevated from quantitatively minor membrane constituents to very important players in cellular functioning and responses. Seven phosphoinositides have been identified, and glycerophosphoinositol and its phosphorylated derivatives were demonstrated to be important in modulating cell proliferation and G-protein–dependent activities (Berrie et al., 1999; Corda et al., 2002). The detection, identification, and quantification of phosphatidylinositol phosphates (GPInsP) and phosphatidylinositol bisphosphates (GPInsP_2) with different fatty acid composition can be achieved by ESI-MS/MS (Milne et al., 2005; Wenk et al., 2003). However, the determination of the isomeric headgroups remains a challenge. For many years, glycerophosphoinositols have been traditionally analyzed by anion-exchange HPLC using gradients of aqueous mobile phases with high concentrations of buffer solutions and employing radiolabeling techniques as detection and quantitation systems (Alter and Wolf, 1995; Lips et al., 1989; Morris et al., 2000).

This technique has the main disadvantage of being a multi-step procedure that only measures the true levels of these compounds if the radiolabeling is taken to isotopic equilibrium, which is often difficult to achieve. Since no radioactive glycerophosphoinositide standards are commercially available, they are usually synthesized by radiolabeling of cell extracts and deacylation (Hama et al., 2000). This process introduces impurities as well as some practical considerations, since the radiolabeling is not always feasible as in analysis of glycerophosphoinositides from tissues and organs.

Some nonradioactive methods have been used to analyze phosphatidylinositides. GPIns4P and GPIns4,5P_2 were well resolved from each other and from other phospholipids using normal-phase HPLC and evaporative light-scattering detection (Gunnarsson et al., 1997). The three regioisomers of GPInsP_2 were separated by anion-exchange HPLC with NaOH gradients after deacylation, followed by suppressed conductivity detection (Nasuhoglu et al., 2002). Recently, a new method for analysis of intact phosphoinositides by LC-MS using a microbore silica column was reported (Pettitt et al., 2006).

The β-cyclodextrin–bonded column has been successfully employed for the analysis of phosphorylated carbohydrates, allowing the separation of both enantiomeric and positional isomers (Feurle et al., 1998). It has also been used for analysis of glycerophosphoinositol (Dragani et al., 2004). This stationary phase can be operated under water/organic conditions with the use of volatile modifiers at low concentrations, which is suitable for ESI-MS.

Here we describe an LC-MS method for separation, identification and quantification of the four GroPInsP_2 and GroPInsP_3 lipid species from cell extracts by using β-cyclodextrin–bonded column with on-line negative ESI-MS detection and internal standard calibration.

The HPLC system consisted of two Shimadzu (Shimadzu Scientific Instruments, Inc., Columbia, MD) LC-10 ADVP binary pumps and an SCL-10AVP system controller. A 20-μl sample was injected via a CTC

HTC PAL autosampler (Leap Technologies, Carrboro, NC). The separation was performed on a Nucleodex β-OH column, 5 μm (200 × 4.0 mm i.d.) from Macherey-Nagel (Düren, Germany). The mobile phase included 50-mM ammonium formate aqueous solution (solvent A) and acetonitrile (solvent B). The following gradient was used: 0 to 1 min, B = 80%; 1 to 15 min, B = 80 to 70%; 15 to 30 min, B = 70 to 66%; 30 to 70 min, B = 66 to 61%; 70 to 75 min, B = 61 to 60%; 75 to 76 min, B = 60 to 80%; and 76 to 90, B = 80%. The mobile phase was delivered at a flow rate of 0.3 ml/min.

The MS spectrum was acquired in negative mode on a 4000 Q-Trap LC-MS/MS system, fitted with turbo spray ion source from Applied Biosystems (Foster City, CA). The turbo spray source was operated at 200° with an ion voltage of −4500 V, and nitrogen as CUR and nebulizer gas. The CUR was 10 l/hr. The flow rate of ion source gas 1 and gas 2 was 20 and 40 l/hr, respectively. The declustering potential (DP) was −110 V, and the collision energy (CE) was −5 V. Settings include channel electron multiplier (CEM): 2200 V; scan type: enhanced MS (EMS), unit resolution for Q1; scan rate: 1000 amu/s; scan range from m/z 300 to 590; duration: 90 min; step size: 0.08 amu; Q3 entry barrier: 8 V; and pause between mass ranges: 5.007 ms. A representative extracted ion current (XIC) is shown on Fig. 2.7 depicting the separation of the three GroPInsP_2 and GroPInsP_3 isomers.

Calibration curves were constructed by mixing deacylated GPInsP_n standards working solutions with equal volume of 100-μM CMP. They were then analyzed by LC-MS. Calibration curves for deacylated GPIns4,5P_2 and GPIns3,5P_2 are shown on Fig. 2.8A and B. They are created by using the relationship between "peak area of GroPInsPn/peak area of CMP" and "amount of GroPInsPn/amount of CMP." As an example, the method was applied to the analysis of RAW 264.7 cell extracts after stimulation with platelet activating factor (PAF) for 15 min. The extracts were deacylated and analyzed by LC-MS using CMP as an internal standard. The XIC traces from this analysis are shown on Fig. 2.9A and B.

7. Computational Analysis of Mass Spectral Data

Complex mixtures of phospholipids from biological extracts generate mass spectra with hundreds to thousands of peaks associated with the molecular ions present in the sample, which includes any molecules that acquire a charge in the desired instrument mode over the applicable m/z range. In lipid extracts, these ions include phospholipids, solvent contaminants, fragments of larger macromolecules, and multiply charged ions, among other components. In addition, in the absence of chromatographic separation, any

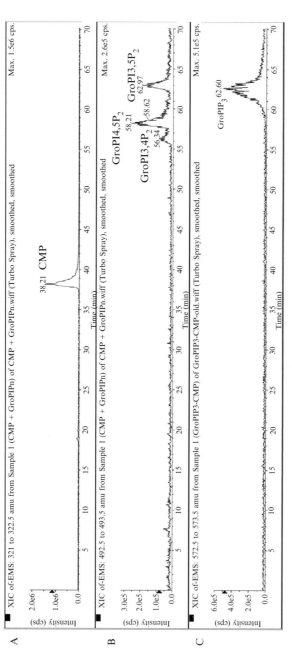

Figure 2.7 Liquid chromatography–mass spectrometry spectra of GroPInsP_2 and GroPInsP_3 standards with internal standard cytidine-5′-monophosphate (CMP). (A) Extracted ion chromatograms (XICs) of cytidine-5′-monophosphate (m/z 322); (B) Extracted ion chromatograms of GroPInsP_2 (m/z 493); (C) Extracted ion chromatograms of GroPInsP_3 (m/z 573).

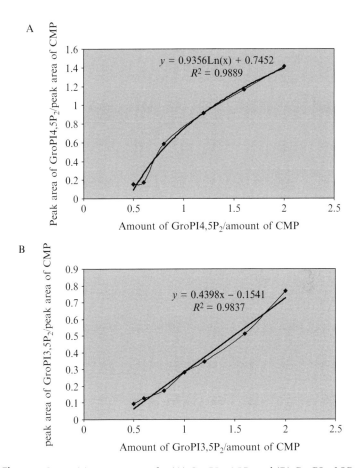

Figure 2.8 Calibration curves for (A) GroPIns4,5P_2 and (B) GroPIns3,5P_2.

particular m/z peak may contain ions from several different isobaric phospholipids or isotopic ions from molecular ions at $m/z-1$, $m/z-2$, which may be appropriately corrected for isotopic distribution if the molecular composition is known.

7.1. Direct infusion (intra-source separation)

For direct infusion of samples in the MS, an initial qualitative screen to identify particular m/z values whose peak ions are consistently different between two conditions is readily developed. For direct infusion, we have created code to find all the peaks in each MS scan and to statistically compare these peaks across conditions (Forrester et al., 2004). For direct infusion phospholipid spectra, alignment of the same peaks across different samples is

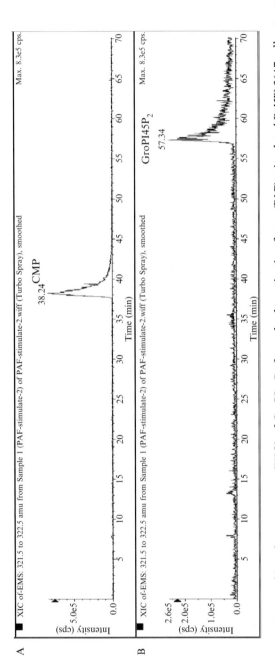

Figure 2.9 Extracted ion chromatograms (XIC) of GroPInsPn from platelet activating factor (PAF)–stimulated RAW 264.7 cell extracts. (A) Extracted ion chromatograms of cytidine-5′-monophosphate (CMP) (m/z 322). (B) Extracted ion chromatograms of GroPInsP_2 (m/z 493).

relatively simple because there is usually only one peak per unit m/z, so that peak intensities from the same nominal m/z are directly comparable across samples. The comparison of peak ion intensities (or areas) may be done in several ways, with a desired normalization and using standard statistical tests. The visualization of these changes is relevant for interpretation and useful for suggesting specific m/z peak values which were different under the condition(s) being compared. Addition of internal standards to uniform background cell extracts can be used to validate this method, and titrations may be used to confirm the monotonic behavior of the peak ion intensity on molecular concentration (i.e., increasing concentration of a molecule generates an increase in peak intensity at the relevant m/z). Due to nonlinearities (e.g., ion competition), however, the potential for false changes must be recognized as a true increase, or decrease in a molecule may consistently affect the ionization of other molecules during the infusion process. Note that this is true whether or not one normalizes to single or multiple internal standards. A general rule is that direct infusion spectra should only be compared (peak by peak) if the total spectra are sufficiently similar. Similarity can be defined both by the pattern the spectra create and by the total ion count (or mean ion count). A correlation coefficient can be used to quantitatively assess global similarity of spectra and a filter applied to limit the comparison of spectra to those with a sufficient number of peaks above the noise. If this is the case, one may then be reasonably confident that those isolated specific peaks, which differ in a statistically significant manner due to the condition, reflect true changes in molecular abundance of contributing ions at that m/z. Since a set of "control" spectra are generally used for any comparison, only those changes associated with the condition should repeatedly and reliably be indicated by statistical testing. This implies that the protocol for the control and condition should be identical (same "random" error sources from biological variability, experimenter, instrumentation, extraction methods, etc.) except for the effect of the condition. Any effects that are systematically different between the controls and condition will likely be reflected in the spectra. Randomizing the infusion of samples (between condition and controls) and any other protocol steps are suggested to minimize systematic influences except for the condition being tested. Since the comparison is the same for every m/z, the qualitative screen can be used to assess changes in peak ions at every m/z.

A single typical direct infusion spectra is binned in 0.07 m/z units and spans a range of m/z 350 to 1200 (12,142 bins). Each sample is scanned once per second for approximately 60 s (user-defined) so that a single spectrum actually consists of 60 scans. The vendor-supplied software typically time-averages these scans for graphical display. The ASCII data for each MS file can be obtained from the vendor files (.Raw files) by using XConvert.exe located in the \bin folder of Xcalibur. This program allows batch processing of a large number of .Raw files. The format of these ASCII files

(~60 MB) is cumbersome and contains mostly redundant text so that it is useful to write a parsing program to store the relevant time-m/z-ion intensity information in separate files. We wrote a program in Fortran (G77) to batch process large numbers of these ASCII files. We have implemented a filter on the 60 scans (for each bin), which eliminates aberrant scans (infusion problems) by rank-ordering the ion counts and removing a user-defined percentile (e.g., the upper and lower tenth percentile) prior to time-averaging. This tends to have little effect on maxima, but creates well-defined minima in the spectra. Following time-averaging, a peak finding (or maxima) algorithm may then be applied to each file. It is also possible to forgo this step and directly compare all m/z binned intensity values (12,142) from each spectrum. For the phospholipid spectra, we currently determine the maximum for each nominal m/z by finding the maximum between adjacent minima. For example, for nominal m/z 885, the code searches for a minimum intensity between 884.5 and 885.0 and another minimum between 885.5 and 886.0. An exclusion limit on the minima spacing (e.g., 0.3 m/z units) can prevent erroneous assignments. The maximum intensity and m/z value between these two minima are then stored (e.g., peak m/z 885.5, peak intensity 6.37×10^7). For more complex spectra, one can smooth the time-averaged spectra and use a low-high-low pattern to determine peaks. We currently find the maxima for every nominal m/z (850 maxima for m/z 350–1199) and use a characteristic noise level from the spectra to filter appropriately by S/N level when necessary. However, it is generally useful to compare all m/z values since, in some comparisons, the maximum intensity for a particular m/z could be at the noise level, but the condition may generate a sizable peak (or vice versa). In this formulation, each unique mass spectra corresponds to a vector of intensity values (e.g., 850) associated with each unit m/z value. The subsequent normalization and statistical comparison of these intensities across spectra are then relatively transparent; we wrote algorithms in a statistical package (SPlus) to batch-import these m/z-peak data files, to perform the relevant statistical tests, and to output the results comparing the appropriate conditions. In practice, this method provides a reliable way to qualitatively screen for potential phospholipid changes due to some perturbation or condition, and we created "arrays" to highlight the m/z values that appear to show statistically significant increasing or decreasing peak intensities due to the condition. Unlike gene expression arrays, the magnitude of peak intensity changes due to signaling events. For example, in mass spectra for phospholipids are not generally dramatic (a two "fold" peak change is quite large), since many structural lipids are already present in the control populations, and the cell likely does not tolerate very large changes in most species molecular abundance. To confirm direct infusion results, to minimize ion competition, to separate molecular classes of lipids, and to quantify the phospholipid molecular species in extracts, LC-MS is employed.

7.2. LC-MS data analysis

The batch processing of full-scan LC-MS data requires significantly more computational processing than the direct infusion data, primarily due to the (typically) 60 min runs per sample, with approximately one scan per second. The MS data are contained in the vendor (Analyst, Applied Biosystems) .wiff files and must be converted into ASCII prior to further processing. We use a converter mzStar.exe from the Institute for Systems Biology (ISB), which converts the .wiff files into mzXML format (Pedrioli et al., 2004). We convert the mzXML files to .mgf (Mascot Generic Format, ASCII files) and header files (ASCII) using mzXML2other.exe (ISB), which requires compilation on a cygwin, X-Windows terminal emulating a Unix shell. Each step is batchable (processes multiple files without user input), and the .mgf sample files are approximately 100 MB ASCII each. The .mgf files contain sequentially ordered scans (the retention time of each scan is located in the header file) and the m/z bins and their associated ion intensity values. We wrote code in Fortran (G77) to batch process these .mgf files creating two forms of output: (1) smoothed (time-averaged in $\sim$10-s increments) XICs of retention time/intensity values for every nominal m/z value, and (2) integrated, aligned (by m/z and retention time) peaks across the samples. The former (1) is used to graphically check selected XICs across multiple samples for correct peak identification, integration windows selection, and peak alignment. The method of analysis is certainly not unique; however, we have also experimented with using three-dimensional (3D; m/z-retention time-intensity space) kernel smoothing and peak finding, but found that computational time and appropriate alignment checking across many samples ($\sim$30) was far more difficult to implement. By generating XICs for each m/z, the problem of 3D peak finding and integration is reduced to the more tractable two-dimensional (2D) problem in retention time-intensity space. Specifically, to create an XIC, a selected m/z window must be chosen (e.g., m/z 809.5 to 810.5) and the ion counts summed (or averaged) in this window to represent the ion counts for that particular retention time ($\sim$10 s time window). Recall that the short time-averaging (user-selectable in the code) is essentially a smoother and natural compression of the data. This time-averaging interval should be less than the retention time "width" of any peaks. Otherwise, temporal resolution is lost. Therefore, for any particular nominal m/z, the XICs from n samples are readily compared graphically across samples (intensity as a function of retention time). We currently have a user-selectable range in m/z for creating the XIC and usually select ranges (m/z -0.5, m/z $+0.5$) looping over the range of m/z values (e.g., 350.5 to 1199.5) whose range is dynamically obtained from the spectra files. Since the 3D peak maxima about a nominal m/z (e.g., m/z 419.9 at 23.3 min) may not be symmetric about the XIC window selected, the XICs will indicate ion contributions in nominal m/z 419.5 and nominal m/z 420.5 with peaks ~23.3 minutes in both cases. This can be ameliorated by smoothing, finding the 3D peak max, and dynamically defining the m/z intervals (e.g., by slicing into 2D segments and cutting off the integration in m/z when the maxima falls below some predefined threshold relative to the global 3D maximum); alternatively, the internal standards for the lipid class should have a similar 3D peak m/z location, will be similarly subject to the same asymmetry in the chosen XIC selection range, and should therefore be valid for quantitatively normalizing by area. In Fig. 2.10 we indicate pseudo-code for the LC-MS analysis program. Briefly, the code first loops over all the MS files of interest (~3 GB for 30 files), reading in the header files, which contain the

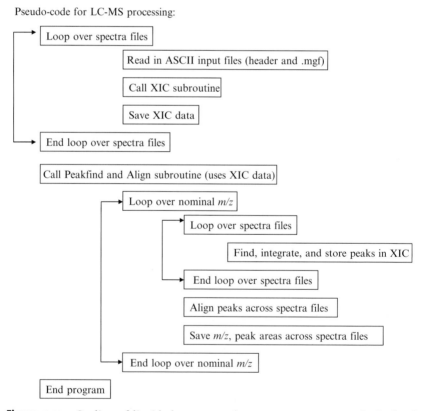

Figure 2.10 Outline of liquid chromatography-mass spectrometry code for batch processing full-scan *wiff* spectra.

number of scans and retention times of each, and the data fields, which contain the scan number, followed by a list of m/z and corresponding ion intensity data. For each file, a subroutine is called, which loops over all the m/z values and stores an equivalent XIC for each m/z, as previously described. Given a list of XICs for each spectra file for every m/z, a peak finding and alignment subroutine are called, which take the saved XIC data as input. The subroutine chooses an m/z and examines the XICs for that m/z across each of the files. Looping over each file (XIC here), candidate peaks are identified using a median filter (user-selectable) and low-high-low pattern. Candidate peaks are integrated taking binned incremental steps (in retention time) away from the maxima (left and right) and applying a slope (user-selectable) and relative magnitude (user-selectable) filter to determine the endpoints of integration. Since the XICs can be noisy "near" the peak, it is advisable to either smooth or have exception handling for using the slope-filter to determine integration endpoints. We typically continue searching for integration endpoints even if the slope filter fails, as long as the ion intensity remains an appreciable fraction of the peak intensity (e.g., >60%, user-selectable). However, one must be careful because this selection becomes relevant for peaks that are not well separated in retention time ("just" resolved). In most cases, it is useful to check graphically the XICs for the m/z to confirm that the peak of interest is well separated, and the integration window determined by the code is appropriate (Fig. 2.11). The set of candidate peaks (amplitudes and areas) need to be aligned across the files so that they can be reliably compared. Experimentally, large retention time shifts (larger than 5 min) are possible across separate spectra files, and it is sometimes useful to prealign the spectra (e.g., optimizing a correlation coefficient between files and a chosen XIC or "average" XIC). Whether or not pre-alignment of XICs across files is used, we decided to align peaks by counting the number of peaks IDd in a retention time window (for that fixed nominal m/z) across the files. If this number is above a threshold (user-selectable) based on the percent of the spectra files, then the candidate peak is considered a "true" peak centered at that relevant retention time ("true" peak time). For example, using a nominal m/z 723.5 from 30 spectra files, 23 files could indicate a peak (or peaks) between 9 and 10 min, 4 files between 10 and 11 min, and 3 files between 8 and 9 min. The "true" time would be the average (e.g., 9.8 min). Then, for each file, the maximum peak intensity that occurs in some retention time window (e.g., ±2 min, user-selectable) about this "true" time is used to define the "true" peak for that file, and the integrated intensity (and peak amplitude) is saved. If no peak is identified for a particular file in this window, and then a zero is reported for that file. The more spectra files one has to compare, the lower the median threshold one can use for peak identification, since "random" candidate peaks (by low-high-low identification) are unlikely to repeat in the same retention time window across a

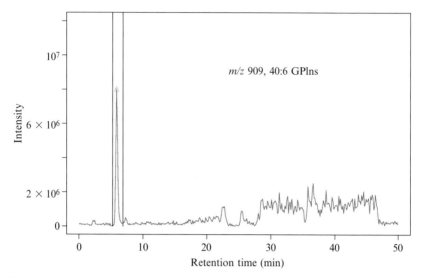

Figure 2.11 Generation of extracted ion chromatograms (XIC; m/z 909.06 to 909.96) from Fortran code as described in the text. The horizontal axis is retention time, and the vertical axis ion intensity. The vertical black lines indicate the region of integration dynamically selected by the code. The red triangle indicates the peak(s) automatically identified by the algorithms described in the text. (See color insert.)

significant percent of the files. Once again, it is useful to graphically display the XICs across files and visually see the repeated peaks and their retention time fluctuations across files (Fig. 2.12). This is especially the case if one computed a large change in a molecular species due to a condition and would like to verify that large retention time shift across files were not erroneously picking up another, different peak in the XIC (at the same m/z). If pre-alignment of XICs is used and the corrected correlation coefficients are near 1, then the retention time windows selected for assigning peaks may be of shorter duration (~30 s to 1 min). For identification and a quantitative assessment of individual phospholipids, multiple chemically defined internal standards per class identify typical retention times, peak profiles (widths), and retention time locations. Using multiple internal standards also allows for the examination of the variation in peak retention time as a function of m/z in addition to differential ionization across the lipid class. In practice, the molecular species in a phospholipid class are readily identified from full-scan data by m/z retention time information in comparison to similar peak profiles and retention times relative to the internal standards. Fragmentation of the peak ions at the relevant m/z retention time provides confirmation of the molecular species whose ions are contributing to the peak in the XIC.

Following the generation of integrated, aligned peak areas across multiple spectra files, accurate quantification of molecular species requires

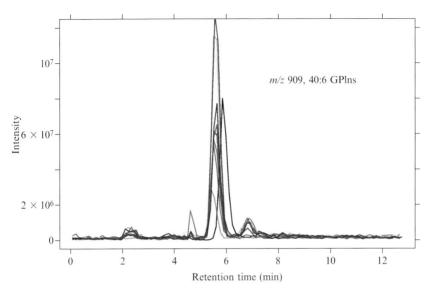

Figure 2.12 A zoomed-in section including the extracted ion chromatogram (XIC) from Fig. 2.11 (*m/z* 909.06 to 909.96), which also includes extracted ion chromatograms generated from nine other full-scan spectra files, indicative of the variation in amplitude and peak location across these files. Peak areas from the 10 files are automatically aligned as described and output in ASCII format for further analysis (normalization and quantification). Many mass-to-charge (*m/z*) values in glycerophospholipid (GPL) full-scan spectra contain extracted ion chromatograms with multiple peaks at different retention times. (See color insert.)

titrations of internal standards for individual species of phospholipids over the presumed physiological range of variation in abundance. To generate standard curves, we proceed as follows: (1) add fixed quantities of multiple fixed odd-chain internal standards (per glycerophospholipid [GPL] class) to every sample; (2) add varying amounts of even-carbon standards (e.g., 10, 50, 100, 500 ng) in triplicate; (3) normalize the peak areas of the even-carbon standards to the average of the fixed odd-carbon areas (per GPL class); and (4) repeat the process several times to estimate experiment to experiment variability in the slopes and intercepts (assuming linearity). An example of standard curves for the GPA class using experimental protocol and code described above is given in Fig. 2.13, and the variation in slope values and intercepts across independent experiments for the even-carbon internal standards are listed in Table 2.4.

For LC-MS with the protocol as described, we have found that the variation in slopes across GPLs within a class is nonlinear (and class-dependent) and has no obvious dependence on double bonds or carbon number. This is unfortunate since only rough slope and intercept estimates are possible for

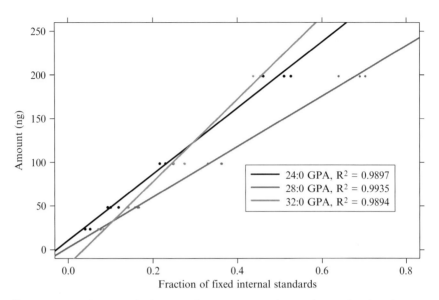

Figure 2.13 Three standard curves for 24:0, 28:0, and 32:0 glycerophosphatidic acid (GPA) (black, green, and red, respectively) generated as described in the text using the automated code. The horizontal axis refers to peak areas normalized to the mean of four fixed odd-carbon internal standards, and the vertical axis is the amount added in nanograms (ng). (See color insert.)

Table 2.4 Variation in glycerosphosphatidic acid (GPA) slopes across independent experiments

GPA	Experiment		
	1^a	2^a	3^a
24:0	254	289	X
28:0	189	184	X
32:0	238	267	266
34:1	172	244	204
34:2	147	179	145
36:0	192	352	330
36:1	206	247	202
36:2	195	288	194
38:4	253	218	230
38:6	299	388	337
40:6	437	403	384

[a] Standard deviations in slope values for individual experiments are approximately 5%.

Table 2.5 Approximate amounts (ng) of individual GPSer molecular species in RAW 264.7 cells (5 × 10⁶ cells)

GPSer[a]	Estimated amount (ng)	
32:1	52	40
32:0	30	24
34:3	41	18
34:2	62	54
34:1	393	441
34:0	163	172
36:4	20	23
36:3	30	30
36:2	249	314
36:1	1119	1319
36:0	464	510
38:6	21	25
38:5	60	70
38:4	188	326
38:3	293	340
38:2	144	12
38:1	60	65
38:0	9	26
40:7	37	47
40:6	575	692
40:5	770	895
40:4	451	538
40:3	200	260
40:2	78	87
40:0	33	41

[a] Shown are duplicate control (unstimulated) samples for a 2-h KDO2 Lipid A stimulation using a 10% serum LIPID MAPS protocol (www.lipidmaps.org).

molecular species for which internal standards are not available. These estimated slopes for molecular species that lack titrations are based on the variation in slopes for other members of the same GPL class. One may use the range of slopes or regression estimates (e.g., in m/z) across the GPL class to estimate the unknown slopes with appropriately large errors in the quantification of these molecular species. In Table 2.5, we list estimated amounts of different GPSer in RAW 264.7 cells following the LC-MS protocol described in the previous section. Once the molecular identity at a nominal m/z and retention time is known, isotopic corrections for m/z +2 contributions from the molecular species at a nominal m/z are readily computed, and normalized peak areas may be appropriately corrected and quantified within a GPL class. External normalization (e.g., cell counts, DNA) is useful for comparing across samples as the fluctuation in individual GPLs may be significant due to extrinsic factors. Addition of internal standards prior to extraction may be used to estimate extraction efficiency per class and potential differential extraction efficiencies within a GPL class.

ACKNOWLEDGMENTS

This work was supported in part by the NIH Large Scale Collaborative Initiative LIPID MAPS (U54 GM069338). The authors thank Andrew Goodman and Michelle Armstrong for excellent technical assistance with various aspects of this work. Dr. Brown is the Ingram Associate Professor of Cancer Research in Pharmacology.

REFERENCES

Alter, C. A., and Wolf, B. A. (1995). Identification of phosphatidylinositol 3,4,5-trisphosphate in pancreatic islets and insulin-secreting β-cells. *Biochem. Biophys. Res. Commun.* **208,** 190–197.

Berrie, C. P., Iurisci, C., and Corda, D. (1999). Membrane transport and *in vitro* metabolism of the Ras cascade messenger, glycerophosphoinositol-4-phosphate. *Eur. J. Biochem.* **266,** 413–419.

Bligh, E. G., and Dyer, W. J. (1959). A rapid method of total lipid extraction and purification. *Can. J. Biochem. Physiol.* **37,** 911–917.

Bloor, W. R. (1928). The determination of small amounts of lipid in blood plasma. *J. Biol. Chem.* **77,** 53–73.

Brügger, B., Erben, G., Sandhoff, R., Wieland, F. T., and Lehmann, W. D. (1997). Quantitative analysis of biological membrane lipids at the low picomole level by nano-electrospray ionization tandem mass spectrometry. *Proc. Natl. Acad. Sci. USA* **94,** 2339–2344.

Christie, W. W. (2003). "Lipid analysis: Isolation, separation, identification, and structural analysis of lipids." Oily Press, Bridgewater England.

Clarke, N. G., and Dawson, R. M. C. (1981). Alkaline O leads to N-transacylation. *Biochem. J.* **195,** 301–306.

Corda, D., Iurisci, C., and Berrie, C. P. (2002). Biological activities and metabolism of the lysophosphoinositides and glycerophosphoinositols. *Biochim. Biophys. Acta* **1582,** 52–69.

Cunningham, T. W., Lips, D. L., Bansal, V. S., Caldwell, K. K., Mitchell, C. A., and Majerus, P. W. (1990). Pathway for the formation of D-3 phosphate containing inositol phospholipids in intact human platelets. *J. Biol. Chem.* **265,** 21676–21683.

DeLong, C. J., Baker, P. R. S., Samuel, M., Cui, Z., and Thomas, M. J. (2001). Molecular species composition of rat liver phospholipids by ESI-MS/MS: The effect of chromatography. *J. Lipid Res.* **42,** 1959–1968.

Dowhan, W. (1997). Molecular basis for membrane phospholipid diversity: Why are there so many lipids? *Annu. Rev. Biochem.* **66,** 199–232.

Dowhan, W., and Bogdanov, M. (2002). Functional roles of lipids in membranes. *In* "New Comprehensive Biochemistry, Biochemistry of Lipids, Lipoproteins, and Membranes" (D. E. Vance and J. E. Vance, eds.), Vol. 36, pp. 1–35. Elsevier Science B.V., Amsterdam.

Downes, C. P., Gray, A., and Lucocq, J. M. (2005). Probing phosphoinositide functions in signaling and membrane trafficking. *Trends Cell. Biol.* **15,** 259–268.

Dragani, L. K., Berrie, C. P., Corda, D., and Rotilio, D. (2004). Analysis of glycerophosphoinositol by liquid chromatography-electrospray ionization tandem mass spectrometry using a β-cyclodextrin-bonded column. *J. Chromatogr. B Analyt. Technol. Biomed. Life Sci.* **802**, 283–289.
Ekroos, K., Chernushevich, I. V., Simons, K., and Schevchenko, A. (2002). Quantitative profiling of phospholipids by multiple precursor ion scanning on a hybrid quadrupole time-of-flight mass spectrometer. *Anal. Chem.* **74**, 941–949.
Fahy, E., Subramaniam, S., Brown, H. A., Glass, C. K., Merrill, A. H., Jr., Murphy, R. C., Raetz, C. R. H., Russell, D. W., Seyama, Y., Shaw, W., Shimizu, T., Spener, F., *et al.* (2005). A comprehensive classification system for lipids. *J. Lipid Res.* **46**, 839–862.
Fenn, J. B., Mann, M., Meng, C. K., Wong, S. F., and Whitehouse, C. M. (1989). Electrospray ionization for mass spectrometry of large molecules. *Science* **246**, 64–71.
Fenwick, G. R., Eagles, J., and Self, R. (1983). Fast atom bombardment mass spectrometry of intact phospholipids and related compounds. *Biomed. Mass Spectrom.* **10**, 382–386.
Feurle, J., Jomaa, H., Wilhelm, M., Gutsche, B., and Herderich, M. (1998). Analysis of phosphorylated carbohydrates by high-performance liquid chromatography-electrospray ionization tandem mass spectrometry utilizing a β-cyclodextrin bonded stationary phase. *J. Chromatogr. A* **803**, 111–119.
Folch, J., Ascoli, I., Lees, M., Meath, J. A., and LeBaron, F. N. (1951). Preparation of lipide extracts from brain tissue. *J. Biol. Chem.* **191**, 833–841.
Forrester, J. S., Milne, S. B., Ivanova, P. T., and Brown, H. A. (2004). Computational lipidomics: A multiplexed analysis of dynamic changes in membrane lipid composition during signal transduction. *Mol. Pharm.* **65**, 813–821.
Fridriksson, E. K., Shipkova, P. A., Sheets, E. D., Holowka, D., Baird, B., and McLafferty, F. W. (1999). Quantitative analysis of phospholipids in functionally important membrane domains from RBL-2H3 mast cells using tandem high-resolution mass spectrometry. *Biochemistry* **38**, 8056–8063.
Gunnarsson, T., Ekblad, L., Karlsson, A., Michelsen, P., Odham, G., and Jergil, B. (1997). Separation of polyphosphoinositides using normal-phase high-performance liquid chromatography and evaporative light scattering detection or electrospray mass spectrometry. *Anal. Biochem.* **254**, 293–296.
Hama, H., Takemoto, J. Y., and DeWald, D. B. (2000). Analysis of phosphoinositides in protein trafficking. *Methods* **20**, 465–473.
Han, X., and Gross, R. W. (1994). Electrospray ionization mass spectroscopic analysis of human erythrocyte plasma membrane phospholipids. *Proc. Natl. Acad. Sci. USA* **91**, 10635–10639.
Han, X., and Gross, R. W. (1995). Structural determination of picomole amounts of phospholipids via electrospray ionization tandem mass spectrometry. *J. Am. Soc. Mass Spectrom.* **6**, 1202–1210.
Han, X., and Gross, R. W. (2003). Global analyses of cellular lipidomes directly from crude extracts of biological samples by ESI mass spectrometry: A bridge to lipidomics. *J. Lipid Res.* **44**, 1071–1079.
Hirschmann, H. (1960). The nature of substrate asymmetry in stereoselective reactions. *J. Biol. Chem.* **235**, 2762–2767.
Hsu, F. F., and Turk, J. (2000a). Charge-driven fragmentation processes in diacyl glycerophosphatidic acids upon low-energy collisional activation: A mechanistic proposal. *J. Am. Soc. Mass Spectrom.* **11**, 797–803.
Hsu, F. F., and Turk, J. (2000b). Charge-remote and charge-driven fragmentation processes in diacyl glycerophosphoethanolamine upon low-energy collisional activation: A mechanistic proposal. *J. Am. Soc. Mass Spectrom.* **11**, 892–899.

Hsu, F. F., and Turk, J. (2000c). Characterization of phosphatidylinositol, phosphatidylinositol-4-phosphate, and phosphatidylinositol-4, 5-bisphosphate by electrospray ionization tandem mass spectrometry: A mechanistic study. *J. Am. Soc. Mass Spectrom.* **11,** 986–999.

Hsu, F. F., and Turk, J. (2001). Studies on phosphatidylglycerol with triple quadrupole tandem mass spectrometry with electrospray ionization: Fragmentation processes and structural characterization. *J. Am. Soc. Mass Spectrom.* **12,** 1036–1043.

IUPAC-IUB Commission on Biochemical Nomenclature (CBN) (1967). The nomenclature of lipids. *Eur. J. Biochem.* **2,** 127–131.

Ivanova, P. T., Milne, S. B., Forrester, J. S., and Brown, H. A. (2004). Lipid arrays: New tools in the understanding of membrane dynamics and lipid signaling. *Mol. Interv.* **4,** 84–94.

Ivanova, P. T., Cerda, B. A., Horn, D. M., Cohen, J. S., McLafferty, F. W., and Brown, H. A. (2001). Electrospray ionization mass spectrometry analysis of changes in phospholipids in RBL-2H3 mastocytoma cells during degranulation. *Proc. Natl. Acad. Sci. USA* **98,** 7152–7157.

Kerwin, J. L., Tuininga, A. R., and Ericsson, L. H. (1994). Identification of molecular species of glycerophospholipids and sphingomyelin using electrospray mass spectrometry. *J. Lipid Res.* **35,** 1102–1114.

Larsen, Å, Uran, S., Jacobsen, P. B., and Skotland, T. (2001). Collision-induced dissociation of glycerophospholipids using electrospray ion-trap mass spectrometry. *Rapid Commun. Mass Spectrom.* **15,** 2393–2398.

Lesnefsky, E. J., Stoll, M. S. K., Minkler, P. E., and Hoppel, C. L. (2000). Separation and quantitation of phospholipids and lysophospholipids by high-performance liquid chromatography. *Anal. Biochem.* **285,** 246–254.

Lips, D. L., Majerus, P. W., Gorga, F. R., Young, A. T., and Benjamin, T. L. (1989). Phosphatidylinositol 3-phosphate is present in normal and transformed fibroblasts and is resistant to hydrolysis by bovine brain phospholipase C II. *J. Biol. Chem.* **264,** 8759–8763.

Lytle, C. A., Gan, Y. D., and White, D. C. (2000). Electrospray ionization/mass spectrometry compatible reversed-phase separation of phospholipids: Piperidine as a post column modifier for negative ion detection. *J. Microbiol. Methods* **41,** 227–234.

Michell, R. H., Heath, V. L., Lemmon, M. A., and Dove, S. K. (2006). Phosphatidylinositol 3,5-bisphosphate: Metabolism and cellular functions. *Trends Biochem. Sci.* **31,** 52–63.

Milne, S. B., Ivanova, P. T., Forrester, J. S., and Brown, H. A. (2006). LIPIDOMICS: An analysis of cellular lipids by ESI-MS. *Methods* **39,** 92–103.

Milne, S. B., Forrester, J. S., Ivanova, P. T., Armstrong, M. D., and Brown, H. A. (2003). Multiplexed lipid arrays of anti-immunoglobulin M-induced changes in the glycerophospholipid composition of WEHI-231 cells. *AfCS Res. Reports* **1,** 1–11. Accessible at www.signaling-gateway.org/reports/v1/DA0011/DA0011.htm.

Milne, S. B., Ivanova, P. T., DeCamp, D., Hsueh, R. C., and Brown, H. A. (2005). A targeted mass spectrometric analysis of phosphatidylinositol phosphate species. *J. Lipid Res.* **46,** 1796–1802.

Morris, J. B., Hinchliffe, K.A, Ciruela, A., Letcher, A. J., and Irvine, R. F. (2000). Thrombin stimulation of platelets causes an increase in phosphatidylinositol 5-phosphate revealed by mass assay. *FEBS Lett.* **475,** 57–60.

Murphy, R. C., and Harrison, K. A. (1994). Fast atom bombardment mass spectrometry of phospholipids. *Mass Spectrom. Rev.* **13,** 57–75.

Murphy, R. C., Fiedler, J., and Hevko, J. (2001). Analysis of nonvolatile lipids by mass spectrometry. *Chem. Rev.* **101,** 479–526.

Nasuhoglu, C., Feng, S., Mao, J., Yamamoto, M., Yin, H. L., Earnest, S., Barylko, B., Albanesi, J. P., and Hilgemann, D. W. (2002). Nonradioactive analysis of phosphatidylinositides and other anionic phospholipids by anion-exchange high-performance liquid chromatography with suppressed conductivity detection. *Anal. Biochem.* **301,** 243–254.

Pedrioli, P. G., Eng, J. K., Hubley, R., Vogelzang, M., Deutsch, E. W., Raught, B., Pratt, B., Nilsson, E., Angeletti, R. H., Apweiler, R., Cheung, K., Costello, C. E., *et al.* (2004). A common open representation of mass spectrometry data and its application to proteomics research. *Nat. Biotechnol.* **22,** 1459–1466.

Pettitt, T. R., Dove, S. K., Lubben, A., Calaminus, S. D. J., and Wakelam, M. J. O. (2006). Analysis of intact phosphoinositides in biological samples. *J. Lipid Res.* **47,** 1588–1596.

Pulfer, M., and Murphy, R. C. (2003). Electrospray mass spectrometry of phospholipids. *Mass Spectrom. Rev.* **22,** 332–364.

Ramanadham, S., Hsu, F. F., Bohrer, A., Nowatzke, W., Ma, Z., and Turk, J. (1998). Electrospray ionization mass spectrometric analyses of phospholipids from rat and human pancreatic islets and subcellular membranes: Comparison to other tissues and implications for membrane fusion in insulin exocytosis. *Biochemistry* **37,** 4553–4567.

Rouzer, C.A, Ivanova, P. T., Byrne, M. O., Milne, S. B., Marnett, L. J., and Brown, H. A. (2006). Lipid profiling reveals arachidonate deficiency in RAW 264.7 cells: Structural and functional implications. *Biochemistry* **45,** 14795–14808.

Schwudke, D., Oegema, J., Burton, L., Entchev, E., Hannich, J. T., Ejsing, C. S., Kurzhchalia, T., and Schevchenko, A. (2006). Lipid profiling by multiple precursor and neutral loss scanning driven by the data-dependent acquisition. *Anal. Chem.* **78,** 585–595.

Smith, P. B. W., Snyder, A. P., and Harden, C. S. (1995). Characterization of bacterial phospholipids by electrospray ionization tandem mass spectrometry. *Anal. Chem.* **67,** 1824–1830.

Taguchi, R., Hayakawa, J., Takeuchi, Y., and Ishida, M. (2000). Two-dimensional analysis of phospholipids by capillary liquid chromatography/electrospray ionization mass spectrometry. *J. Mass Spectrom.* **35,** 953–966.

Wang, C., Xie, S., Yang, J., Yang, Q., and Xu, G. (2004). Structural identification of human blood phospholipids using liquid chromatography/quadrupole-linear ion trap mass spectrometry. *Anal. Chim. Acta* **525,** 1–10.

Watson, A. D. (2006). Lipidomics—a global approach to lipid analysis in biological systems. *J. Lipid Res.* **47,** 2101–2111.

Wenk, M. R. (2005). The emerging field of lipidomics. *Nat. Rev. Drug Discov.* **4,** 594–610.

Wenk, M. R., Lucast, L., DiPaolo, G., Romanelli, A. J., Suchy, S. F., Nussbaum, R. L., Cline, G. W., Shulman, G. I., McMurray, W., and DeCamilli, P. (2003). Phosphoinositide profiling in complex lipid mixtures using electrospray ionization mass spectrometry. *Nat. Biotechnol.* **21,** 813–817.

CHAPTER THREE

Detection and Quantitation of Eicosanoids via High Performance Liquid Chromatography-Electrospray Ionization-Mass Spectrometry

Raymond Deems, Matthew W. Buczynski, Rebecca Bowers-Gentry, Richard Harkewicz, and Edward A. Dennis

Contents

1. Introduction	60
2. Methods	62
2.1. Sample collection	62
2.2. Eicosanoid isolation	62
2.3. Reverse-phase liquid chromatography	62
2.4. Chiral chromatography	63
2.5. Mass spectrometry	63
2.6. Quantitation	63
3. Results and Discussion	74
3.1. MRM transition selection	74
3.2. Stereoisomer detection	77
3.3. Lower limit of detection	78
3.4. Recoveries	79
3.5. Miscellany	79
3.6. Summary	80
Acknowledgments	80
References	81

Abstract

Eicosanoids constitute a large class of biologically active arachidonic acid (AA) metabolites that play important roles in numerous physiological processes. Eicosanoids are produced by several distinct routes, including the cyclooxygenase, lipoxygenase, and P450 enzymatic pathways, as well as by nonenzymatic processes. In order to completely understand the eicosanoid response of a cell

Departments of Chemistry, Biochemistry, and Pharmacology, University of California, San Diego, La Jolla, California

Methods in Enzymology, Volume 432
ISSN 0076-6879, DOI: 10.1016/S0076-6879(07)32003-X
© 2007 Elsevier Inc.
All rights reserved.

or tissue to a given stimulus, measuring the complete profile of eicosanoids produced is important. Since the eicosanoids are products of a single species, AA, and represent, for the most part, the addition of various oxygen species, the hundreds of eicosanoids have very similar structures, chemistries, and physical properties. The identification and quantitation of all eicosanoids in a single biological sample are a challenging task, one that high-performance liquid chromatography-mass spectrometry (LC-MS) is well suited to handle. We have developed a LC-MS/MS procedure for isolating, identifying, and quantitating a broad spectrum of eicosanoids in a single biological sample. We currently can measure over 60 eicosanoids in a 16-min LC-MS/MS analysis. Our method employs stable isotope dilution internal standards to quantitate these specific eicosanoids. In the course of setting up the LC-MS system, we have established a library that includes relative chromatographic retention times and tandem mass spectrometry data for the most common eicosanoids. This library is available to the scientific community on the website www.lipidmaps.org.

1. INTRODUCTION

The eicosanoids comprise a broad class of AA metabolites that mediate a wide variety of important physiological functions. Most of the AA found in cells is esterified to the *sn*-2 position of neutral lipids or phospholipids (Schaloske and Dennis, 2006; Six and Dennis, 2000). Upon activation, phospholipase A_2s release the AA from cellular phospholipids. The free AA can then be converted by a score of enzymes in three distinct oxidation pathways (cyclooxygenase [Simmons *et al.*, 2004; Smith *et al.*, 2000], lipoxygenase [Funk, 2001; Peters-Golden and Brock, 2003; Spokas *et al.*, 1999], and cytochrome P450 [Sacerdoti *et al.*, 2003]) into hundreds of bioactive species. These pathways exhibit a fair amount of redundancy, as some eicosanoid species can be produced by more than one enzyme. Likewise, some enzymes are capable of producing more than one eicosanoid product. In addition to these reactions, AA and its metabolites can also undergo nonenzymatic oxidation and dehydration reactions to produce many of these same compounds as well as other metabolic species, including stereoisomers of the enzymatic products. Thus, eicosanoid biosynthesis is a complex network of interacting pathways and interconnected metabolites. It is quite possible that perturbing one arm of this system could produce changes and compensations in the other arms. In order to fully understand how a given cell or tissue responds to a stimulus or how a drug targeted for one eicosanoid might affect the distribution of the other eicosanoids, one must determine the entire eicosanoid spectrum.

Enzyme-linked immunosorbent assays have long been the primary means of quantitating eicosanoids (Reinke, 1992; Shono *et al.*, 1988). This method requires specific antibodies for each eicosanoid to be quantitated; however, relatively few eicosanoids have commercially available antibodies. This significantly limits the number of eicosanoids that can be detected and quantitated. This technique is also expensive and inefficient in that only a single eicosanoid can be determined with each assay. Thus, it is not amenable to analyzing a large number of different eicosanoids. Gas chromatography–MS (GC–MS) methods were developed that greatly improved upon these limitations (Baranowski and Pacha, 2002) and allowed the simultaneous analysis of multiple eicosanoids. To volatilize the eicosanoids for GC, they must first be chemically derivatized. A wide variety of derivatization methods are available; however, a single derivatization method is not suitable for all eicosanoids. Furthermore, Murphy *et al.* (2005) have found that some eicosanoids are not suited for GC-MS analysis. These volatilization issues were overcome with the development of electrospray ionization (ESI), which allows the eicosanoids to be analyzed by MS directly from an aqueous sample. The eicosanoid carboxylate moiety readily ionizes in the ESI source.

The similarities in eicosanoid structure and chemical characteristics requires that both high performance LC and collision-induced decomposition (CID) be employed, in conjunction with ESI-MS, to isolate and unambiguously identify the individual eicosanoid species. LC isolates the eicosanoids based on their chemical and physical characteristics, while CID produces characteristic precursor/product transitions that can be employed in multi-reaction monitoring (MRM) mode on the MS. ESI-MRM was first employed in this field by Margalit *et al.* (1996) to quantitate 14 eicosanoids directly from a biological sample. In 2002, the resolving power of LC was coupled with the sensitivity of ESI-MRM to study five eicosanoids from LPS (lipopolysaccharide)-stimulated synovial cells (Takabatake *et al.*, 2002). Recently, Kita *et al.* (2005) developed a high-throughput method for the detection of 18 different eicosanoids from biological samples.

We present here a protocol for identifying and quantitating a large number of eicosanoids in a single ESI-based LC-MS/MS run without requiring derivitization. This technique employs a solid-phase extraction procedure to isolate eicosanoids, an LC method to separate species, and a MS-CID technique to unambiguously identify a large number of eicosanoids. To accurately quantitate eicosanoids using these procedures, we have employed the well-established stable isotope dilution method (Hall and Murphy, 1998). We have included deuterated eicosanoids as internal standards to track and measure losses during sample preparation and to account for the various response issues associated with mass spectral analysis.

We currently can identify over 60 discrete chemical species of eicosanoid in a single 16-min run and can quantitate a significant fraction of them (Buczynski *et al.*, 2007; Harkewicz *et al.*, 2007). We present the methods for the extraction and analysis of eicosanoids from media and cells produced during cell culture. We then describe the protocols for these methods and conclude with a discussion of some characteristics of this procedure.

2. Methods

2.1. Sample collection

The following procedure was developed for the isolation of eicosanoids from six-well cell culture plates containing 2.0 ml of media. We have also adapted this method to other sample types and sample volumes by scaling our procedure as needed. The media was removed and 100 μl of a mixture of internal standards (containing 10 ng/100 μl of each standard in EtOH) was added followed by 100 μl of EtOH to bring the total concentration of EtOH to 10% by volume. Samples were centrifuged for 5 min at 3000 rpm to remove cellular debris. The eicosanoids were then isolated via solid-phase extraction.

When intracellular eicosanoids were analyzed, adherent cells were scraped into 500 μl of MeOH, and then 1000 μl of phosphate-buffered saline (PBS) and 100 μl of internal standards were added. Scraping cells in aqueous solutions was shown to activate eicosanoid production, whereas doing so in MeOH effectively stopped the reactions and lysed the cells. These samples were then processed the same as the media.

2.2. Eicosanoid isolation

Eicosanoids were extracted using Strata® X SPE columns (Phenomenex, Torrance, CA). Columns were washed with 2 ml of MeOH followed by 2 ml of H_2O. After applying the sample, the columns were washed with 1 ml of 10% MeOH, and the eicosanoids were then eluted with 1 ml of MeOH. The eluant was dried under vacuum and redissolved in 100 μl of solvent A (water-acetonitrile-formic acid [63:37:0.02; v/v/v]) for LC-MS/MS analysis.

2.3. Reverse-phase liquid chromatography

The analysis of eicosanoids was performed by LC-MS/MS. Eicosanoids were separated by reverse-phase LC on a C18 column (2.1 × 250 mm; Grace-Vydac, Deerfield, IL) at a flow rate of 300 μl/min at 25°. All samples were loaded via a Pal auto-sampler (Leap Technologies, Carrboro, NC) that maintained the samples at 4° to minimize degradation of eicosanoids while queued for analysis. The column was equilibrated in Solvent A, and samples (dissolved in Solvent A) were injected using a 50-μl injection loop and eluted with a linear gradient from 0 to 20% solvent B (acetonitrile-isopropyl alcohol [50:50; v/v]) between 0 and 6 min; solvent B was increased to 55% from 6 to 6.5 min and held until 10 min; solvent B was increased to 100% from 10 to 12 min and held until 13 min; and then, solvent B was dropped to 0% by 13.5 min and held until 16 min.

2.4. Chiral chromatography

When it was required to isolate isomeric eicosanoids, normal-phase chiral liquid chromatography was carried out using the same pumping system described above for reverse-phase chromatography. Separation was carried out on a 4.6 × 250 mm Chiral Technologies (West Chester, PA) derivatized amylose column (Chiralpak® AD-H) equipped with a guard column (Chiralpak® AD-H guard column) held at 35°. Buffer A was hexane/anhydrous ethanol/water/formic acid: 96/4/0.08/0.02, v/v; buffer B was 100% anhydrous ethanol. This small amount of water in buffer A is miscible in the hexane/anhydrous ethanol mix and was found to be vital for satisfactory chiral separation and peak shape. Gradient elution was achieved using 100/0:A/B at 0 min; linearly ramped to 90/10:A/B by 13 min; linearly ramped to 75/25:A/B by 15 min and held until 25 min; and then linearly ramped back to 100/0:A:B by 27 min and held there until 42 min to achieve column re-equilibration. The chiral chromatography effluent was coupled to a mass spectrometer for further analysis.

2.5. Mass spectrometry

All MS analyses were performed using an Applied Biosystems (Foster City, CA) 4000 QTRAP hybrid, triple-quadrupole, linear ion trap mass spectrometer equipped with a Turbo V ion source and operated in MRM mode. For all experiments, the Turbo V ion source was operated in negative electrospray mode (chiral chromatography utilized the ion source in chemical ionization mode, as shown later) and the QTRAP was set as follows: $CUR = 10$ psi, $GS1 = 30$ psi, $GS2 = 30$ psi, $IS = -4500$ V, $CAD = HIGH$, $TEM = 525°$, $ihe = ON$, $EP = -10$ V, and $CXP = -10$ V. The voltage used for CID (-15 to -35 V) and the declustering potentials (-30 to -100 V) varied according to molecular species and were maximized for each eicosanoid.

The Turbo V ion source was operated in atmospheric pressure chemical ionization (APCI) mode when employing chiral chromatography using the following settings: $CUR = 10$ psi, $GS1 = 45$ psi, $GS2 = 60$ psi, $NC = -3.0$ μA, $CAD = HIGH$, $TEM = 400°$, $ihe = ON$, $DP = -60$ V, $EP = -15$ V, and $CXP = -10$ V.

2.6. Quantitation

Eicosanoid quantitation was performed by the stable isotope dilution method previously described by Hall and Murphy (Hall and Murphy, 1998). For each eicosanoid to be quantitated, an internal standard was selected that had a different precursor ion mass than the target analyte, but was chemically and structurally as similar to the target analyte as possible. This is ideally achieved by using a deuterated analog of the analyte.

We employed these standards whenever they were commercially available. In other cases, we employed a deuterated analog that was the closest to the desired analog in characteristics. For example, 15d-$\Delta^{12,14}$ PGJ$_2$ (d4) was employed as the internal standard for PGJ$_2$, 15d-$\Delta^{12,14}$ PGJ$_2$, and 15d-$\Delta^{12,14}$ PGD$_2$. Table 3.1 lists the internal standards that we are currently employing (boxes in gray) and indicates which internal standard is used with which analyte. Presently, eight deuterated internal standards are used to quantitate 16 eicosanoids. An aliquot of the internal standard (10 ng std/100 μl of ethanol) was added to either the media or cell extracts immediately following its isolation. The samples were then processed as previously detailed.

The primary standards contained an accurately known amount of each eicosanoid (non-deuterated) to be quantitated and an accurate aliquot of the internal standards. The concentration of the primary standards must be known with high accuracy. This can be accomplished in one of several ways. In some cases, they are commercially available. Cayman Chemicals, for example, offers a "Quanta-PAK" version of many eicosanoids that contains a deuterated internal standard and a vial containing an accurately determined amount of the non-deuterated primary standard. Some of the eicosanoids (e.g., some HETEs and leukotrienes) have significant ultraviolet (UV) absorption that can be employed to determine the concentration of the standard. The amount of standard can also be determined gravimetrically if a microbalance if available.

The set of primary standards was then prepared by adding accurately determined amounts of the given analyte (non-deuterated) to 100 μl of the same internal standard used to spike the samples. (Note: the concentration of the internal standard does not need to be accurate, but it is crucial that an accurately known volume of the exact same internal standard is added to the sample and to the primary standards.) A typical standard curve consisted of 0.3, 1, 3, 10, 30, and 100 ng of primary standard per 100 μl of internal standard containing 10 ng of each internal standard. The internal standard and the primary standard samples were run before and after each set of unknown samples, and 10 μl of each was loaded onto the column.

A linear standard curve was generated where the ratio of analyte standard peak area to internal standard peak area in the primary standards was plotted versus the amount of primary standard (ng). Figure 3.1 shows examples of three typical standard curves. Linear regression analysis was used to calculate the slope and intercept of the standard curves that were then used to calculate the unknowns. R^2 values for these curves of greater than 0.99 were routinely obtained. The ratio of the unknown analyte peak area to internal standard peak area in the sample was then compared to the appropriate standard curve to calculate the amount of analyte in the sample. Since, in some cases, the deuterated standards contained a small percent of non-deuterated analyte, the LC-MS/MS of the internal standard was analyzed to determine the amount of non-deuterated analyte present. In this case, the

Table 3.1 Eicosanoid library

Eicosanoid[a]	Systematic name	[M−H] (m/z)	Production (m/z)	LC retention time[b] (min)	Internal standard	Recovery[c]	Limit of detection (pg on column)
AA	5Z,8Z,11Z,14Z-eicosatetraenoic acid	303	259	12.4	AA-d$_8$	E	50
AA-d$_8$	5Z,8Z,11Z,14Z-eicosatetraenoic acid (5,6,8,9,11,12,14,15-d8)	311	267	12.4		ND	ND
AA-EA	N-(5Z,8Z,11Z,14Z-eicosatetraenoyl)-ethanolamine	346	259	10.6		E	10
5(S)6(R)DiHETE	5S,6R–dihydroxy-7E,9E,11Z,14Z-eicosatetraenoic acid	335	163	9.0		M	5
5(S)6(S)DiHETE	5S,6S-dihydroxy-7E,9E,11E,14Z-eicosatetraenoic acid	335	163	9.0		E	1
5(S)15(S)DiHETE	5S,15S-dihydroxy-6E,8Z,11Z,13E-eicosatetraenoic acid	335	201	7.8		E	5
8(S)15(S)DiHETE	8S,15S-dihydroxy-5Z,9E,11Z,13E-eicosatetraenoic acid	335		ND[d]		ND	ND
±5,6-DiHETrE	5,6-dihydroxy-8Z,11Z,14Z-eicosatrienoic acid	337	145	9.0		E	1
±8,9-DiHETrE	8,9-dihydroxy-5Z,11Z,14Z-eicosatrienoic acid	337	127	8.8		E	1

(continued)

Table 3.1 (continued)

Eicosanoid[a]	Systematic name	[M−H] (m/z)	Production (m/z)	LC retention time[b] (min)	Internal standard	Recovery[c]	Limit of detection (pg on column)
±11,12-DiHETrE	11,12-dihydroxy-5Z,8Z,14Z-eicosatrienoic acid	337	167	8.7		E	1
±14,15-DiHETrE	14,15-dihydroxy-5Z,8Z,11Z-eicosatrienoic acid	337	207	8.6		E	1
±5,6-EpETrE	5,6-epoxy-8Z,11Z,14Z-eicosatrienoic acid	319	191	10.1		E	5
±8,9-EpETrE	8,9-epoxy-5Z,11Z,14Z-eicosatrienoic acid	319	127	10.0		E	5
±11,12-EpETrE	11,12-epoxy-5Z,8Z,14Z-eicosatrienoic acid	319	167	9.8		E	10
±14,15-EpETrE	14,15-epoxy-5Z,8Z,11Z-eicosatrienoic acid	319	139	9.7		E	50
5(R)HETE	5R-hydroxy-6E,8Z,11Z,14Z-eicosatetraenoic acid	319	115	9.64	5(S)HETE-d$_8$	ND	ND
5(S)HETE	5S-hydroxy-6E,8Z,11Z,14Z-eicosatetraenoic acid	319	115	9.64	5(S)HETE-d$_8$	E	1
5(S)HETE-d$_8$	5S-hydroxy-6E,8Z,11Z,14Z eicosatetraenoic acid (5,6,8,9,11,12,14,15-d8)	327	116	9.60		ND	ND
8(R)HETE	8R-hydroxy-5Z,9E,11Z,14Z-eicosatetraenoic acid	319	155	9.43		ND	ND
8(S)HETE	8S-hydroxy-5Z,9E,11Z,14Z-eicosatetraenoic acid	319	155	9.43		E	1
9-HETE	9-hydroxy-5Z,7E,11Z,14Z-eicosatetraenoic acid	319	151	9.49		E	1
11(R)HETE	11R-hydroxy-5Z,8Z,12E,14Z-eicosatetraenoic acid	319	167	9.31	5(S)HETE-d$_8$	ND	ND
11(S)HETE	11S-hydroxy-5Z,8Z,12E,14Z-eicosatetraenoic acid	319	167	9.31	5(S)HETE-d$_8$	E	1

12(R)HETE	12R-hydroxy-5Z,8Z,10E,14Z-eicosatetraenoic acid	319	179	9.38		ND	ND
12(S)HETE	12S-hydroxy-5Z,8Z,10E,14Z-eicosatetraenoic acid	319	179	9.38		E	1
15(R)HETE	15R-hydroxy-5Z,8Z,11Z,13E-eicosatetraenoic acid	319	175	9.19	5(S)HETE-d$_8$	ND	ND
15(S)HETE	15S-hydroxy-5Z,8Z,11Z,13E-eicosatetraenoic acid	319	175	9.19	5(S)HETE-d$_8$	E	1
20-HETE	20-hydroxy-5Z,8Z,11Z,14Z-eicosatetraenoic acid	319	245	8.98		E	1
12(S)HHTrE	12S-hydroxy-5Z,8E,10E-heptadecatrienoic acid	279	163	8.7		E	50
5(S)HpETE	5S-hydroperoxy-6E,8Z,11Z,14Z-eicosatetraenoic acid	335	155	9.7		E	5
12(S)HpETE	12S-hydroperoxy-5Z,8Z,10E,14Z-eicosatetraenoic acid	335	153	9.4		E	1
15(S)HpETE	15S-hydroperoxy-5Z,8Z,11Z,13E-eicosatetraenoic acid	335	113	9.2		E	1
LTB$_4$	5S,12R-dihydroxy-6Z,8E,10E,14Z-eicosatetraenoic acid	335	195	8.2		M	5
6 trans LTB$_4$	5S,12R-dihydroxy-6E,8E,10E,14Z-eicosatetraenoic acid	335	195	7.8		E	1
6 trans 12 epi LTB$_4$	5S,12S-dihydroxy-6E,8E,10E,14Z-eicosatetraenoic acid	335	195	8.0		E	5

(*continued*)

Table 3.1 (continued)

Eicosanoid[a]	Systematic name	[M−H] (m/z)	Production (m/z)	LC retention time[b] (min)	Internal standard	Recovery[c]	Limit of detection (pg on column)
LTC$_4$	5S-hydroxy,6R-(S-glutathionyl), 7E,9E,11Z,14Z-eicosatetraenoic acid	624	272	8.8		P	1
11-trans LTC$_4$	5S-hydroxy,6R-(S-glutathionyl), 7E,9E,11E,14Z-eicosatetraenoic acid	624	272	9.2		M	1
LTE$_4$	5S-hydroxy,6R-(S-cysteinyl),7E,9E,11Z,14Z-eicosatetraenoic acid	438	235	10.1		M	5
11-trans LTE$_4$	5S-hydroxy,6R-(S-cysteinyl),7E,9E,11E,14Z-eicosatetraenoic acid	438	235	10.4		M	1
5(S)6(R)15(S) LXA$_4$	5S,6R,15S-trihydroxy-7E,9E,11Z,13E-eicosatetraenoic acid	351	115	5.2		M	1
5(S)6(S)15(S) LXA$_4$	5S,6S,15S-trihydroxy-7E,9E,11Z,13E-eicosatetraenoic acid	351		ND[d]		ND	ND
5(S)14(R)15(S) LXB$_4$	5S,14R,15S-trihydroxy-6E,8Z,10E,12E-eicosatetraenoic acid	351		ND[d]		ND	ND
5-OxoETE	5-oxo 6E,8Z,11Z,14Z-eicosatetraenoic acid	317	203	9.8		E	5
12-OxoETE	12-oxo-5Z,8Z,10E,14Z-eicosatetraenoic acid	317	153	9.4		E	1
15-OxoETE	15-oxo-5Z,8Z,11Z,13E-eicosatetraenoic acid	317		ND[d]		ND	ND

PGA$_2$	9-oxo-15S-hydroxy-5Z,10Z,13E-prostatrienoic acid	333		ND^d	ND	ND
dhk-PGA$_2$	9,15-dioxo-5Z,10-prostadienoic acid	333		ND^d	ND	ND
PGB$_2$	15S-hydroxy-9-oxo-5Z,8(12),13E-prostatrienoic acid	333	175	6.6	E	5
PGD$_2$	9S,15S-dihydroxy-11-oxo-5Z,13E-prostadienoic acid	351	189	4.6	M	5
PGD$_2$-d$_4$	9S,15S-dihydroxy-11-oxo-5Z,13E-prostadienoic acid (3,3,4,4-d4)	355	193	4.6	PGD$_2$-d$_4$	ND
PGD$_2$-EA	N-(9S,15S-dihydroxy-11-oxo-5Z,13E-prostadienoyl)-ethanolamine	394	271	3.3		E	10
15d-$\Delta^{12,14}$PGD$_2$	9S-hydroxy-11-oxo-5Z,12E,14E-prostatrienoic acid	333	271	8.2	15d-$\Delta^{12,14}$PGJ$_2$-d4	E	1
dhk-PGD$_2$	11,15-dioxo-9S-hydroxy-5Z-prostenoic acid	351	207	5.9	dhk-PGD$_2$-d4	P	1
dhk-PGD$_2$-d4	11,15-dioxo-9S-hydroxy-5Z-prostenoic acid (3,3,4,4-d4)	355	211	5.9		ND	ND
6-keto PGE$_1$	6,9-dioxo-11R,15S-dihydroxy-13E-prostenoic acid	367	143	3.1		P	5
PGE$_2$	9-oxo-11R,15S-dihydroxy-5Z,13E-prostadienoic acid	351	189	4.3	PGE$_2$-d$_4$	M	10
PGE$_2$-d$_4$	11R,15S-dihydroxy-9-oxo-5Z,13E-prostadienoic acid (3,3,4,4-d4)	355	193	4.3		ND	ND
PGE$_2$-EA	N-(11R,15S-dihydroxy-9-oxo-5Z,13E-prostadienoyl)-ethanolamine	394	203	3.0		M	10000

(*continued*)

Table 3.1 (continued)

Eicosanoid[a]	Systematic name	[M−H] (m/z)	Production (m/z)	LC retention time[b] (min)	Internal standard	Recovery[c]	Limit of detection (pg on column)
bicyclo-PGE$_2$	9,15-dioxo-5Z-prostaenoic acid-cyclo[11S,16]	333	175	7.4		E	5
dhk-PGE$_2$	9,15-dioxo-11R-hydroxy-5Z-prostenoic acid	351	207	5.3	dhk-PGD$_2$-d4	M	1
19(R)-hydroxy PGE$_2$	9-oxo-11R,15S,19R-trihydroxy-5Z,13E-prostadienoic acid	367	287	2.4		M	1
20-hydroxy PGE$_2$	9-oxo-11R,15S,20-trihydroxy-5Z,13E-prostadienoic acid	367	287	2.4		M	5
15-keto PGE$_2$	9,15-dioxo-11R-hydroxy-5Z,13E-prostadienoic acid	349	161	4.7		M	1
tetranor PGEM	11R-hydroxy-9,15-dioxo-2,3,4,5-tetranor-prostan-1,20 dioic acid	327	291	2.3		P[e]	50
6,15-diketo 13,14-dihydro PGF1α	6,15-dioxo-9S,11R-dihydroxy-13E-prostenoic acid	369	267	3.6		P	50
6-keto PGF$_{1α}$	6-oxo-9S,11R,15S-trihydroxy-13E-prostenoic acid	369	207	2.9	6-keto PGF$_{1α}$-d4	E	10
6-keto PGF$_{1α}$-d4	6-oxo-9S,11R,15S-trihydroxy-13E-prostenoic acid (3,3,4,4-d4)	373	211	2.9		ND	ND
PGF$_{2α}$	9S,11R,15S-trihydroxy-5Z,13E-prostadienoic acid	353	193	4.0	PGF$_{2α}$-d$_4$	M	10
PGF$_{2α}$-d$_4$	9S,11R,15S-trihydroxy-5Z,13E-prostadienoic acid (3,3,4,4-d4)	357	197	4.0		ND	ND
PGF$_{2α}$-EA	N-(9S,11R,15S-trihydroxy-5Z,13E-prostadienoyl)-ethanolamine	396	334	3.0		M	10

Name	Systematic name				IS		
11β-PGF$_{2α}$	9S,11S,15S-trihydroxy-5Z,13E-prostadienoic acid	353	193	3.7		M	1
dhk-PGF$_{2α}$	9S,11S-dihydroxy-15-oxo-5Z-prostenoic acid	353	209	5.2	dhk-PGF$_{2α}$-d4	M	5
dhk-PGF$_{2α}$-d4	9S,11S-dihydroxy-15-oxo-5Z-prostenoic acid (3,3,4,4-d4)	357	213	5.2		ND	ND
2,3 dinor-11β PGF2α	9S,11S,13S-trihydroxy-2,3-dinor-5Z,13E-prostadienoic acid	325	145	3.0		M	1
20-hydroxy PGF$_{2α}$	9S,11S,15S,20-tetrahydroxy-5Z,13E-prostadienoic acid	369	193	2.3		M	50
15-keto PGF$_{2α}$	9S,11R-dihydroxy-15-oxo-5Z,13E-prostadienoic acid	351	217	4.5		M	1
PGF$_{2β}$	9R,11R,15S-trihydroxy-5Z,13E-prostadienoic acid	353		ND^d		ND	ND
tetranor PGFM	9S,11R-dihydroxy-15-oxo-2,3,4,5-tetranor-prostan-1,20-dioic acid	329	293	2.1		P^e	10
PGG$_2$	9S,11R-epidioxy-15S-hydroperoxy-5Z,13E-prostadienoic acid	367		ND^d		ND	ND
PGH$_2$	9S,11R-epidioxy-15S-hydroxy-5Z,13E-prostadienoic acid	351		ND^d		ND	ND
PGJ$_2$	11-oxo-15S-hydroxy-5Z,9Z,13E-prostatrienoic acid	333	189	6.5	15d-$\Delta^{12,14}$PGJ$_2$-d4	E	1
Δ^{12}PGJ$_2$	11-oxo-15S-hydroxy-5Z,9Z,12E-prostatrienoic acid	333		ND^d		ND	ND
15d-$\Delta^{12,14}$PGJ$_2$	11-oxo-5Z,9Z,12E,14Z-prostatetraenoic acid	315	271	8.8	15d-$\Delta^{12,14}$PGJ$_2$-d4	E	5

(*continued*)

Table 3.1 (continued)

Eicosanoid[a]	Systematic name	[M−H] (m/z)	Production (m/z)	LC retention time[b] (min)	Internal standard	Recovery[c]	Limit of detection (pg on column)
15d-Δ[12,14]PGJ$_2$-d4	11-oxo-5Z,9Z,12E,14Z-prostatetraenoic acid (3,3,4,4-d4)	319	275	8.8		ND	ND
PGK$_2$	9,11-dioxo-15S-hydroxy-5Z,13E-prostadienoic acid	349	205	4.4		ND	ND
TXB$_2$	9S,11,15S-trihydroxy-thromboxa-5Z,13E-dien-1-oic acid	369	169	3.6	TXB$_2$-d4	M	10
TXB$_2$-d4	9S,11,15S-trihydroxy-thromboxa-5Z,13E-dien-1-oic acid (3,3,4,4-d4)	373	173	3.6		ND	ND
11-dehydro TXB$_2$	9S,15S-dihydroxy-11-oxo-thromboxa-5Z,13E-dien-1-oic acid	367		ND[d]		ND	ND
2,3-dinor TXB$_2$	9S,11,15S-trihydroxy-2,3-dinor-thromboxa-5Z,13E-dien-1-oic acid	341		ND[d]		ND	ND

[a] Gray boxes indicate internal standards.
[b] The retention times are given to indicate the relative elution position of the compounds with the understanding that the absolute values are not significant.
[c] Recoveries were grouped as P for poor (<25%), M for moderate (25–75%), E for excellent (<76%), and ND for not determined.
[d] These compounds were not analyzed quantitatively for this table, but the MS/MS spectra and LC retention times are available in the LIPID MAPS Eicosanoid Library at www.lipidmaps.org or in Harkewicz et al., 2007.
[e] Tetranor PGFM and tetranor PGEM did not bind to the Strata-X SPE column and thus could not be detected by this method.

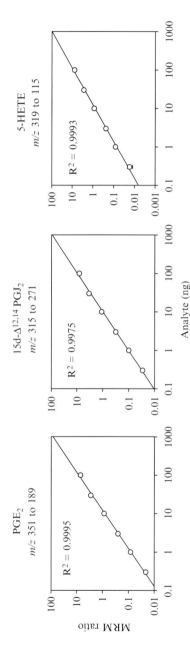

Figure 3.1 Eicosanoid standard curves. Three typical standard curves are shown. The standard solutions are prepared as described in the "Methods" section. The solutions contained 10 ng of an internal standard for each analyte and 0.3 to 100 ng of the analyte in 140 μl of which 10 μl were analyzed. The multi-reaction monitoring (MRM) transitions employed to monitor the analytes are listed in the figure. The internal standards and transitions were: PGE_2 (d4) m/z 355–193, 15d-$\Delta^{12,14}$ PGJ_2 (d4) m/z 319–275, and 5-HETEs (d8) m/z 327–116. The data were presented in log scale, but the linear regression analysis to determine parameters was done on the original non-log data.

non-deuterated contaminant was subtracted from each analysis. The dynamic range that can be covered is limited on the low end by the amount of non-deuterated analyte in the internal standards and the sensitivity of the mass spectrometer. The upper limit is restricted by ion suppression and detector saturation issues.

3. Results and Discussion

3.1. MRM transition selection

To date, we have compiled a library of over 60 eicosanoids that we can detect and quantitate with these methods. These compounds are listed in Table 3.1 with the precursor and product ions used in the MRM analysis. We have also included in this table several compounds for which we have MS/MS spectra but for which we have not selected MRM transitions. We have not included any recovery or limit of detection data for these compounds.

We have compiled the MS/MS spectra of each of these compounds. We have published these data (Harkewicz et al., 2007). These data also can be accessed on the LIPID MAPS web page at http://www.lipidmaps.org. In addition to MS/MS spectra, the web visitor can obtain chemical structures for standards in both GIF and ChemDraw® formats, specific details regarding LC and MS parameters employed in our analysis, structures of dominant fragment ions (including literature references, when available, for fragment assignments), and retention times for a stated set of chromatographic conditions. Lastly, a web link to Cayman Chemical provides useful information and references on specific eicosanoids.

The product ions employed here for the MRM detection were selected to yield the best discrimination from other eicosanoids that co-elute in the vicinity of the analyte and to yield the highest signal. By balancing LC retention time and product ion selection, we were able to successfully distinguish the large majority of the eicosanoids listed. Various product ions can be selected to obtain greater sensitivity if conflicting eicosanoids are not present in a given set of samples.

The MS/MS spectra of most eicosanoids show numerous product ions. While a similar eicosanoid may have the same product ions, their relative intensities usually vary. The ratio of intensities of these product ions can be used to distinguish these species. In this case, multiple MRM transitions can then be analyzed, and the ratio of product ions found in the unknown can be compared with either an MS/MS library spectra or a pure standard run under the same conditions. This would aid in confirming the identity of a chromatographic peak.

Figure 3.2 shows the chromatograms of a few selected eicosanoids to illustrate several points about this procedure. All of the panels in Fig. 3.2 were generated by overlaying the individual MRM chromatograms for the various MRM pairs onto a single plot.

The sharpest peaks have a width at half height of 6 s and the baseline peak width on the order of 18 s. The Applied Biosystems 4000 QTrap can handle over 100 MRM pairs in a single scan. The dwell time employed when we are scanning 60 to 70 analytes is 25 ms. This produces a cycle time of 2 s per scan. This translates to at least nine data points per peak for the narrowest peaks, which is sufficient to accurately define the peak shape for quantitation. The Applied Biosystems 4000 QTrap software allows the user to break the MRM pairs into sets, and these sets can be run in series during the course of a single analysis so that only a fraction of the MRM pairs are being scanned during any time period. Running fewer MRM pairs in each scan allows the dwell time to be increased. Although this would not increase the absolute intensity of the peaks, it would increase the time averaging for each data point, thus decreasing the noise levels and increasing the signal-to-noise (S/N) ratio.

Figure 3.2A displays the retention times of several typical prostaglandins. Most of the prostaglandins show baseline separation from all other prostaglandins, and the analysis and identification of the peaks is fairly clear cut.

Figure 3.2B displays the region of the chromatogram that contains the HETEs, DiHETEs, EpETrEs, and DiHETrEs. Some prostaglandins and the leukotrienes also elute in this region; however, their molecular ions m/z are different from the hydroxy eicosanoids so that they do not appear in these scans. Even so, Fig. 3.2B clearly indicates that significant care must be taken when assigning the peaks in this region to the correct eicosanoid. While the differences in the retention times of the HETEs are reproducible, for the HETEs that elute between 8- and 12-HETE, they are not very large. The absolute value of the retention time for a given eicosanoid fluctuates due to batch variability in the LC solvents, and these shifts can be larger than the differences between the retention times of two neighboring eicosanoids. For this reason, a complete set of standards should be run both before and after each set of unknowns to accurately determine eicosanoid retention times, especially when assigning peak identities. If these HETEs are present in the samples, the gradient in this region of the chromatogram can be altered to spread out this region and improve the resolution.

Figure 3.2C shows the chromatograms of LTB_4, LTC_4, 11-trans LTC_4, LTE_4, and 11-trans LTE_4. This panel also demonstrates one of the significant powers of the MRM method. Examination of Fig. 3.2A and B shows that the region from 8 to 11 min is very crowded, and that a significant number of eicosanoids elute in this region. Yet, when the MRM transitions are different, the chromatograms are very clean, and the assignments are easily made. This panel also highlights an important precaution that must be considered with the MRM method. While the chromatogram of a particular analyte may appear to be very clean because no other compounds exhibit the same MRM transition, a tremendous amount of material may be eluting in the same place, but with different MRM transitions.

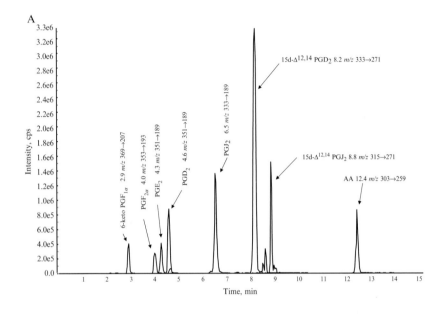

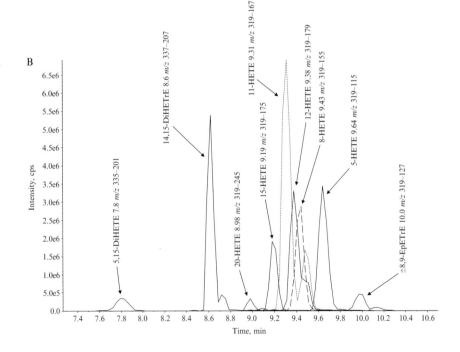

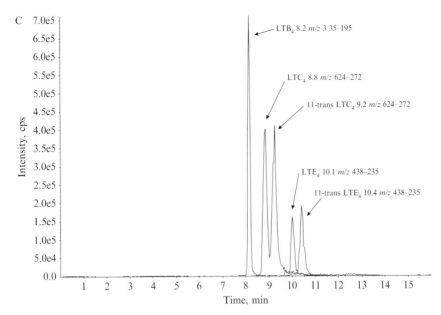

Figure 3.2 High performance liquid chromatography (HPLC) eicosanoid chromatography on reverse-phase C18. The chromatography profiles of selected eicosanoid standards run on reverse-phase C18 HPLC (see "Methods" section). In each panel, the individual chromatograms produced by a given multi-reaction monitoring (MRM) pair have been overlaid. Each label lists the eicosanoid, retention time, and multi-reaction monitoring transition that produced a given chromatogram. (A) Representative sample of prostaglandins and arachidonic acid (AA). (B) Representative sample of the hydroxy-eicosanoids, including examples of HETEs, diHETEs, diHETrEs, and EpETrEs. (C) Representative sample of leukotrienes.

This material can still cause significant ion suppression of the analyte being examined. For example, we have found that PGE_2 and PGD_2 appear to co-elute with material co-extracted from Dulbecco's Modification of Eagle's Medium (DMEM) cell culture media, significantly diminishing the MS sensitivity of these eicosanoids (see "Recoveries" section). The proper use of internal standards can compensate for some of these issues during quantitation.

3.2. Stereoisomer detection

Table 3.1 contains eight pairs of isomers that have identical retention time and MS/MS spectra, and thus cannot be distinguished with the C18 column. To identify and quantitate these isomers normal phase chiral chromatography is required.

3.3. Lower limit of detection

Table 3.1 also contains an estimate of the lower limit of detection (LOD) for the eicosanoid when employing our standard procedures. A set of standards was made and then serially diluted to produce a set of standards such that 1 to 5000 pg would be loaded onto the column. A signal was judged to be significant if the signal area was three times the noise at the three-standard-deviation level. Most limits were between 1 and 50 pg loaded onto the column. Only PGE_2-EA had an LOD greater than 50 ng. The detection limits that we obtained were achieved with a 2.1 × 250 mm Grace-Vydac reverse-phase C18 column with 5 µm particle size. Decreasing the column diameter, the particle size, and the flow rate can increase sensitivity, as can changing the column packing. We have analyzed only one isomer of any isomeric analytes and only non-deuterated analytes. Therefore, we have listed the LOD and recovery values for the other isomers and for the deuterated standards as not determined (ND) in Table 3.1. However, their recoveries and LODs should be very similar to the corresponding isomer that was analyzed or the equivalent non-deuterated analyte.

Figure 3.3 shows an example of the analysis of a sample that was close to our lower limit of detection. It is the chromatogram showing the analysis of PGE_2 and PGD_2 in a sample of spinal fluid from a rat that had been

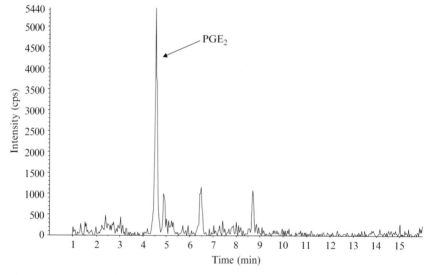

Figure 3.3 PGE_2 analysis of rat spinal fluid. Spinal fluid was collected from a rat whose paw had been injected with carrageenan to induce an inflammatory pain state. A sample of spinal fluid was removed (42 µl), processed as described in the "Methods" section and analyzed by liquid chromatography–mass spectrometry–mass spectrometry. Eighty percent of the sample was analyzed. The PGE_2 peak corresponds to 52 pg of PGE_2 being loaded onto the column. The signal-to-noise (S/N) ratio was 16-fold over a three-standard-deviation noise level.

subjected to the carrageenan model of inflammatory pain. This represents 52 pg of PGE_2 load onto the column and correlates with 150 pg of PGE_2 present in the original 42-μl spinal fluid sample.

3.4. Recoveries

Our primary goal was to analyze as many eicosanoids in a single run as possible. We chose to focus on maximizing the recovery of AA, prostaglandins, and HETEs. Table 3.1 shows the recoveries that this system achieved for most of the analytes that we detect. These recoveries were determined by adding a known amount of each analyte, contained in a standard mix, to 2.0 ml of water or DMEM, and then isolating the eicosanoids via the standard sample preparation method previously outlined. We also did an "add back" experiment where DMEM alone was extracted by the same method; however, instead of adding an aliquot of the standard eicosanoid mixture to the DMEM sample before extraction, the eicosanoids were added to the post-extraction methanol column effluent. Losses in the "add back" samples would indicate that materials in the media are being co-extracted with the eicosanoids, and that these contaminants affect the eicosanoid MS response. All three sets of samples were analyzed with our LC-MS/MS procedure. The eicosanoid peak intensities of these samples were compared to those of standards that had not been through the isolation procedure, but instead were directly analyzed by LC-MS/MS.

Table 3.1 reports the relative recoveries of standards that were extracted from DMEM. The mono- and di-hydroxy eicosanoids had excellent recoveries at between 75 and 100%. The leukotrienes and prostaglandins had only moderate recoveries in the range of 50%. In most cases, the recoveries of these compounds from water were in the 80 to 100% range. The "add back" experiment showed that, for the prostaglandins, most of the losses occurred in the "add back" experiment, implying that some component from the media is being extracted that decreases the MS response to the prostaglandins. However, the "add back" levels for the cysteinyl leukotrienes were 100% within experimental error, suggesting that the losses occurred during the extraction process, and that media components are not affecting leukotriene detection. Tetranor PGEM and PGFM do not bind to the Strata-X SPE columns under these conditions and could not be detected by this protocol. The PGK_2 had very large errors, and extraction could not be measured. Again, the yield of a given class of compounds could be improved by altering the conditions or column type, but often at the expense of other analytes.

3.5. Miscellany

The process of culturing cells can affect the eicosanoid levels in other ways as well. For example, PGD_2 can undergo dehydration to form PGJ_2, 15d-$\Delta^{12,14}$ PGJ_2, and 15d-$\Delta^{12,14}$ PGD_2. This dehydration has been reported to be accelerated in serum albumin (Fitzpatrick and Wynalda, 1983; Maxey et al., 2000). This decomposition will not be compensated for by the internal standards since they are not added until after the cell incubations. We have also found that when 10% serum is present during the cell culture, no LTC_4 could be detected; however, LTE_4, which is a breakdown product of LTC_4, was detected. When the same experiments are run in serum-free media, significant levels of the LTC_4 were detected, but very little LTE_4. Presumably, the serum is catalyzing the conversion of LTC_4 to LTE_4. Clearly, the types of recovery experiments outlined above must be conducted whenever applying this system to a new type of sample (e.g. media, serum, or tissues) or when changing the conditions of an already-tested sample. Care must also be taken to determine the effects that the addition of any agents (e.g., inhibitors, activators, or drugs) have on the extractions and on the quantitation of any other analytes. LC-MS/MS should also be done on the agents being added to determine if they have any MRM transitions that could be mistaken for one of the standard analytes.

Traditionally, lipids solutions were routinely acidified before subjecting them to liquid extractions. We compared our recoveries with and without acidifying the media before application to the SPE columns. We found that there were no significant differences; therefore, we do not routinely acidify the media. It is possible that eicosanoids other than those we employed in our testing could benefit by acidification. In addition, some eicosanoid species are susceptible to air oxidation and/or adhere to vessels and should not be taken to dryness. To guard against these losses, 20 μl of a 50/50 solution of glycerol/ethanol can be added to the methanol column elution just before drying down the samples on the Speedvac. The glycerol remains, and the eicosanoids are concentrated into the glycerol, which can then be taken up in the LC equilibration buffer for LC-MS/MS analysis.

3.6. Summary

The method outlined above is a sensitive, accurate one for identifying and quantitating a large number of eicosanoids in a single LC-MS/MS run. This method can detect 1 pg (on column) of many eicosanoids, which is similar to the sensitivity of EIA analysis. Its primary advantage is that a large number of different species can be measured in a single analysis, while EIA analysis requires specific antibodies that are often not commercially available.

ACKNOWLEDGMENTS

We would like to thank Dr. Robert C. Murphy (University of Colorado School of Medicine, Denver, CO) for his invaluable help and advice on the use of LC-MS/MS as a quantitative tool to analyze eicosanoids. This work was supported by the LIPID

MAPS Large Scale Collaborative Grant number GM069338 from the U.S. National Institutes of Health.

REFERENCES

Baranowski, R., and Pacha, K. (2002). Gas chromatographic determination of prostaglandins. *Mini. Rev. Med. Chem.* **2**, 135–144.
Buczynski, M. W., Stephens, D. L., Bowers-Gentry, R. C., Grkovich, A., Deems, R. A., and Dennis, E. A. (2007). Multiple agonist induced changes in eicosanoid metabolites correlated with gene expression in macrophages. *J. Biol. Chem.* Epub in advance of print.
Fitzpatrick, F. A., and Wynalda, M. A. (1983). Albumin-catalyzed metabolism of prostaglandin D2: Identification of products formed *in vitro*. *J. Biol. Chem.* **258**, 11713–11718.
Funk, C. D. (2001). Prostaglandins and leukotrienes: Advances in eicosanoid biology. *Science* **294**, 1871–1875.
Hall, L. M., and Murphy, R. C. (1998). Electrospray mass spectrometric analysis of 5-hydroperoxy and 5-hydroxyeicosatetraenoic acids generated by lipid peroxidation of red blood cell ghost phospholipids. *J. Am. Soc. Mass Spectrom.* **9**, 527–532.
Harkewicz, R., Fahy, E., Andreyev, A., and Dennis, E. A. (2007). Arachidonate-derived dihomoprostaglandin production observed in endotoxin-stimulated macrophage-like cells. *J. Biol. Chem.* **282**, 2899–2910.
Kita, Y., Takahashi, T., Uozumi, N., and Shimizu, T. (2005). A multiplex quantitation method for eicosanoids and platelet-activating factor using column-switching reversed-phase liquid chromatography-tandem mass spectrometry. *Anal. Biochem.* **342**, 134–143.
Margalit, A., Duffin, K. L., and Isakson, P. C. (1996). Rapid quantitation of a large scope of eicosanoids in two models of inflammation: Development of an electrospray and tandem mass spectrometry method and application to biological studies. *Anal. Biochem.* **235**, 73–81.
Maxey, K. M., Hessler, E., MacDonald, J., and Hitchingham, L. (2000). The nature and composition of 15-deoxy-Delta(12,14)PGJ(2). *Prostaglandins Other Lipid Mediat.* **62**, 15–21.
Murphy, R. C., Barkley, R. M., Zemski, B. K., Hankin, J., Harrison, K., Johnson, C., Krank, J., McAnoy, A., Uhlson, C., and Zarini, S. (2005). Electrospray ionization and tandem mass spectrometry of eicosanoids. *Anal. Biochem.* **346**, 1–42.
Peters-Golden, M., and Brock, T. G. (2003). 5-lipoxygenase and FLAP. *Prostaglandins Leukot. Essent. Fatty Acids* **69**, 99–109.
Reinke, M. (1992). Monitoring thromboxane in body fluids: A specific ELISA for 11-dehydrothromboxane B2 using a monoclonal antibody. *Am. J. Physiol.* **262**, E658–E662.
Sacerdoti, D., Gatta, A., and McGiff, J. C. (2003). Role of cytochrome P450-dependent arachidonic acid metabolites in liver physiology and pathophysiology. *Prostaglandins Other Lipid Mediat.* **72**, 51–71.
Schaloske, R. H., and Dennis, E. A. (2006). The phospholipase A$_2$ superfamily and its group numbering system. *Biochim. Biophys. Acta* **1761**, 1246–1259.
Shono, F., Yokota, K., Horie, K., Yamamoto, S., Yamashita, K., Watanabe, K., and Miyazaki, H. (1988). A heterologous enzyme immunoassay of prostaglandin E2 using a stable enzyme-labeled hapten mimic. *Anal. Biochem.* **168**, 284–291.
Simmons, D. L., Botting, R. M., and Hla, T. (2004). Cyclooxygenase isozymes: The biology of prostaglandin synthesis and inhibition. *Pharmacol. Rev.* **56**, 387–437.
Six, D. A., and Dennis, E. A. (2000). The expanding superfamily of phospholipase A$_2$ enzymes: Classification and characterization. *Biochim. Biophys. Acta* **1488**, 1–19.

Smith, W. L., DeWitt, D. L., and Garavito, R. M. (2000). Cyclooxygenases: Structural, cellular, and molecular biology. *Annu. Rev. Biochem.* **69,** 145–182.

Spokas, E. G., Rokach, J., and Wong, P. Y. (1999). Leukotrienes, lipoxins, and hydroxyeicosatetraenoic acids. *Methods Mol. Biol.* **120,** 213–247.

Takabatake, M., Hishinuma, T., Suzuki, N., Chiba, S., Tsukamoto, H., Nakamura, H., Saga, T., Tomioka, Y., Kurose, A., Sawai, T., and Mizugaki, M. (2002). Simultaneous quantification of prostaglandins in human synovial cell-cultured medium using liquid chromatography/tandem mass spectrometry. *Prostaglandins Leukot. Essent. Fatty Acids* **67,** 51–56.

CHAPTER FOUR

STRUCTURE-SPECIFIC, QUANTITATIVE METHODS FOR ANALYSIS OF SPHINGOLIPIDS BY LIQUID CHROMATOGRAPHY–TANDEM MASS SPECTROMETRY: "INSIDE-OUT" SPHINGOLIPIDOMICS

M. Cameron Sullards, Jeremy C. Allegood, Samuel Kelly, Elaine Wang, Christopher A. Haynes, Hyejung Park, Yanfeng Chen, and Alfred H. Merrill, Jr.

Contents

1. Introduction: An Overview of Sphingolipid Structures and Nomenclature — 86
2. Analysis of Sphingolipids by Mass Spectrometry — 89
   2.1. Electron ionization-mass spectrometry — 89
   2.2. Fast atom bombardment and liquid secondary ionization mass spectrometry — 90
   2.3. Electrospray ionization — 90
   2.4. Liquid chromatography, electrospray ionization, mass spectrometry and tandem mass spectrometry — 91
   2.5. MALDI, mass spectrometry, and tandem mass spectrometry — 91
3. Analysis of Sphingolipids by "OMIC" Approaches — 92
4. Materials and Methods — 95
5. Materials — 98
   5.1. Biological samples — 98
6. Extraction — 98
   6.1. Preparation of samples for LC-MS/MS — 99
7. Identification of the Molecular Species by Tandem Mass Spectrometry — 100
   7.1. Materials for infusion and LC-ESI-MS/MS — 100

7.2. Sphingolipid subspecies characterization prior
to quantitative LC-MS/MS analysis 100
8. Quantitation by LC-ESI-MS/MS Using Multireaction Monitoring 101
8.1. Analysis of sphingoid bases and sphingoid base
1-phosphates in positive ion mode 102
9. Analysis of (Dihydro)Ceramides, (Dihydro)Sphingomyelins, and
(Dihydro)Monohexosyl-Ceramides in Positive Ion Mode 104
9.1. Modifications for analysis of glucosylceramides
and galactosylceramides 105
9.2. Analysis of ceramide 1-phosphates by positive and negative
ion modes 106
9.3. Analytical methods for additional analytes 106
10. Other Methods 109
Acknowledgments 111
References 112

Abstract

Due to the large number of highly bioactive subspecies, elucidation of the roles of sphingolipids in cell structure, signaling, and function is beginning to require that one perform structure-specific and quantitative (i.e., "sphingolipidomic") analysis of all individual subspecies, or at least of those are relevant to the biologic system of interest. As part of the LIPID MAPS Consortium, methods have been developed and validated for the extraction, liquid chromatographic (LC) separation, and identification and quantitation by electrospray ionization (ESI), tandem mass spectrometry (MS/MS) using an internal standard cocktail that encompasses the signaling metabolites (e.g., ceramides, ceramide 1-phosphates, sphingoid bases, and sphingoid base 1-phosphates) as well as more complex species (sphingomyelins, mono- and di-hexosylceramides). The number of species that can be analyzed is growing rapidly with the addition of sulfatides and other complex sphingolipids as more internal standards become available. This review describes these methods as well as summarizes others from the published literature.

Sphingolipids are an amazingly complex family of compounds that are found in all eukaryotes as well as some prokaryotes and viruses. The size of the sphingolipidome (i.e., all of the individual molecular species of sphingolipids) is not known, but must be immense considering mammals have over 400 headgroup variants (for a listing, see http://www.sphingomap.org), each of which is comprised of at least a few—and, in some cases, dozens—of lipid backbones. No methods have yet been developed that can encompass so many different compounds in a structurally specific and quantitative manner. Nonetheless, it is possible to analyze useful subsets of the sphingolipidome, such as the backbone sphingolipids involved in signaling (sphingoid bases, sphingoid base 1-phosphates, ceramides, and ceramide 1-phosphates) and metabolites at important branchpoints, such as the partitioning of ceramide into sphingomyelins, glucosylceramides, galactosylceramides, and ceramide 1-phosphate versus turnover to the backbone sphingoid base. This review describes methodology that has been developed as part of the LIPID MAPS Consortium (www.lipidmaps.org) as well as other methods that can be used for sphingolipidomic analysis to the extent that such is currently feasible. The focus of this review is primarily mammalian sphingolipids; hence, if readers are interested in methods to study other organisms, they should consult the excellent review by Stephen Levery in another volume of *Methods in Enzymology* (Levery, 2005), which covers additional species found in plants, fungi, and other organisms.

It should be noted from the start that although many analytical challenges remain in the development of methods to analyze the full "sphingolipidome," the major impediment to progress is the limited availability of reliable internal standards for most of the compounds of interest. Because it is an intrinsic feature of mass spectrometry that ion yields tend to vary considerably among different compounds, sources, methods, and instruments, an analysis that purports to be quantitative will not be conclusive unless enough internal standards have been added to correct for these variables. Ideally, there should be some way of standardizing every compound in the unknown mixture; however, that is difficult, if not impossible, to do because the compounds are not available, and the inclusion of so many internal standards generates a spectrum that may be too complex to interpret. Therefore, a few representative internal standards are usually added, and any known differences in the ion yields of the analytes of interest versus the spiked standard are factored into the calculations. Identification of appropriate internal standards has been a major focus of the LIPID MAPS Consortium, and the methods described in this review are based on the development of a certified (i.e., compositionally and quantitatively defined by the supplier) internal standard cocktail that is now commercially available (Avanti Polar Lipids, Alabaster, AL).

For practical and philosophical reasons, an internal standard cocktail was chosen over the process of an investigator adding individual standards for only the analytes of interest. On the practical level, addition of a single cocktail minimizes pipetting errors as well as keeping track of whether each internal standard is still usable (e.g., has it degraded while in solution?). Philosophically, the internal standard cocktail was chosen because an underlying premise of systems analysis asserts that, due to the high relevancy of unexpected interrelationships involving more distant components, one can only understand a biological system when factors outside the primary focus of the experiment have also been examined. Indeed, the first payoffs of "omics" and systems approaches involve the discoveries of interesting compounds in unexpected places when a "sphingolipidomic" analytical method was being used as routine practice instead of a simpler method that would have only measured the compound initially thought to be important (Zheng *et al.*, 2006). Thus, routine addition of a broad internal standard cocktail at the outset of any analysis maximizes the opportunity for such discoveries, both at the time the original measurements are made and when one decides to return to the samples later, which can fortunately be done for many sphingolipids because they remain relatively stable in storage.

1. INTRODUCTION: AN OVERVIEW OF SPHINGOLIPID STRUCTURES AND NOMENCLATURE

An important first step in "omic" research is to have structurally precise (yet convenient) ways of describing the compounds that have been measured (e.g., instead of reporting the amount of unspecified "ceramides" in a biological sample, report the individual subspecies). For sphingolipids, this means specifying the nature of the sphingoid base backbones (sometimes also called "long-chain bases"), the amide-linked fatty acids (found in most complex sphingolipids), headgroups, and any other modifications. A proposed nomenclature (Fahy et al., 2005) is described below, but it should always be borne in mind that it is often not possible to unassailably prove some aspects of the structure of a compound (i.e., the stereochemistry, positions of double bonds, etc.) using MS alone. Nonetheless, in cases where nature has biased the structures toward specific species, it is appropriate to propose structures based on prior experience (as has been done for genes, proteins, glycans, etc.), while ever mindful of these assumptions.

A substantial number of compounds in nature have sphingolipid-like structures (such as fumonisins and myriocin); however, those that are formally regarded to be sphingolipids have a lipid backbone comprised of a 1,3-dihydroxy, 2-amino alkane, or alkene as summarized in Fig. 4.1. The most prevalent sphingoid base in mammalian cells is D-*erythro*-sphingosine, (2S,3R,4E)-2-aminooctadec-4-ene-1,3-diol, which is also called E-sphing-4-enine. A convenient abbreviation for sphingoid bases is to assign a letter reflecting the number of hydroxyls ("d" referring to the two (di-) hydroxyl groups and "t" for three, tri-) followed by numbers reflecting the length of the alkyl chain (18 carbons for sphingosine) and the presence of double bonds. If the location and stereochemistry of the double bond are known, it can be depicted in various ways (e.g., 4E-18:1, 4-*trans*-d18:1, or d18:1$^{trans\Delta 4}$ for sphingosine), although this is usually implicit for mammalian sphingolipids, and the usual abbreviation for sphingosine is "d18:1." This scheme can quite easily depict other sphingoid bases that vary in alkyl chain length (e.g., the d20:1 found in brain gangliosides) and the number and positions of double bonds (e.g., the 4-*trans*-,14-*cis*-d18:2 found in brain and plasma), the presence of additional hydroxyl groups (e.g., t18:0 for 4-hydroxysphinganine, which is sometimes called phytosphingosine or the full chemical name [2S,3R,4R]-2-amino-octadecane-1,3,4-triol), and other features.

Derivatives of sphingoid bases that retain the free amino group include the sphingoid base 1-phosphates, sphingosylphosphocholine (lysosphingomyelin), sphingosyl-hexoses (such as galactosylsphingosine, which is also called psychosine), and N-methyl- derivatives (Zheng et al., 2006). N-acyl-derivatives are often categorically called "ceramides," although this name has begun to

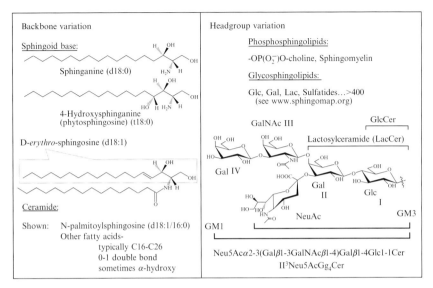

Figure 4.1 Examples of structures of sphingolipids and nomenclature and abbreviations used.

specify N-acylsphingosines, so N-acylsphinganines are called dihydroceramides, and N-acyl-4D-hydroxysphinganines are often called "phytoceramides" or "4-hydroxydihydroceramides" (with the hydroxyl specified to distinguish them from α-hydroxyceramides, where the hydroxyl group is on the fatty acid). These are the major species in most mammalian sphingolipids, and, in instances where other sphingoid bases are present, the names connote the added features of the backbone, if they have been determined (which is often not the case). The amide-linked fatty acids are typically saturated or monounsaturated with chain lengths from 14 to 26 carbon atoms (or even longer in the special case of skin), and sometimes have a hydroxyl group on the α- or ω-carbon atom. The abbreviated nomenclature for ceramides lists the features of the fatty acid after those of the sphingoid base; for example, N-palmitoylsphingosine can be abbreviated d18:1;16:0 or d18:1/16:0, and N-α-hydroxypalmitoylsphingosine is d18:1;h16:0 or d18:1/h16:0.

The so-called "complex" sphingolipids are comprised of a "ceramide" backbone with a headgroup in phospho- or glycosyl-linkage to the 1-hydroxyl (an enzyme that adds a fatty acid to the 1-hydroxyl has also been found, but its physiologic function is not yet clear) (Shayman and Abe, 2000). The major phosphosphingolipids of mammals are ceramide phosphates and sphingomyelins (ceramide phosphocholines); however, since other organisms have additional categories of phosphosphingolipids (e.g., ceramide phosphoethanolamines and phosphoinositols), these broader categories are nonetheless encountered by mammals in food, and thus some microflora might be detected in some tissues if the methods are sufficiently sensitive. The major glycosphingolipids of mammalian organisms have glucose (Glc) or galactose (Gal) attached in β-glycosidic linkage, to which additional carbohydrates and functional groups may be attached. Glycosphingolipids are classified several ways, with the most general being to subdivide them into neutral glycosphingolipids, which contain one or more uncharged sugars such as Glc, Gal, N-acetylglucosamine (GlcNAc), N-acetylgalactosamine (GalNAc), and fucose (Fuc) versus acidic glycosphingolipids, which additionally contain ionized functional groups, phosphate, sulfate, or charged sugar residues such as sialic acid—a generic term that encompasses N-acetylneuraminic acid, NeuAc, and a large family of structurally related compounds that vary with the species (e.g., mice also have N-glcolyl-neuraminic acid, NeuGc).

The galactosylceramide (GalCer) branch is relatively small—about a half-dozen headgroup species—of which the sulfatides (sulfated GalCer and Gal-GalCer) are the most prevalent. In contrast, hundreds of glycolipids exist downstream from glucosylceramide (GlcCer), and additional information about their structures can be obtained from websites such as www.glycoforum.gr.jp and www.sphingomap.org. The nomenclature for these compounds has been systematized (Chester, 1998), and is based on five major "root" glycans from which most of the more complex glycosphingolipids derive (ganglio-, Gg; globo-, Gb, and isoglobo-, iGb; lacto-,

Nomenclature for classification of glycosphingolipids					
Root name (Abbrev)	Carbohydrate in the "root" structure				Other subcategories
	IV	III	II	I	
Ganglio (Gg)	Galβ1–3GalNAcβ1–4Galβ1–4Glcβ1-Cer				
Lacto (Lc)	Galβ1–3GlcNAcβ1–3Galβ1–4Glcβ1-Cer				
Neolacto (nLc)	Galβ1–4GlcNAcβ1–3Galβ1–4Glcβ1-Cer				
Globo (Gb)	GalNAcβ1–3Galα1–4Galβ1–4Glcβ 1-Cer				
Isoglobo (iGb)	GalNAcβ1–3Galα1–3Galβ1–4Glcβ1-Cer				
Mollu (Mu)	GalNAcβ1–2Manα1–3Manβ1–4Glcβ1-Cer				
Arthro (At)	GalNAcβ1–4GlcNAcβ1–3Manβ1–4Glcβ1-Cer				
Ganglioside (Gn) (n = # of neuraminic acids; M(ono), D(di-), T(ri-), etc.)					Contain N-acetyl (NeuAc) or N-glycolyl- (NeuGc) neuraminic acid (Sialic acid)
Galactose series Sulfatide					Galβ-1Cer contain sulfate

Figure 4.2 Classification of glycosphingolipids.

Lc, and neolacto-, nLc), as shown in Fig. 4.2. Use of these root names helps simplify the nomenclature somewhat. For example, the compound shown in Fig. 4.1 can be described as Neu5Acα2–3(Galβ1–3GalNAcβ1–4)Galβ1–4Glc1–1Cer or as II^{3}Neu5AcGg$_4$Cer; however, in many cases, this is still so cumbersome that historic names for some of the more common compounds are often used (i.e., the same compound in Fig. 4.1 is called ganglioside GM1a). The Svennerhold system uses G to depict that the compound is a ganglioside, with the number of sialic acid residues denoted with a capital letter (i.e., *Mono-*, *Di-* or *Tri-*) plus a number reflecting the subspecies within that category, such as ganglioside GM1 versus GM3, etc. For a few glycosphingolipids, historically assigned names as antigens and blood group structures are also in common usage (e.g., Lexis x and sialyl Lewis x). It does not appear that a nomenclature has yet been chosen for sphingolipids that are covalently attached to protein (e.g., ω-hydroxy–ceramides and –glucosylceramides are attached to surface proteins of skin; inositol-phosphoceramides are used as membrane anchors for some fungal proteins).

2. Analysis of Sphingolipids by Mass Spectrometry

Due to the structural complexity of sphingolipids, MS has been used for their analysis for decades; however, extension of the technology to quantitative analysis is a newer development.

2.1. Electron ionization-mass spectrometry

Electron ionization MS (EI-MS) was initially used to elucidate the structures of ceramides (Samuelsson and Samuelsson, 1968, 1969a) and neutral glycosphingolipids (Samuelsson and Samuelsson, 1969b; Sweeley and Dawson, 1969). These early experiments permitted the analysis of sphingolipids as intact molecular species and yielded diagnostic fragmentations that could distinguish isomeric structures (Hammarstrom and Samuelsson, 1970; Samuelsson and Samuelsson, 1970). Because these molecules were either relatively large or polar, they required derivatization to trimethylsilyl (64) or permethyl ethers to reduce the polarity and increase volatility for efficient transfer to the gas phase. However, EI induced extensive fragmentation owing to its relatively high-energy imparted during ionization, which in some cases prevented observation of intact higher mass molecular ions. Additionally, resolution of complex mixtures with varying headgroups, sphingoid bases, and N-acyl combinations was particularly challenging using these methods, as resulting spectra displayed multiple fragments from multiple precursors.

2.2. Fast atom bombardment and liquid secondary ionization mass spectrometry

Fast atom bombardment (FAB) and lipid secondary ionization mass spectrometry (LSIMS) are ionization techniques in which nonvolatile liquids are ionized off a probe tip by bombardment with a beam of either accelerated atoms or ions. FAB and LSIMS are considered to be much "softer" ionization techniques than EI because they eliminate the need for derivatization and yield intact molecular ions with less fragmentation, thus making them more amenable to analysis of somewhat complex mixtures of monohexosylceramides (Hamanaka et al., 1989), sphingoid bases (Hara and Taketomi, 1983), and sphingomyelins (Hayashi et al., 1989).

Structural information can also be obtained from these methods of MS analysis, especially if MS/MS analyses are employed to select an ion of interest, collisionally dissociate it, and detect the resultant product ions. When either $(M + H)^+$ or $(M - H)^-$ precursor ions fragment, they do so at specific positions to yield product ions distinctive for sphingolipid headgroups, sphingoid bases, and N-acyl fatty acids (Adams and Ann, 1993). These pathways of fragmentation can be influenced by the inclusion of alkali metal ions as $(M + Me^+)^+$ where $Me^+ = Li^+, Na^+, K^+, Rb^+$, or Cs^+ (Ann and Adams, 1992).

A limitation of FAB and LSIMS is that both require a matrix to ionize the analyte. These matrices induce a significant amount of background chemical noise, which can limit sensitivity, especially for lower molecular mass species, such as the free sphingoid bases. Methods such as Dynamic FAB (Suzuki et al., 1989, 1990) partially solved this problem with continuous application of solvent and matrix to the probe tip; however, the ~100-fold reduction in background noise was not enough to completely eliminate background chemical noise from the matrix at low analyte concentrations.

2.3. Electrospray ionization

The electrospray ionization method allows solutions containing sphingolipids to be infused directly into the ion source of a mass spectrometer through a hollow metal needle that is held at a high positive or negative potential. At the ending tip of the needle, highly charged droplets form and are drawn into the orifice of the MS by both a potential and an atmospheric pressure difference. During transition from atmospheric pressure to vacuum, the solvent evaporates, whereas the analyte of interest remains as an ionized species in the gas phase. ESI is a much softer ionization technique, and yields primarily intact ions with little or no fragmentation when the ionization conditions have been optimized.

2.4. Liquid chromatography, electrospray ionization, mass spectrometry and tandem mass spectrometry

Since ESI already involves infusion of liquids into the MS, it is a logical extension to use ESI-MS to analyze the eluates from liquid chromatography. From the MS perspective, this has the advantage that the LC can be used for desalting of the compounds of interest as well as to enhance the sensitivity by chromatographic focusing of the sample in a small volume, in addition to whatever desirable separations have been achieved as a result of the LC (e.g., to separate isomers that might not be distinguished by MS alone). LC also reduces the complexity of the eluent at any given elution time, which greatly decreases ionization suppression effects from other species, further improving quantitative accuracy and sensitivity.

LC-MS/MS has been used to identify, quantify, and determine the structures of free sphingoid bases, free sphingoid base phosphates, ceramides, monohexosylceramides (both galactosylceramides and glucosylceramides), lactosylceramides, sphingomyelins (Merrill et al., 2005; Sullards, 2000; Sullards and Merrill, 2001), and more complex glycosphingolipids (Kushi et al., 1989; Suzuki et al., 1991). Complete chromatographic resolution of all individual species is not required in LC-MS/MS because the mass spectrometer is able to differentiate between many components present in a mixture by their mass and structure (e.g., all of the N-acyl chain-length variants of Cer) (Bielawski et al., 2006; Merrill et al., 2005; Pettus et al., 2003; Sullards and Merrill, 2001). Sphingolipid separations have mainly been distributed between two types of LC: reversed phase (Kaga et al., 2005; Lee et al., 2003; Merrill et al., 2005) for separations based on the length and saturation of N-acyl chains (e.g., to separate sphingosine and sphinganine) and normal phase (Merrill et al., 2005; Pacetti et al., 2005; Pettus et al., 2004) to separate compounds primarily by their headgroup constituents (e.g., resolving ceramide from sphingomyelin, etc.). The advantages that are derived from LC separation prior to MS may not be immediately obvious. For example, sphingosine and sphinganine have easily resolved precursor and product ions and can be analyzed without LC; however, sphingosine is typically much more abundant in biological samples than sphinganine, and the (M + 2) isotope of sphingosine, which has the same precursor mass-to-charge ratio (m/z) as sphinganine, interferes with the quantitation of sphinganine unless the sphingoid bases have been separated by LC (Merrill et al., 2005).

2.5. MALDI, mass spectrometry, and tandem mass spectrometry

Matrix-assisted laser desorption/ionization (MALDI) has been used to identify ceramides, monohexosylceramides, and sphingomyelins (Fujiwaki et al., 1999) as well as complex glycosphingolipids such as lactosylceramides and gangliosides (Suzuki et al., 2006). Samples are prepared for MALDI by mixing a solution containing the analyte with a solution containing a MALDI matrix compound, which is typically a small substituted organic acid that contains a moiety that can absorb the photonic energy of the laser. Absorption of the laser energy causes the matrix to become vibrationally excited and volatilize, taking analyte molecules with it into the gas phase and inducing charge exchange reactions between the excited matrix compound and the analyte. This causes the analyte to become ionized, typically as a singly charged species even if the analyte has multiple potentially ionizable moieties. Therefore, one advantage of MALDI ionization is that compounds that might otherwise give complex spectra due to multiple ionization states will show fewer peaks in MALDI. Choice of the matrix is key to successful generation of intact molecular species with minimal fragmentation and has been thoroughly reviewed (David, 2006). A complication in MALDI is that the matrix produces background chemical noise at lower m/z values that can preclude analysis of smaller compounds. MALDI ionization is typically used in conjunction with time-of-flight (TOF) mass analyzers because the laser pulse and resulting ion plume provide a discreet event that is compatible with TOF mass analysis.

3. ANALYSIS OF SPHINGOLIPIDS BY "OMIC" APPROACHES

The large numbers of compounds and the wide range of chemical properties—from essentially water-insoluble for ceramides to highly water-soluble sphingosine 1-phosphate and some glycosphingolipids—pose special challenges in the development of methods that can analyze a significant fraction of the sphingolipidome. First, one must have a method that extracts all of the compounds from the biological sample of interest. Then, the compounds must be detected in ways that distinguish even compounds that are closely related structurally (and in some cases, are isobars or isomers, such as GlcCer and GalCer). Ultimately, the methods should be able to establish the amounts of each compound, or at least to be able to detect changes or differences.

Some laboratories have approached this by so-called "shotgun" methodologies where mixtures of compounds in relatively crude extracts are ionized by ESI or MALDI, and the identities of the compounds of interest in the complex mass spectra are discerned by methods such as fragmentation followed by subsequent MS/MS (see "Other Methods" section). This approach is a sensible one for an initial survey of an extract from a biological material; however, many dangers are inherent in "shotgun" lipidomics, not the least of which is its inability to discern differences between isomeric species (such as GlcCer and GalCer); the frequent occurrence of other compounds in crude extracts may affect ionization and, hence, not only complicate quantitative analysis but also lead to the impression that some compounds are not present when they actually are. Modifications of the shotgun approach that use a pre-MS fractionation method such as LC greatly improve its reliability.

An alternative approach is to develop LC-MS/MS methods for a more modest number of compounds in which the structural identification and quantitation are more robust from the outset, then add more species as experience (and the availability of the appropriate internal standards) allows. This approach was taken by the Sphingolipid Core of the LIPID MAPS Consortium under the descriptor "inside-out sphingolipidomics," which refers to a focus on methodologies that establish the nature and amounts of the lipid backbones (sphingoid bases and ceramides) followed by the more complex derivatives. Thus, the categories of compounds that are currently covered by these methods follow the *de novo* biosynthetic pathway (as well as turnover of the lipid backbones), as summarized in Fig. 4.3.

LC-MS/MS was chosen because it provides: (1) a high level of specificity with regard to differentiation of complex molecular species via retention time, molecular mass, and structure; (2) levels of sensitivity that are orders of magnitude higher than that of classical techniques—often ~1 fmole or less, as is needed for analysis of small samples ($\sim 10^3$ to 10^6 cells); and (3) signal responses that can be more readily correlated with the amounts of the analytes across a wide dynamic range, which is important for the "omics" analysis to be useful for systems biology studies. In spite of all these positive attributes, one still needs to be careful in the choice of sample handling and extraction protocols, validation of the internal standards, and other aspects for LC-MS/MS to fulfill its promise as a quantitative MS analysis technique. The separation that is achieved by LC can have the desired effect of removing ionization-suppressing contaminants, and possibly resolving isomeric species that could compromise the analysis; however, it might also make it difficult to obtain uniform ionization of unknowns and standards if they elute from the LC under different solvent conditions. Thus, in the selection of an LC method, it is desirable to have as many as possible of the analytes elute in the same fraction as the standards and, when this is not possible, to have enough internal standards to assess any variation in ion yield that might be affected by the solvent composition where the analytes elute.

Fortunately, sphingolipids tend to be readily ionized, and many fragment to produce abundant and distinctive product ions that are indicative of the headgroup, sphingoid base, or fatty acid moieties (Fig. 4.4). Free sphingoid bases and most complex sphingolipids ionize readily via ESI in the positive ion mode to form $(M+H)^+$ ions, and sphingoid base-1-phosphates, ceramide-1-phosphates, sphingomyelins, sulfatides, and gangliosides form abundant $(M-H)^-$, $(M-15)^-$, and $(M-nH)^{n-}$ species, respectively, when ionized in the negative ion mode. Each also undergoes structure-specific fragmentation that can be used for identification in MS/MS mode as a precursor–product pair or by neutral loss.

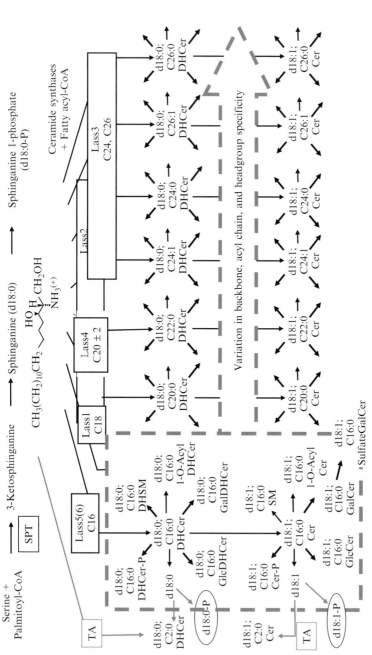

Figure 4.3 Partial metabolic pathway for sphingolipid biosynthesis (and turnover of ceramide) summarizing the compounds analyzed by the sphingolipidomic method described in this review, with structures of selected species and numerically designated enzymatic reactions. CerS/Lass refers to the gene family for ceramide synthases that have the shown selectivity for fatty acyl-CoAs to produce the compounds shown. Also shown is the transacylase that can form N-acetyl-sphingoid bases (C2-ceramide); a separate enzyme produces the 1-O-acylceramides. For simplicity, the downstream metabolites are only shown for the C16:0 subspecies; however, the same complexity will be found for the other subspecies. For other abbreviations, see Fig. 4.1.

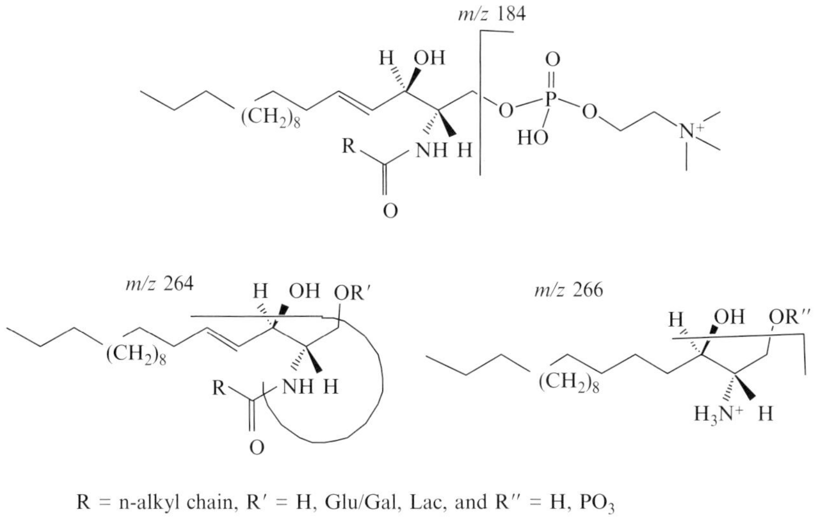

"Inside-Out" Sphingolipidomics   95

R = n-alkyl chain, R' = H, Glu/Gal, Lac, and R" = H, PO_3

Figure 4.4 Fragmentation of sphingolipids observed in the positive ion mode. Fragmentation of long-chain bases, long-chain base phosphates, ceramides, and monohexosylceramides involves dehydration at the 3-position, dehydration at the 1-position, or cleavage of the 1-position moiety with charge retention on the sphingoid base. Sphingomyelin similarly cleaves at the 1-position; however, the charge is retained on the phosphoryl choline headgroup yielding the m/z 184 ion.

Much of the power of LC-MS/MS lies in the improbability that many biomolecules will have the same characteristics on LC, ionize to yield the same precursor m/z, and give the same fragment ion(s). This said, in LC-MS/MS, there is a trade-off in the amount of time that one can spend resolving the categories of compounds of interest by LC (i.e., one tries to keep the time for the LC as short as possible to allow more samples to be run) versus the multireaction monitoring (MRM) pairs that can be monitored during the time when compounds of interest are eluting from the column.

The sphingolipidomic method developed for the LIPID MAPS Consortium uses LC-MS/MS and has been used to identify and quantify a wide range of sphingolipids by optimizing the extraction yields (Fig. 4.5), LC, and MS parameters (Fig. 4.6), as described below for the published methods (Merrill *et al.*, 2005; Sullards and Merrill, 2001) with minor modifications, along with a brief description of the newer methods that are in development for sulfatides and other species.

4. MATERIALS AND METHODS

Many methods have been developed for analysis of sphingolipids from the first intermediates of the *de novo* biosynthetic pathway (i.e., 3-ketosphinganine, sphinganine and sphinganine 1-phosphate, and

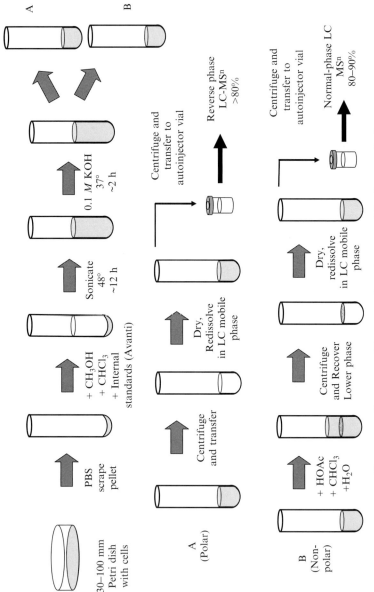

Figure 4-5 Summary of the extraction scheme used for these methods.

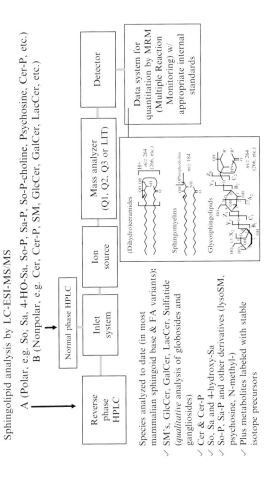

Figure 4.6 Summary of the work flow and metabolites analyzed by LC-ESI-MS/MS of sphingolipids by the methods described in the text.

dihydroceramides) through the metabolites of ceramide (ceramide phosphate, sphingomyelin, GlcCer, GalCer, and sphingosine and sphingosine 1-phosphate).

5. Materials

5.1. Biological samples

The biological sample can range from cells in culture (typically 1×10^6 cells, but depending on the analyte of interest, several orders of magnitude fewer cells might be sufficient). The same protocol is also effective with many other biological samples, such as 1 to 10 mg of tissue homogenates at 10% wet weight per volume in phosphate-buffered saline (PBS), small volumes (1 to 10 μl) of blood, urine, and the like. However, the recovery and optimization of the extraction volume, time, and other aspects should be tested for each new sample type. Tissue culture medium (typically 0.1 to 0.5 ml) can also be analyzed, but should be lyophilized first (in the test tubes described in the next paragraph) before extraction to reduce the aqueous volume. When the samples are prepared, aliquots or duplicate samples should be assayed for the normalizing parameter (μg DNA, mg protein, cell count, etc.) to be used. In our experience, it is best to do these assays first because, if they fail, the effort invested in the lipid analysis will be wasted.

The samples should be placed in Pyrex 13 × 100–mm borosilicate, screw-capped glass test tubes with Teflon caps (the Corning number for these tubes with Teflon-lined caps is 9826–13). It is vital that these specific tubes be used because transfer of samples from one container to another is often a source of variability and frustration. The samples should be stored frozen at $-80°$ until extracted. If it is necessary to ship the samples, this should be done in a shipping container that keeps the vials frozen, separated, and "right-side-up," such as Exakt-Pak containers (catalog number for containers holding up to 20 vials, MD8000V20; for 40 vials, MD8010V40) (http://www.exaktpak.com/).

6. Extraction

The steps of this extraction scheme have been summarized in Fig. 4.5.

1. Start with the samples in 13 × 100–mm screw-capped glass test tubes with Teflon caps. Most samples (washed cells, tissue homogenates, etc.) will already have an approximate aqueous volume of ∼0.1 ml; if not (e.g., if the sample has been lyophilized), bring to this volume with water.

2. Add 0.5 ml of methanol, then 0.25 ml of chloroform and the internal standard cocktail (which contains 0.5 nmol each of the C12:0 fatty acid homologs of SM, Cer, GlcCer, LacCer, and ceramide 1-phosphate; C25-ceramide; and d17:1 sphingosine, d17:0 sphinganine, d17:1 sphingosine 1-phosphate, and d17:0 sphinganine-1-phosphate) (Avanti Polar Lipids, Alabaster, AL).
3. Sonicate as needed to disperse the sample, then incubate overnight at 48° in a heating block.
4. Cool and add 75 μl of 1 M KOH in methanol, sonicate, and incubate 2 hr at 37°. (This step reduces clogging of the LC columns, but can be eliminated for some samples, if desired.)
5. Transfer half of the extract (which will be used for analysis of the more polar sphingolipids, such as sphingoid bases and sphingoid base 1-phosphates, by reverse phase LC-ESI-MS/MS) (0.4 ml) to a new test tube (this will be used in the next section).
6. To the half of the extract that remains in the original test tube(s), add 3 μl of glacial acetic acid to bring the pH to ~ neutral, then add 1 ml of chloroform and 2 ml of distilled, deionized water, and mix with a rocking motion to avoid formation of an emulsion.
7. Centrifuge in a tabletop centrifuge to separate the layers, then carefully remove the upper layer with a Pasteur pipette, leaving the interface (with some water).
8. Evaporate the lower phase to dryness under reduced pressure (using a stream of N_2 or a speed vac) and save this fraction for normal phase LC-ESI-MS/MS of the complex sphingolipids.

6.1. Preparation of samples for LC-MS/MS

For the sphingoid base fraction (Step 5 in "Extraction" section), use a tabletop centrifuge to remove insoluble material, then transfer to a new glass test tube and evaporate the solvent under N_2 or reduced pressure (speed vac). Add 300 μl of the appropriate mobile phase (as shown later) for reverse-phase LC-ESI-MS/MS, sonicate, then transfer to 1.5-ml microfuge tubes (organic solvent resistant) and centrifuge for several minutes or until clear. Transfer 100 μl of the clear supernatant into a 200-μl glass autoinjector sample vial for LC-ESI-MS/MS analysis.

Prepare the extract for normal-phase LC-ESI-MS/MS (Step 8 in "Extraction" section) in the same manner, but using the appropriate mobile phase.

Note: For some samples, slightly higher recoveries and/or more uniform recoveries across different sphingolipid subspecies are obtained if the residue from these redissolving steps are treated again and the extracts pooled.

7. IDENTIFICATION OF THE MOLECULAR SPECIES BY TANDEM MASS SPECTROMETRY

The first step of the analysis is to identify the molecular subspecies of each category of sphingolipid that is present by analysis of their unique fragmentation products using precursor ion or neutral loss scans. Once these subspecies have been identified, the investigator can build an MRM protocol for quantitative analysis of those analytes for which both the ionization and dissociation (collision energy) conditions have been optimized for individual molecular species. These parameters will vary from instrument to instrument, and thus are left to the investigator to perform based on their own circumstances (examples of parameters have been given in Merrill et al., 2005). A general scheme for the instrumentation is given in Fig. 4.6.

7.1. Materials for infusion and LC-ESI-MS/MS

Positive-ion infusion spray solution: $CH_3OH/HCOOH$ (99:1) (v:v) containing 5-mM ammonium formate.

Reverse-phase LC:

1. The recommended reverse-phase LC column is a Supelco 2.1 mm i.d. × 5 cm Discovery C18 column (Supelco, Bellefont, PA).
2. The following are solvents for reverse phase LC:

   a. Reverse-phase LC solution A: $CH_3OH/H_2O/HCOOH$ (74:25:1) (v:v:v) with 5-mM ammonium formate
   b. Reverse-phase LC solution B: $CH_3OH/HCOOH$ (99:1) (v:v) with 5-mM ammonium formate

Normal-phase LC:

1. The recommended normal-phase column is a Supelco 2.1 mm i.d. × 5 cm LC-NH$_2$ column, with the exception for the LC method for separation of GlcCer and GalCer, which uses a Supelco 2.1 mm i.d. × 25 cm LC-Si column.
2. The following are solvents for normal-phase LC:
   Normal-phase solvent A: $CH_3CN/CH_3OH/CH_3COOH$ (97:2:1) (v:v:v) with 5-mM ammonium acetate
   Normal-phase solvent B: $CH_3OH:CH_3COOH$ (99:1) (v:v) with 5-mM ammonium acetate

7.2. Sphingolipid subspecies characterization prior to quantitative LC-MS/MS analysis

1. Dilute a 50-μl aliquot (one-sixth of complex fraction) of the reconstituted sample for complex sphingolipid analysis (as described in Section 6.1) to a final volume of 1 ml with positive ion infusion spray solution.

2. Infuse this solution into the ion source with a 1-ml syringe at 0.6 ml/hr.
3. Perform precursor ion scan(s) for predetermined unique molecular decomposition products (e.g., fragment ions of m/z 264 and 266 for the d18:1 and d18:0 sphingoid base backbones, respectively, and m/z 184 for SM, etc.).
4. Identify the sphingolipid precursor and product ion pairs for species present, noting potential isobaric combinations of sphingoid base and N-acyl fatty acids.

Note: At this point, it is usually most time-efficient to reanalyze the sample by LC-ESI-MS/MS using the LC method described below since the final optimization of the ionization parameters will need to be done with the same solvent composition as the compounds of interest elute from LC.

5. Using the "parts list" of species (Fig. 4.4) for each observed subspecies, determine the optimal ionization conditions and collision energy for each species. Typically, collision energy will need to be increased within each sphingolipid class as the N-acyl chain or sphingoid base increases in length.

8. Quantitation by LC-ESI-MS/MS Using Multireaction Monitoring

The optimized ionization and fragmentation conditions for each analyte of interest, combined with the LC elution position, are used to construct the MRM analysis protocols. In MRM, the mass spectrometer is programmed to monitor specific, individually optimized precursor and product ion pairs (with respect to instrument parameters for highest yield of the precursor and product ions of interest) in specific LC timeframes. The signal generated by each ion transition in most cases uniquely identifies a particular molecular species by retention time, mass, and structure, although there are still rare occasions where another compound adds to the signal; hence, whenever possible, samples should be examined for this possibility by another technique, such as further analysis of the ions of interest by tandem MS or by an orthogonal technique, such as comparing the results of other types of chromatographic separation, ionization (e.g., positive versus negative ion mode), and fragmentation. To determine the relationships between the signal response for the spiked internal standard (which can be very close to 1:1 for some instruments, such as the ABI 4000 QTrap when the MRM parameters have been optimized) and a correction factor (if needed), one typically compares the internal standard with the analyte of interest (e.g., the 17 carbon homolog of sphingosine [d17:1] versus naturally occurring sphingosine [d18:1]; in the case of more complex sphingolipids, C12-chain–length internal standard vs at least two chain length variants, such as the C16- and C24-chain–length versions of the analyte, which are usually available commercially [from Avanti Polar Lipids, Alabaser, AL, Matreya, Pleasant Gap, PA, among others]). This comparison has been performed for all of the species covered by the standard cocktail mix, and a very high equivalence exists between the ion yields of most of the naturally occurring analytes and the selected standard when analyzed using the ABI 4000 QTrap; however, some categories show more deviation using the API 3000 triple quadrupole instrument, which illustrates the need to determine these relationships for the particular instrument being used for the analysis.

8.1. Analysis of sphingoid bases and sphingoid base 1-phosphates in positive ion mode

1. Portions of the reconstituted samples (prepared above) are diluted with reverse-phase LC solutions A and B for a final A:B ratio of 80:20 (v:v). (If it is desirable to concentrate the sample, the solvent can be removed by evaporation and then the residue redissolved in this solvent mixture.) The samples are centrifuged to remove any precipitate, and then placed in autosampler vials and loaded into the autosampler.
2. The LC column (reverse-phase C18) is equilibrated with reverse-phase LC solution A and B (80:20) for 0.5 min at flow rate of 1.0 ml/min.
3. The sample (typically 10 to 50 μl) is injected (then the needle is programmed to wash with at least 5 ml of reverse-phase solution A to prevent potential sample carryover) and the following elution protocol is followed (at a flow rate of 1.0 ml per minute):

   a. Wash the column for 0.6 min with reverse-phase solutions A and B (80:20, v:v).
   b. Apply a linear gradient to 100% of reverse-phase solution B over 0.6 min.
   c. Wash with 100% reverse-phase solution B for 0.3 min.
   d. Apply a linear gradient to reverse-phase solution A:B (80:20) over 0.3 min.
   e. Re-equilibrate the column with a 80:20 mix of reverse-phase solutions A and B for 0.5 min.

4. Determine the areas under the peaks for internal standards and analytes of interest using extracted ion chromatograms.
5. Quantify analytes relative to internal standard spike.

During the LC run, the ions of interest are followed as described by Merrill *et al.* (2005), with the salient points summarized here with comparisons between two types of mass analyzers, a triple quadrupole MS/MS (the API 3000), and a hybrid quadrupole-linear ion trap (ABI 4000 QTrap, Applied Biosystems, Foster City, CA). Positive mode analysis of the $(M + H)^+$ ions of long-chain bases such as sphingosine (d18:1), sphinganine (d18:0), 4-D-hydroxysphinganine (t18:0), the 17-carbon homologs (d17:1 and d17:0), and the 20-carbon homologs (d20:1 and d20:0) by MS/MS reveals that they fragment via single and double dehydration to product ions of m/z 282/264, 284/266 (sphinganine also fragments to a headgroup ion with m/z 60 that can be followed to avoid overlap with sphingosine isotopes), 300/282/264, 268/250, 270/252, 310/292, and 312/294, respectively. The ratios between single- and double-dehydration products varies by collision energy and type of mass spectrometer. Signal response for the single-dehydration products is greater than double-dehydration products in the ABI 4000 QTrap (even given higher collision energies), whereas in the API 3000 triple quadrupole, the ratio is reversed, with a much stronger double-dehydration signal than for single dehydration. Sphingoid base 1-phosphates derivatives undergo a similar dehydration and cleavage of the headgroup to yield the same m/z product ions as the double-dehydrated product ions described above; however, this does not interfere with their analysis in the same run because they are distinguished both by their precursor ion mass and retention times on LC.

Ionization parameters for sphingoid bases and sphingoid base 1-phosphates are similar for sphingoid bases containing 17 to 20 carbons. (In the API 3000 triple quadrupole ionization, settings are identical; however small eV differences are required in the ABI 4000 QTrap to achieve optimal signal response.) The main variable in these analyses is collision energy, which increases with sphingoid base chain length. Modifications to sphingoid bases such as N-methylation, additional sites of unsaturation, and hydroxyl addition will also require increasing collision energy and can be detected by shifts in precursor and product ion m/z as well as alteration of reverse-phase LC retention. Although one must check if there are changes in ion yield due to differences in the elution solvent, possible ion suppressing compounds in crude extracts, or other changes when compounds in a subspecies series elute slightly differently from the LC, small shifts in retention provide a useful verification of the identities of the species when product ions are otherwise identical.

Notes on 3-ketosphingoid base analysis: 3-Ketosphinganine (3kSa) can be analyzed by reverse-phase LC with the other sphingoid bases. It has the same precursor m/z as sphingosine, and, although there are differences in the degree of dehydration in the fragments (sphingosine fragments primarily to m/z 264 whereas 3-ketosphinganine fragments to m/z 282), the most reliable way to differentiate these compounds is by examining the differences in elution on reverse-phase LC (3-ketosphinganine elutes after sphingosine). Interestingly, one also finds 3-keto-sphingoid bases as part of 3-ketodihydroceramides, which are identifiable by earlier elution on normal-phase LC as peak pairs with normal ceramide (with 3-ketodihydroceramide eluting first).

9. ANALYSIS OF (DIHYDRO)CERAMIDES, (DIHYDRO)SPHINGOMYELINS, AND (DIHYDRO)MONOHEXOSYL-CERAMIDES IN POSITIVE ION MODE

1. Portions of the reconstituted samples (as prepared in a previous section) are diluted with normal-phase solution A. (If it is desirable to concentrate the sample, the solvent can be removed by evaporation and the residue redissolved in this solvent mixture.) The samples are centrifuged to remove any precipitate and then placed in autosampler vials and loaded into the autosampler.
2. The normal-phase LC-NH$_2$ column (Supelco 2.1 mm i.d. × 5 cm LC-NH$_2$) is equilibrated for 0.5 min with normal-phase solution A (98:2, v:v) at 1.5 ml/min.
3. The sample (typically 10 to 50 μl) is injected (then the needle is programmed to wash with at least 5 ml of normal-phase solution A to prevent potential sample carryover), and the following elution protocol is followed (at a flow rate of 1.5 ml/min):

   a. Wash the column 0.5 min with normal-phase solution A.
   b. Apply a linear gradient to 10% normal-phase solution B over 0.2 min.
   c. Hold at A:B (90:10, v:v) for 0.5 min.
   d. Gradient over 0.4 min to normal-phase solutions A and B (82:18, v:v).
   e. Hold at A:B (82:18, v:v) for 0.6 min.
   f. Apply a linear gradient to 100% normal-phase solution B over 0.4 min.
   g. Re-equilibrate the column with normal-phase solution A for 0.5 min.

   Note: By this protocol, one monitors the specific precursor and product ion transitions for ceramides for the first 0.8 min, followed by monohexosylceramides for 1.15 min and then sphingomyelins for 1.05 min; however, due to age or brand of column, the elution of monohexosylceramides may shift, requiring time adjustment as needed.

   Note regarding chromatography in general: It is useful to bear in mind that, when converting methods between two different size columns, it is advisable to convert flow rates, hold times, and gradients based on column volume. Pre-gradient holds should be at least five to six column volumes worth of solvent; for maximum resolution during a gradient, solvent change per column volume should be 2 to 4%. For the columns described in this review, the lengths described correspond to the column volumes in parentheses: 2.1 × 50 mm (0.11 ml), 2.1 × 150 mm (0.33 ml), and 2.1 × 250 mm (0.55 ml). Also, as columns are changed, check the recommended flow rates for that column length.
4. Determine the areas under the peaks for internal standards and analytes of interest using extracted ion chromatograms.
5. Quantify analytes relative to internal standard spike.

Product ion analysis of the $(M + H)^+$ ions of ceramides reveals cleavage of the amide bond and dehydration of the sphingoid base to form highly abundant, structurally specific fragment ions. These product ions yield information regarding the number of carbon atoms in the chain, degree of hydroxylation, unsaturation, or other structural modifications of the long-chain base (e.g., sphingosine, m/z 264; sphinganine, m/z 266; and 4-hydroxysphinganine, m/z 264—which is the same as for sphingosine-based ceramides, but these can be distinguished by differences in LC elution times). With this knowledge about the sphingoid base composition and the original precursor m/z, the identity of the fatty acids can be deduced.

Ionization parameters for complex sphingolipids are linked to their headgroup classes with size and charge of headgroup playing key roles. Accordingly, differences in N-acyl chain length somewhat shift the required collision energy (in our experience, ~2.5 eV every two to four carbons).

Product ion scans of the $(M + H)^+$ ions of GlcCer, LacCer, and more complex glycolipids reveal that these ions undergo dissociation by two pathways: cleavage at the glycosidic linkage(s) at low collision energies with loss of the carbohydrate headgroup as a neutral species with charge remaining on the ceramide moiety and cleavage of both the sugar head-group and the fatty acid acyl chain at higher energies with charge retention on the dehydrated sphingoid base.

Low energy dissociation to remove the headgroup can be useful for analyzing glycosidic linkages in neutral glycolipids and gangliosides; how-ever, for quantitation of individual sphingoid base and N-acyl fatty acid combinations of monohexosylceramides, high energy dissociations to the sphingoid base-specific product ion are preferable. Choosing this fragmentation pathway allows distinction of isobaric d18:1/C18:0 versus d20:1/C16:0 glucosylceramides, which would have the same low energy deglycosylated fragment m/z 566 but high-energy conjugated carbocation fragments that differ (m/z 264 versus m/z 292, respectively).

Sphingolipids containing phosphodiester-linked headgroups, such as in sphingomyelin (SM), fragment very differently: the $(M + H)^+$ species fragments at the phosphate-ceramide bond, with charge retention on the phospho-headgroup to yield highly abundant ions of m/z 184. Ceramide phosphoethanolamines (CPE) also fragment at the phosphate-ceramide bond, but the headgroup is lost as a neutral species of mass 141 u.

9.1. Modifications for analysis of glucosylceramides and galactosylceramides

Because GlcCer and GalCer elute in the same fractions by the above method, biological samples that contain both of these monohexosylceramides must be analyzed by a separate method.

1. Reconstitute the samples in normal-phase solution A and load into the autosampler.
2. Pre-equilibrate a normal phase LC-Si column (Supelco 2.1 mm i.d. × 25 cm LC-Si) for 1.0 min with a normal-phase solution A at 1.5 ml per min.
3. Inject 50 μl of each reconstituted extract and continue to elute with normal-phase solution A at 1.5 ml/min for 8 min. GlcCer elutes at 2.56 min and GalCer at 3.12 min using this isocratic normal-phase system.
   Note: GlcCer and GalCer are isobaric species, so only one MRM transition is needed per shared sphingoid base and fatty acid combination.
4. Determine the areas under the peaks for internal standards and analytes of interest using extracted ion chromatograms and quantify analytes relative to internal standard spike.

9.2. Analysis of ceramide 1-phosphates by positive and negative ion modes

Ceramide 1-phosphates may be examined by the reverse-phase LC method described above for sphingoid bases and sphingoid base 1-phosphates by extending the final wash with 100% reverse-phase LC solution B until the ceramide 1-phosphates elute (C12-ceramide phosphate elutes at the beginning of the change to this solvent, followed by longer chain-length species).

An orthogonal approach (normal-phase LC negative mode following phosphate as the product ion in the MRM) can also be employed using either silica (LC-Si) or diol (LC-Diol) columns (Supelco) in conjunction with elution solvents that contain triethylammonium acetate (TEAA) and begin with acetonitrile and gradient to methanol and water (formic acid and ammonium formate buffer systems can also be used because they allow CerP to ionize well in negative mode).

In positive mode, the major fragments from ceramide 1-phosphate arise from cleavage of the amide bond and dehydration of the sphingoid base to form m/z 264 (for sphingosines), 266 (for sphinganines), and so on. As with the other complex sphingolipids, to maintain ionization efficiency, collision energy should be increased slightly with chain length. In negative mode, the major product ion from the precursor $(M - H)^-$ ions is phosphate. This can be used for quantitation; however, the fidelity of the MRM pair for ceramide 1-phosphate should be confirmed since the product is no longer structure-specific for the sphingoid base and N-acyl chain combinations.

9.3. Analytical methods for additional analytes

Methods for more complex glycosphingolipids are still under development. Sulfatides are readily analyzed in negative ion mode and yield primarily $(M - H)^-$ ions that fragment to the sulfate group (m/z 96.9) (lower-abundance fragment ions representing the lipid backbone can also be observed). For LC, either reverse- (C18) or normal- (LC-Si or LC-Diol) phase columns can be used, with methanol/water gradients to methanol (with 5 mM ammonium acetate and 0.01% ammonium hydroxide to promote ionization) for reverse phase. For normal phase, acetonitrile, methanol, and water combinations (with the same mobile phase modifiers or triethylammonium acetate) achieve satisfactory LC resolution.

Gangliosides also ionize readily in the negative ion mode in methanol. For gangliosides analyzed by the triple quadrupole MS/MS, the $(M-2H)^{2-}$ ions fragment to yield highly abundant $C1\beta$-H_2O ions of m/z 290, which reflect the N-acetyl neuraminic acid moiety (Figs. 4.7 and 4.8). The enhanced product ion scan feature that is available using the ion trapping function of the 4000 QTrap provides more structural information because it yields better sensitivity and more abundant high mass product ions (Fig. 4.8). Additionally, cleavage at the glycosidic bonds produces characteristic Y_n-type ions and through ring cleavages, such as $^{2,4}X_{2\alpha}$ and $^{2,4}X_{3\alpha}$ (m/z 1351.9 and 1512.8, respectively), which are useful for determination of glycosidic bond linkage.

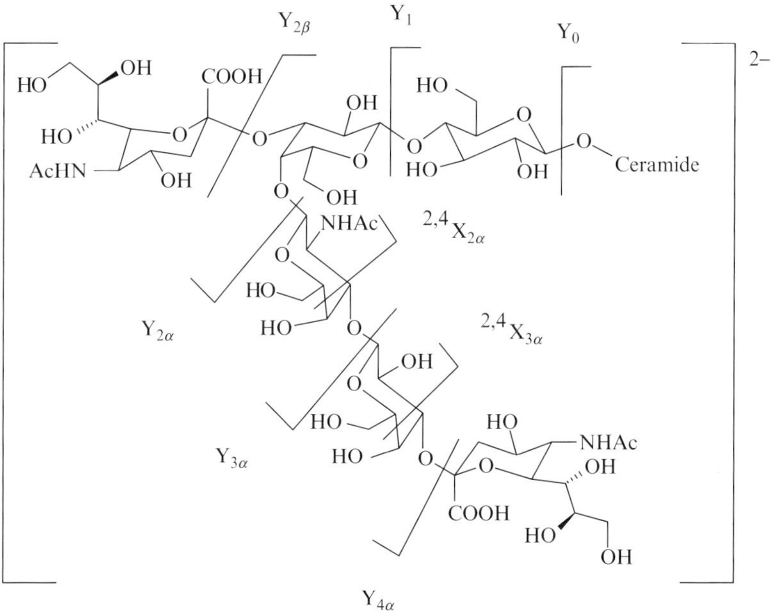

Figure 4.7 Structure of ganglioside GD1a and major cleavage sites and nomenclature.

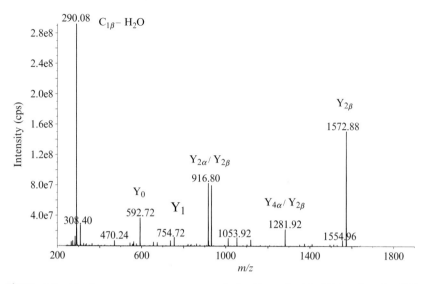

Figure 4.8 Tandem mass spectrometry analysis of ganglioside GD1a using the 4000 QTrap in enhanced product-ion scanning (EPI) mode (~55 fmol consumed). See Fig. 4.7 for the cleavages represented by these labels.

An MS/MS/MS (MS^3) analysis is performed much the same manner as a product ion scan. In this case the first mass analyzer (Q1) is set to pass the precursor ion of interest, which is transmitted to Q2, where it collides with a neutral gas (N$_2$ or Ar) and dissociates to various fragment ions. Rather than mass analyzing the resulting product ions, the linear ion trap (LIT) is set to trap and hold a 2-m/z-unit–wide window centered on the product ion of interest. The selected m/z is irradiated with a single wavelength amplitude frequency to induce further fragmentation to secondary product ions, which are then scanned out of the LIT to the detector. The resulting MS^3 spectrum shows the fragmentation pattern of the selected product ion, and yields additional structural details regarding the primary product ion.

MS^3 analysis provides critical structural information about higher order sphingolipids (such as gangliosides) that is not provided in the MS/MS spectrum. Typically, MS/MS data of these ions do not reveal any information about the components of the ceramide backbone. MS^3 analyses of the Y$_0$ product ions (m/z 592.6), which comprise the core lipid part of the molecule, will determine the composition of the ceramide. In the example shown in Fig. 4.9, the highly abundant S, T, U, and V + 16 ions (m/z 324, 308, 282, and 283, respectively) reveal that the fatty acid is C18:0, and the complementary P and Q ions (m/z 265 and 291, respectively) are characteristic of a d20:1 sphingoid base. Thus,

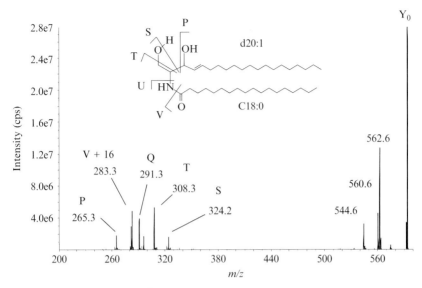

Figure 4.9 MS/MS/MS (MS^3) analysis of the Y$_0$ product ions (m/z 592.7) from Fig. 4.8 to reveal the highly abundant S, T, U, and V + 16 ions (m/z 324, 308, 282, and 283, respectively) that establish that the fatty acid is C18:0, and the complementary P and Q ions (m/z 265 and 291, respectively) that are characteristic of a d20:1 sphingoid base.

MS^3 scans provide an additional level of structural analysis yielding critical information regarding sphingoid base, fatty acid, and headgroups in glycosphingolipids.

10. OTHER METHODS

Two-dimensional (2D) ESI-MS has been introduced for the identification and quantitation of large families of sphingolipids using a triple quadrupole instrument (Han and Gross, 2003; Han et al., 2004). Samples are infused and analyzed by MS and MS/MS in both positive and negative ESI conditions with the addition of aqueous LiOH and/or LiCl to aid formation of charged molecular adducts. As noted earlier in this chapter, limitations of this approach include the possibility of ionization suppression and the inability, in many instances, to distinguish isomeric and isobaric species; nonetheless, it is a relatively easy and rapid way to profile many species, including phosphatidylserines, phosphatidylethanolamines, diacyl- and triacylglycerols, phosphatidylinositols, phosphatidlycholines, sphingomyelins, some cerebrosides, cardiolipins, ceramides, free fatty acids, and phosphatidic acids (Han et al., 2004).

Nanospray in conjunction with high-resolution MS and MS/MS is an extension of the 2D lipid profiling technique (Ejsing et al., 2006; Ekroos et al., 2002; Schwudke et al., 2006) wherein very small volumes of sample are infused into a Q-TOF MS/MS via a chip containing a high-density array of nanospray nozzles. This allows data collection for extended periods of time during which numerous product ion scans across a chosen mass range can be conducted. The resulting product ion data are then queried for structure-specific fragment ions or neutral losses corresponding to the head-group, fatty acid, or sphingoid base combinations of the sphingolipid species, similar to the 2D technique. The detection of spurious peaks is reduced by using a smaller product ion selection window, which is possible using a high-resolution mass analyzer (TOF instead of quadrupole).

Chip-based nanospray techniques have several additional advantages (Ekroos et al., 2002; Schwudke et al., 2006). Since flow rates are on the order of nl/min, very little sample is consumed and background chemical noise is greatly reduced, which enhances overall sensitivity so that minor lipid species are more readily detected. Each chip contains a high-density array of nanospray nozzles that allows automation for higher reproducibility and sample throughput; furthermore, if the nature of the samples causes problems of carry-over and cross contamination, this can be solved by allocation of one spray nozzle per sample. Newer chip-based nanospray systems can be connected directly to LC columns as well as arranged so there is both analysis of a portion of the eluate and simultaneous fraction-collection of the remainder. When used in this manner, one gains the advantages of LC discussed earlier.

Accurate mass and ultra-high-resolution MS of sphingolipids has been performed by Fourier transform MS (FTMS) coupled to nanoESI (McFarland et al., 2005; Vukelic et al., 2005) or MALDI (Ivleva et al., 2004; O'Connor and Costello, 2001). The former was primarily applied to the identification and structure determination of glycosphingolipids; however, the study introduced some interesting new alternative fragmentation techniques for lipids such as infrared multiphoton dissociation (IRMPD), electron capture dissociation (ECD), and electron detachment dissociation that may also be applicable to "omics" studies. MALDI FTMS has been used to analyze glycolipids directly from TLC plates (Ivleva et al., 2004), and profile lipid species present in intact cells (Jones et al., 2004) and plant tissues (Jones et al., 2005). Ultra-high-resolution and accurate mass measurement enables the grouping of various classes of phospholipids (Jones et al., 2004) and lipid or carbohydrate fragment ions using mass defect plots (McFarland et al., 2005).

Ion mobility MS is another method that can separate lipids (Jackson et al., 2005) with the same m/z based on differences in their mobility through a gas buffer and applied electric field. Since the mobility of a given ion is affected by its three-dimensional shape (ions with a greater cross-section migrate more slowly than do more compact ions) in ion mobility MS, this has been able to distinguish glycans (Clowers et al., 2005) and may be useful for some glycolipids.

Direct tissue imaging MS is rapidly emerging as a useful technology for lipid analysis in tissue sphingolipidomics because one does not need to extract the lipids from the sample, and one also obtains information about where the lipid is localized in the tissue. Ionization of the lipids in the sample is either achieved by MALDI (after MALDI matrix material has been deposited on the sample) or secondary ion MS (SIMS). For MALDI, the laser beam is rastered across the tissue sample, collecting spectra at discrete points; SIMS is analogous, but utilizes a tightly focused beam of primary ions to impinge on the surface and generate a secondary stream of ions (Borner et al., 2006; Jackson et al., 2007; Roy et al., 2006; Sjovall et al., 2004; Woods and Jackson, 2006). Two additional techniques, desorption ESI (DESI) and nanoSIMS, may also be used for cellular and subcellular imaging. The former uses charged droplets of solvent from an electrospray to generate intact secondary molecular ions (Wiseman et al., 2006), and the latter uses a tightly focused beam of Cs^+ ions to scan the sample and produces primarily mono and diatomic secondary ions (Kraft et al., 2006). Examples exist for the use of these technologies: MALDI-TOF has detected sphingomyelins, monohexosylceramides, sulfatides, and gangliosides (Jackson et al., 2007; Woods and Jackson, 2006) in different brain regions; and SIMS TOF analysis of brain slices (Borner et al., 2006) revealed that chain-length variants of galactosylceramide are differentially localized in white matter, with C18-subspecies being associated with cholesterol-rich regions, whereas C24-subspecies are found primarily in areas that are also enriched in Na^+ and K^+. The limitations of these approaches to date include: (1) difficulties in obtaining ions from all of the analytes of interest (and in some cases, the energetics of the ionization process make it difficult to obtain parent ions for all of the compounds that are ionized); (2) variable ionization that compromises quantitative comparisons; and (3) the inability to distinguish isomeric and isobaric compounds. Nonetheless, since in many cases the location of lipids is vital to their function, the development of MS methods that provide quantitative and positional information is an important direction for new methods development.

ACKNOWLEDGMENTS

The work that has been described in this review was supported primarily by funds from the LIPID MAPS Consortium grant (U54 GM069338) and in part from an NIH Integrated Technology Resource grant for Medical Glycomics (PA-02-132).

REFERENCES

Adams, J., and Ann, Q. (1993). Structure determination of sphingolipids by mass spectrometry. *Mass Spectrom. Rev.* **12,** 51–85.
Ann, Q., and Adams, J. (1992). Structure-specific collision-induced fragmentation of ceramides cationized with alkali-metal ions. *Anal. Chem.* **65,** 7–13.
Bielawski, J., Szulc, Z. M., Hannun, Y. A., and Bielawska, A. (2006). Simultaneous quantitative analysis of bioactive sphingolipids by high-performance liquid chromatography-tandem mass spectrometry. *Methods* **39,** 82–91.
Borner, K., Nygren, H., Hagenhoff, B., Malmberg, P., Tallarek, E., and Mansson, J. E. (2006). Distribution of cholesterol and galactosylceramide in rat cerebellar white matter. *Biochem. Biophys. Acta* **1761,** 335–344.
Chester, M. A. (1998). IUPAC-IUB Joint Commission on Biochemical Nomenclature (JCBN). Nomenclature of glycolipids—recommendations, 1997. *Eur. J. Biochem.* **257,** 293–298.
Clowers, B. H., Dwivedi, P., Steiner, W. E., Hill, H. H., Jr., and Bendiak, B. (2005). Separation of sodiated isobaric disaccharides and trisaccharides using electrospray ionization-atmospheric pressure ion mobility-time of flight mass spectrometry. *J. Am. Soc. Mass Spectrom.* **16,** 660–669.
David, J. H. (2006). Analysis of carbohydrates and glycoconjugates by matrix-assisted laser desorption/ionization mass spectrometry: An update covering the period 1999–2000. *Mass Spectrom. Rev.* **25,** 595–662.
Ejsing, C. S., Moehring, T., Bahr, U., Duchoslav, E., Karas, M., Simons, K., and Shevchenko, A. (2006). Collision-induced dissociation pathways of yeast sphingolipids and their molecular profiling in total lipid extracts: A study by quadrupole TOF and linear ion trap-orbitrap mass spectrometry. *J. Mass Spectrom.* **41,** 372–389.
Ekroos, K., Chernushevich, I. V., Simons, K., and Shevchenko, A. (2002). Quantitative profiling of phospholipids by multiple precursor ion scanning on a hybrid quadrupole time-of-flight mass spectrometer. *Anal. Chem.* **74,** 941–949.
Fahy, E., Subramaniam, S., Brown, H. A., Glass, C. K., Merrill, A. H., Jr., Murphy, R. C., Raetz, C. R., Russell, D. W., Seyama, Y., Shaw, W., Shimizu, T., Spener, F., *et al.* (2005). A comprehensive classification system for lipids. *J. Lipid Res.* **46,** 839–861.
Fujiwara, T., Yamaguchi, S., Sukegawa, K., and Taketomi, T. (1999). Application of delayed extraction matrix-assisted laser desorption ionization time-of-flight mass spectrometry for analysis of sphingolipids in tissues from sphingolipidosis patients. *J. Chromatogr. B Biomed. Sci. Appl.* **731,** 45–52.
Hamanaka, S., Asagami, C., Suzuki, M., Inagaki, F., and Suzuki, A. (1989). Structure determination of glucosyl beta 1-N-(omega-O-linoleoyl)-acylsphingosines of human epidermis. *J. Biochem. (Tokyo)* **105,** 684–690.
Hammarstrom, S., and Samuelsson, B. (1970). On the biosynthesis of cerebrosides from 2-hydroxy acid ceramides: Use of deuterium labeled substrate and multiple ion detector. *Biochim. Biophys. Res. Commun.* **41,** 1027–1035.
Han, X., and Gross, R. W. (2003). Global analyses of cellular lipidomes directly from crude extracts of biological samples by ESI mass spectrometry: A bridge to lipidomics. *J. Lipid Res.* **44,** 1071–1079.
Han, X., Yang, J., Cheng, H., Ye, H., and Gross, R. W. (2004). Toward fingerprinting cellular lipidomes directly from biological samples by two-dimensional electrospray ionization mass spectrometry. *Anal. Biochem.* **330,** 317–331.
Hara, A., and Taketomi, T. (1983). Detection of D-erythro and L-threo sphingosine bases in preparative sphingosylphosphorylcholine and its N-acylated derivatives and some evidence of their different chemical configurations. *J. Biochem. (Tokyo)* **94,** 1715–1718.

Hayashi, A., Matsubara, T., Morita, M., Kinoshita, T., and Nakamura, T. (1989). Structural analysis of choline phospholipids by fast atom bombardment mass spectrometry and tandem mass spectrometry. *J. Biochem. (Tokyo)* **106**, 264–269.

Ivleva, V. B., Elkin, Y. N., Budnik, B. A., Moyer, S. C., O'Connor, P. B., and Costello, C. E. (2004). Coupling thin-layer chromatography with vibrational cooling matrix-assisted laser desorption/ionization Fourier transform mass spectrometry for the analysis of ganglioside mixtures. *Anal. Chem.* **76**, 6484–6491.

Jackson, S. N., Wang, H. Y., and Woods, A. S. (2007). In situ structural characterization of glycerophospholipids and sulfatides in brain tissue using MALDI-MS/MS. *J. Am. Soc. Mass Spectrom.* **18**, 17–26.

Jackson, S. N., Wang, H. Y., Woods, A. S., Ugarov, M., Egan, T., and Schultz, J. A. (2005). Direct tissue analysis of phospholipids in rat brain using MALDI-TOFMS and MALDI-ion mobility-TOFMS. *J. Am. Soc. Mass Spectrom.* **16**, 133–138.

Jones, J. J., Batoy, S. M., and Wilkins, C. L. (2005). A comprehensive and comparative analysis for MALDI FTMS lipid and phospholipid profiles from biological samples. *Comput. Biol. Chem.* **29**, 294–302.

Jones, J. J., Stump, M. J., Fleming, R. C., Lay, J. O., Jr., and Wilkins, C. L. (2004). Strategies and data analysis techniques for lipid and phospholipid chemistry elucidation by intact cell MALDI-FTMS. *J. Am. Soc. Mass Spectrom.* **15**, 1665–1674.

Kaga, N., Kazuno, S., Taka, H., Iwabuchi, K., and Murayama, K. (2005). Isolation and mass spectrometry characterization of molecular species of lactosylceramides using liquid chromatography-electrospray ion trap mass spectrometry. *Anal. Biochem.* **337**, 316–324.

Kraft, M. L., Weber, P. K., Longo, M. L., Hutcheon, I. D., and Boxer, S. G. (2006). Phase separation of lipid membranes analyzed with high-resolution secondary ion mass spectrometry. *Science* **313**, 1948–1951.

Kushi, Y., Rokukawa, C., Numajir, Y., Kato, Y., and Handa, S. (1989). Analysis of underivatized glycosphingolipids by high-performance liquid chromatography/atmospheric pressure ionization mass spectrometry. *Anal. Biochem.* **182**, 405–410.

Lee, M. H., Lee, G. H., and Yoo, J. S. (2003). Analysis of ceramides in cosmetics by reversed-phase liquid chromatography/electrospray ionization mass spectrometry with collision-induced dissociation. *Rapid Commun. Mass Spectrom.* **17**, 64–75.

Levery, S. B. (2005). Glycosphingolipid structural analysis and glycosphingolipidomics. *Methods Enzymol.* **405**, 300–369.

McFarland, M. A., Marshall, A. G., Hendrickson, C. L., Nilsson, C. L., Fredman, P., and Mansson, J. E. (2005). Structural characterization of the GM1 ganglioside by infrared multiphoton dissociation, electron capture dissociation, and electron detachment dissociation electrospray ionization FT-ICR MS/MS. *J. Am. Soc. Mass Spectrom.* **16**, 752–762.

Merrill, A. H., Jr., Sullards, M. C., Allegood, J. C., Kelly, S., and Wang, E. (2005). Sphingolipidomics: High-throughput, structure-specific, and quantitative analysis of sphingolipids by liquid chromatography tandem mass spectrometry. *Methods* **36**, 207–224.

O'Connor, P. B., and Costello, C. E. (2001). A high pressure matrix-assisted laser desorption/ionization Fourier transform mass spectrometry ion source for thermal stabilization of labile biomolecules. *Rapid Commun. Mass Spectrom.* **15**, 1862–1868.

Pacetti, D., Boselli, E., Hulan, H. W., and Frega, N. G. (2005). High performance liquid chromatography-tandem mass spectrometry of phospholipid molecular species in eggs from hens fed diets enriched in seal blubber oil. *J. Chromatogr. A* **1097**, 66–73.

Pettus, B. J., Bielawska, A., Kroesen, B. J., Moeller, P. D., Szulc, Z. M., Hannun, Y. A., and Busman, M. (2003). Observation of different ceramide species from crude cellular extracts by normal-phase high-performance liquid chromatography coupled to atmospheric pressure chemical ionization mass spectrometry. *Rapid Commun. Mass Spectrom.* **17**, 1203–1211.

Pettus, B. J., Kroesen, B. J., Szulc, Z. M., Bielawska, A., Bielawski, J., Hannun, Y. A., and Busman, M. (2004). Quantitative measurement of different ceramide species from crude cellular extracts by normal-phase high-performance liquid chromatography coupled to atmospheric pressure ionization mass spectrometry. *Rapid Commun. Mass Spectrom.* **18,** 577–583.

Roy, S., Touboul, D., Brunelle, A., Germain, D. P., Prognon, P., Laprevote, O., and Chaminade, P. (2006). [Imaging mass spectrometry: A new tool for the analysis of skin biopsy. Application in Fabry's disease]. *Ann. Pharm. Fr.* **64,** 328–334.

Samuelsson, B., and Samuelsson, K. (1968). Gas-liquid chromatographic separation of ceramides as di-O-trimethylsilyl ether derivatives. *Biochim. Biophys. Acta.* **164,** 421–423.

Samuelsson, B., and Samuelsson, K. (1969a). Gas–liquid chromatography-mass spectrometry of synthetic ceramides. *J. Lipid Res.* **10,** 41–46.

Samuelsson, K., and Samuelsson, B. (1969b). Gas-liquid chromatography-mass spectrometry of cerebrosides as trimethylsilyl ether derivatives. *Biochem. Biophys. Res. Commun.* **37,** 15–21.

Samuelsson, K., and Sameulsson, B. (1970). Gas chromatographic and mass spectrometric studies of synthetic and naturally occurring ceramides. *Chem. Phys. Lipids* **5,** 44–79.

Schwudke, D., Oegema, J., Burton, L., Entchev, E., Hannich, J. T., Ejsing, C. S., Kurzchalia, T., and Shevchenko, A. (2006). Lipid profiling by multiple precursor and neutral loss scanning driven by the data-dependent acquisition. *Anal. Chem.* **78,** 585–595.

Shayman, J. A., and Abe, A. (2000). 1-O-acylceramide synthase. *Methods Enzymol.* **311,** 105–117.

Sjovall, P., Lausmaa, J., and Johansson, B. (2004). Mass spectrometric imaging of lipids in brain tissue. *Anal. Chem.* **76,** 4271–4278.

Sullards, M. C. (2000). Analysis of sphingomyelin, glucosylceramide, ceramide, sphingosine, and sphingosine 1-phosphate by tandem mass spectrometry. *Methods Enzymol.* **312,** 32–45.

Sullards, M. C., and Merrill, A. H., Jr. (2001). Analysis of sphingosine 1-phosphate, ceramides, and other bioactive sphingolipids by high-performance liquid chromatography-tandem mass spectrometry. *Sci. STKE* **2001,** PL1.

Suzuki, M., Sekine, M., Yamakawa, T., and Suzuki, A. (1989). High-performance liquid chromatography-mass spectrometry of glycosphingolipids: I. Structural characterization of molecular species of GlcCer and IV3 beta Gal-Gb4Cer. *J. Biochem. (Tokyo)* **105,** 829–833.

Suzuki, Y., Suzuki, M., Ito, E., Goto-Inoue, N., Miseki, K., Iida, J., Yamazaki, Y., Yamada, M., and Suzuki, A. (2006). Convenient structural analysis of glycosphingolipids using MALDI-QIT-TOF mass spectrometry with increased laser power and cooling gas flow. *J. Biochem. (Tokyo)* **139,** 771–777.

Suzuki, M., Yamakawa, T., and Suzuki, A. (1990). High-performance liquid chromatography-mass spectrometry of glycosphingolipids: II. Application to neutral glycolipids and monosialogangliosides. *J. Biochem. (Tokyo)* **108,** 92–98.

Suzuki, M., Yamakawa, T., and Suzuki, A. (1991). A micro method involving micro high-performance liquid chromatography-mass spectrometry for the structural characterization of neutral glycosphingolipids and monosialogangliosides. *J. Biochem. (Tokyo)* **109,** 503–506.

Sweeley, C. C., and Dawson, G. (1969). Determination of glycosphingolipid structures by mass spectrometry. *Biochem. Biophys. Res. Commun.* **37,** 6–14.

Vukelic, Z., Zamfir, A. D., Bindila, L., Froesch, M., Peter-Katalinic, J., Usuki, S., and Yu, R. K. (2005). Screening and sequencing of complex sialylated and sulfated glycosphingolipid mixtures by negative ion electrospray Fourier transform ion cyclotron resonance mass spectrometry. *J. Am. Soc. Mass Spectrom.* **16,** 571–580.

Wiseman, J. M., Ifa, D. R., Song, Q., and Cooks, R. G. (2006). Tissue imaging at atmospheric pressure using desorption electrospray ionization (DESI) mass spectrometry. *Angew. Chem. Int. Ed. Engl.* **45,** 7188–7192.

Woods, A. S., and Jackson, S. N. (2006). Brain tissue lipidomics: Direct probing using matrix-assisted laser desorption/ionization mass spectrometry. *AAPS J.* **8,** E391–E395.

Zheng, W., Kollmeyer, J., Symolon, H., Momin, A., Munter, E., Wang, E., Kelly, S., Allegood, J. C., Liu, Y., Peng, Q., Ramaraju, H., Sullards, M. C., *et al.* (2006). Ceramides and other bioactive sphingolipid backbones in health and disease: Lipidomic analysis, metabolism and roles in membrane structure, dynamics, signaling and autophagy. *Biochim. Biophys. Acta* **1758,** 1864–1884.

CHAPTER FIVE

ANALYSIS OF UBIQUINONES, DOLICHOLS, AND DOLICHOL DIPHOSPHATE-OLIGOSACCHARIDES BY LIQUID CHROMATOGRAPHY-ELECTROSPRAY IONIZATION-MASS SPECTROMETRY

Teresa A. Garrett, Ziqiang Guan, *and* Christian R. H. Raetz

Contents

1. Introduction	118
2. Materials	121
3. Liquid Chromatography-Mass Spectrometry	122
4. Preparation of Lipid Extracts	122
5. LC-MS Detection and Quantification of Coenzyme Q	123
6. LC-MS Detection and Quantification of Dolichol	130
7. LC-MS and LC-MS/MS Characterization of Dolichol Diphosphate-Linked Oligosaccharides	136
Acknowledgment	140
References	140

Abstract

Prenols, a class of lipids formed by the condensation of five carbon isoprenoids, have important roles in numerous metabolic pathways of the eukaryotic cell. Prenols are found in the cell as free alcohols, such as dolichol, or can be attached to vitamins, as with the fat soluble vitamins. In addition, prenols such as farnesyl- and geranylgeranyl-diphosphate are substrates for the transfer of farnesyl and geranylgeranyl units to proteins with important implications for signal transduction within the cell. Dolichol phosphate- and dolichol diphosphate–linked sugars are central to the formation of the lipid-linked branched oligosaccharide, Dol-PP-(GlcNAc)$_2$(Man)$_9$(Glc)$_3$, used for co-translational *en bloc* protein *N*-glycosylation in the lumen of the endoplasmic reticulum. Toward furthering our understanding of the role of prenol lipids in the cell, we have developed a method for the detection and quantification of dolichol and coenzyme

Department of Biochemistry, Duke University Medical Center, Durham, North Carolina

Methods in Enzymology, Volume 432
ISSN 0076-6879, DOI: 10.1016/S0076-6879(07)32005-3

© 2007 Elsevier Inc.
All rights reserved.

Q by liquid chromatography-electrospray ionization-mass spectrometry (LC-ESI-MS). These methods, developed using the mouse macrophage RAW 264.7 tumor cells, are broadly applicable to other cell lines, tissues, bacteria, and yeast. We also present a new MS-based method for the detection and structural characterization of the intact dolichol diphosphate oligosaccharide Dol-PP-(GlcNAc)$_2$(Man)$_9$(Glc)$_3$ from porcine pancreas.

1. INTRODUCTION

Prenols are a class of lipids formed by carbocation-based condensations of the five carbon isoprenoids, isopentenyl diphosphate, and dimethylallyl diphosphate (Fahy et al., 2005) (Fig. 5.1). These substances are derived from mevalonate (Kuzuyama and Seto, 2003) in animals or from methylerythritol phosphate in plants (Rodriguez-Concepcion, 2004). Bacteria generate isopentenyl diphosphate and dimethylallyl diphosphate by one or the other of these pathways and, in a few instances, by both (Hedl et al., 2004; Rohdich et al., 2004).

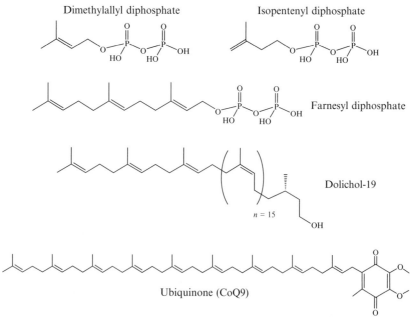

Figure 5.1 Chemical structures of the prenols dolichol and ubiquinone, and their biosynthetic precursors, dimethylallyl diphosphate, isopentenyl diphosphate, and farnesyl diphosphate.

In animal systems, which are the focus of LIPID MAPS, two molecules of isopentenyl diphosphate and one of dimethylallyl diphosphate are condensed by a single enzyme to generate the 15 carbon intermediate, farnesyl diphosphate (Fig. 5.1), in which the stereochemistry of the double bonds is *trans* (Kellogg and Poulter, 1997; Leyes *et al.*, 1999). A separate enzyme elongates farnesyl diphosphate to the 20-carbon geranylgeranyl diphosphate (Ericsson *et al.*, 1998). Subsequently, farensyl- or geranylgeranyl-diphosphate may be further elongated by other prenyl transferases (Kellogg and Poulter, 1997), which incorporate additional isopentenyl units to form the dolichols (Swiezewskaa and Danikiewiczb, 2005), the side chains of the ubiquinones (Turunen *et al.*, 2004), and other substances (Fig. 5.1). Both farnesyl- or geranylgeranyl-diphosphate can also function directly as donor substrates for the addition of farnesyl- or geranylgeranyl- units to proteins, many of which are involved in signal transduction (Gelb *et al.*, 2006).

Mammalian dolichol is a mixture that consists mainly of 17, 18, 19, or 20 isoprene units (Figs. 5.1 and 5.7) (Chojnacki and Dallner, 1988). Small amounts of shorter or longer species may also be detectable. All double bonds except for those of the farnesyl diphosphate primer have the *cis* configuration (Fig. 5.1) because the prenyl transferase responsible for the elongation of farnesyl diphosphate to dolichol orients and condenses its substrates differently than does farnesyl diphosphate synthase (Kellogg and Poulter, 1997). Although free dolichol is the predominant species found in cells, it is initially generated as the diphosphate derivative (Schenk *et al.*, 2001). The diphosphate moiety is subsequently cleaved to yield dolichol phosphate and free dolichol. However, kinases exist that can convert dolichol back to dolichol phosphate (Schenk *et al.*, 2001).

In our experience, dolichol phosphate is much less abundant in animal cells than is dolichol (as shown later). Dolichol phosphate is also rapidly converted to various dolichol phosphate sugars or dolichol diphosphate sugars (Chojnacki and Dallner, 1988; Schenk *et al.*, 2001). In the latter case, dolichol phosphate reacts with UDP-*N*-acetylglucosamine (UDP-GlcNAc) to form dolichol diphosphate-GlcNAc (Kean *et al.*, 1999) to which 13 additional sugars are then added by an important system of glycosyltransferases found in the endoplasmic reticulum (Schenk *et al.*, 2001). The final product is the lipid-linked, branched oligosaccharide Dol-PP-(GlcNAc)$_2$(Man)$_9$(Glc)$_3$ (Fig. 5.2) (Schenk *et al.*, 2001). This conserved intermediate is used for cotranslational *en bloc* protein *N*-glycosylation in the lumen of the endoplasmic reticulum (Hubbard and Ivatt, 1981; Kornfeld and Kornfeld, 1985; Rosner *et al.*, 1982). The sugar composition and glycosidic linkages of Dol-PP-(GlcNAc)$_2$(Man)$_9$(Glc)$_3$ (Fig. 5.2) were determined by classical biochemical and enzymatic methods. However, to our knowledge, the structure of intact Dol-PP-(GlcNAc)$_2$(Man)$_9$(Glc)$_3$ has not been validated by MS or nuclear magnetic resonance (NMR) spectroscopy.

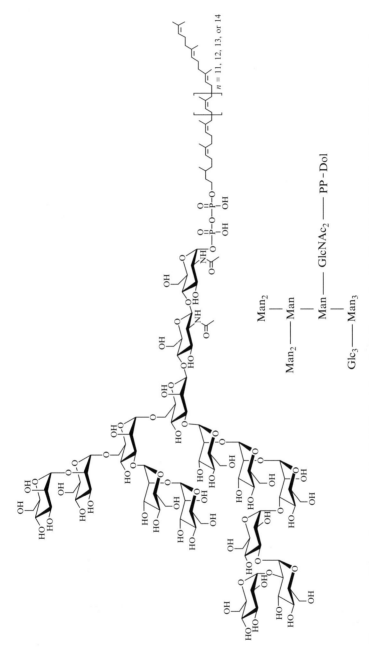

Figure 5.2 Chemical structure of the lipid-linked, branched oligosaccharide, Dol-PP-(GlcNAc)$_2$(Man)$_9$(Glc)$_3$. The inset shows a schematic of the composition and arrangement of the sugars, where GlcNAc is N-acetyl glucosamine, Man is mannose, and Glc is glucose.

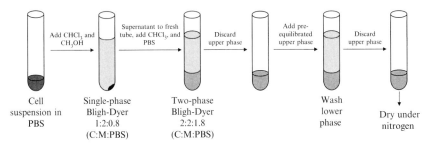

Figure 5.3 Lipid extraction procedure.

Polyprenol chains are also found attached to other small organic molecules in the cell, such as the fat-soluble vitamins A, E, and K (Meganathan, 2001a; Olson, 1964), and the electron transport cofactor, ubiquinone, also called coenzyme Q (Fig. 5.1) (Meganathan, 2001b; Szkopinska, 2000). In the latter case, farnesyl diphosphate is elongated by a different prenyl transferase that incorporates three to seven additional isopentenyl units, depending on the system. In contrast to dolichol, all the double bonds formed during ubiquinone side-chain elongation have the *trans* stereochemistry (Fig. 5.1).

Herein we present new methods for the detection and quantification of dolichol and coenzyme Q in mouse macrophage RAW 264.7 tumor cells (Raetz et al., 2006) using LC-ESI-MS. While the methods reported here were developed using cultured macrophages, we have found that they are broadly applicable to other cell lines and tissues, as well as to yeast and bacteria. We also present a new method for the detection and structural characterization of dolichol diphosphate oligosaccharides, including the first analysis by LC-ESI-MS/MS of intact Dol-PP-(GlcNAc)$_2$(Man)$_9$(Glc)$_3$ from porcine pancreas (Kelleher et al., 2001).

2. MATERIALS

High performance liquid chromatography (HPLC)–grade solvents were from VWR (International Leicestershire, England). Ammonium acetate was from Mallinckrodt (Hazelwood, MO). Zorbax SB-C8 (2.1 × 50 mm, 5 μm) reverse-phase column was from Agilent (Palo Alto, CA). Coenzyme Q6 and Q10 standards were from Sigma-Aldrich (St. Louis, MO). Synthetic nor-dolichol was from Avanti Polar Lipids (Alabaster, AL). Dolichol-diphosphate-oligosaccharide purified from pig pancreas was a gift from Dr. R. Gilmore (University of Massachusetts Medical School, Worcester, MA). All extractions were done in glass tubes equipped with Teflon-lined screw caps using disposable glass pipettes.

3. Liquid Chromatography-Mass Spectrometry

LC-MS analysis was performed using a Shimadzu LC system (comprising a solvent degasser, two LC-10A pumps, and a SCL-10A system controller) coupled to a QSTAR XL quadrupole time-of-flight (TOF), tandem mass spectrometer (Applied Biosystems/MDS Sciex). A Zorbax SB-C8 reversed-phase column (5 μm, 2.1 × 50 mm) was used for all LC-MS analyses. LC was operated at a flow rate of 200 μl/min with a linear gradient as follows: 100% mobile phase A (methanol:acetonitrile:aqueous 1 mM ammonium acetate; 60:20:20) was held for 2 min, then linearly increased to 100% mobile phase B (100% ethanol containing 1 mM ammonium acetate) over 14 min and held at 100% mobile phase B for 4 min. The column was re-equilibrated to 100% mobile-phase A for 2 min prior to the next injection. The post-column split diverted 10% of the LC flow to the ESI source. All MS data were analyzed using Analyst QS software (Applied Biosystems/MDS Sciex). The mass spectrometer was calibrated in both the negative and positive mode using PPG3000 (Applied Biosystems).

4. Preparation of Lipid Extracts

RAW cells, cultured according to (Raetz et al., 2006), were extracted using the method of Bligh and Dyer (1959), as shown in Fig. 5.3. A 150-mm tissue culture plate at about 90% confluence was washed with 10 ml phosphate buffered saline (PBS) and then scraped into 5 ml PBS (137 mM NaCl, 0.027 mM KCl, 0.01 mM Na$_2$HPO$_4$, 0.0018 mM KH$_2$PO$_4$). The cell suspension was centrifuged at 400×g to harvest the cells. This culture plate size typically yields about 100 × 10^6 cells or ~0.1 g of cells (wet weight). The cell pellet is resuspended in 1 ml of PBS and transferred to a 16 × 150-mm-glass centrifuge tube with a Teflon-lined cap.

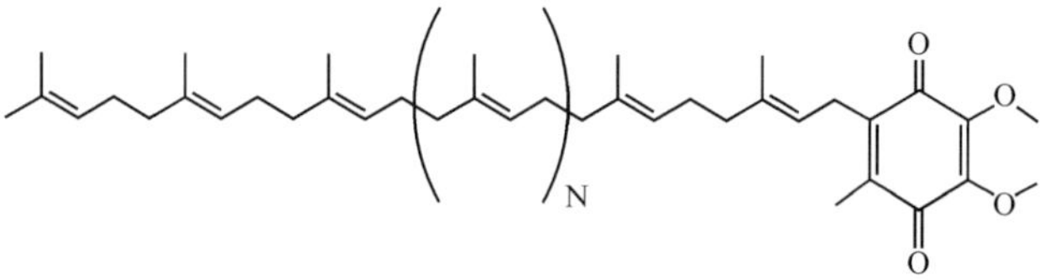

N	CoQ species	Formula [M]	[M+H]$^+$ Observed	[M+H]$^+$ Exact	[M+NH$_4$]$^+$ Observed	[M+NH$_4$]$^+$ Exact	[M+Na]$^+$ Observed	[M+Na]$^+$ Exact
1	CoQ6	C39H58O4	591.434	591.441	608.469	608.467	613.423	613.423
4	CoQ9	C54H82O4	795.647	795.629	812.678	812.655	817.640	817.611
5	CoQ10	C59H90O4	863.713	863.691	880.734	880.718	885.689	885.673

Figure 5.4 Structures and masses of coenzyme Q6, Q9, and Q10.

Next, 1.25 ml of chloroform and 2.5 ml of methanol are added to generate a single-phase Bligh-Dyer mixture (final ratio: 1:2:0.8, C:M:PBS). This mixture is vigorously mixed with a vortex and, if necessary, subjected to sonic irradiation in a bath apparatus for 2 min to break up any chunks of cells. The single-phase extraction mixture is incubated at room temperature for 15 min, then centrifuged at $\sim 500 \times g$ for 10 min in a clinical centrifuge to pellet the cell debris (Fig. 5.3). The supernatant is transferred to a fresh tube; 1.25 ml chloroform and 1.25 ml PBS are added to generate a two-phase system (final ratio; 2:2:1.8, C:M:PBS). After vigorous mixing, the tubes are centrifuged as previously stated to resolve the phases. The upper phase and any interface that may be present is removed and discarded. The interface is generally minimal. Next, the lower phase is washed with 4.75 ml pre-equilibrated neutral upper phase, vortexed, and centrifuged as previously stated to resolve the phases. After removal of the upper phase, the lower phase is dried under a stream of nitrogen, and the dried lipids are stored at $-20°$ until analysis. Generally, from 0.1 g of cells or tissue, about 10 mg of dried lipid is obtained.

Note on scaling up: This method of extraction can be used with larger quantities of cells or tissues. Volumes should be scaled accordingly; however, a three- to five-fold concentration of the initial PBS suspension of cells or tissue should not affect extraction efficiency. With larger volumes, 250 ml Teflon-lined centrifuge bottles (VWR) are useful for centrifugation and separation of the phases.

5. LC-MS Detection and Quantification of Coenzyme Q

Coenzyme Q (CoQ), also known as ubiquinone, is involved in the transport of electrons from complex I (NADH:ubiquinone oxidoreductase) and II (succinate:ubiquinone reductase) to complex III (ubiquinone:cytochrome c oxidoreductase) in the electron transport chain of aerobic organisms (Saraste, 1999). The structure of CoQ is shown in Figs. 5.1 and 5.4. It is a quinone derivative with a long polyprenoid chain that can be 7 to 11 units long depending on the organism (Szkopinska, 2000). In humans, the most abundant CoQ has 10 isoprenoid units. In murine RAW cells, the most abundant CoQ has a side chain consisting of nine units (Olgun *et al.*, 2003).

In order to develop LC-MS methods for the detection of CoQ in total lipid extracts, we needed to (1) determine the linear range for MS response with respect to the amount of standard analyzed; (2) establish that this linear range applied to the MS response when analyzing the standard in total lipid extracts; (3) show that the number of isoprene units on the CoQ did not alter the MS response; and (4) determine the amount of standard to add to our sample to get approximately the same peak area of the extracted ion

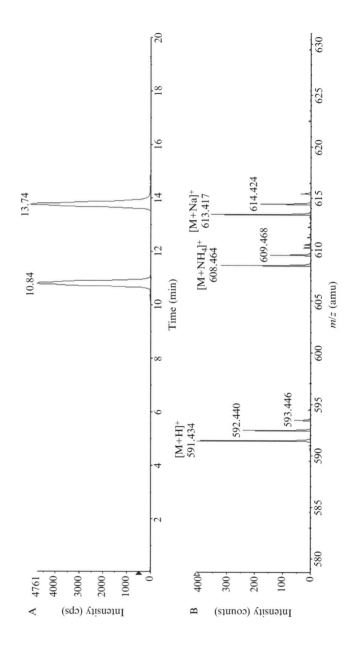

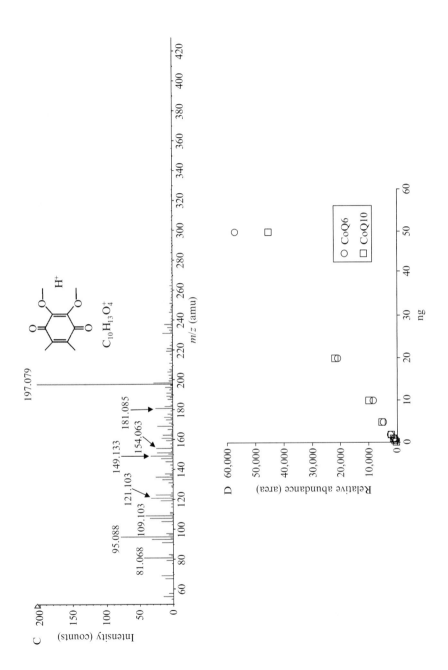

current for the standard and the analyte. We found that we could use a CoQ with a different number of isoprene units as a standard instead of using deuterated CoQ (Gould et al., 2006; Haroldsen et al., 1987).

We obtained two CoQ standards, one with 6 isoprenoid units, CoQ6, and the other with 10 isoprenoid units, CoQ10 (Sigma) (Fig. 5.4). For LC-MS analysis, the standard was re-dissolved in $CHCl_3$ and diluted into DMSO:MeOH (9:1 v/v) at a concentration of 0.1, 0.2, 0.5, 1, 2, or 5 ng/μl. Next, 10 μl of each dilution were injected, in turn, onto a Zorbax SB-C8 (2.1 × 50 mm, 5 mm) column at 200 μl/min and eluted as described above. CoQs were detected in the positive ion mode as $[M+H]^+$, $[M+NH_4]^+$, and $[M+Na]^+$ ions using the following mass spectrometer settings: ESI voltage, +5500 V; declustering potential, 60 V; and focusing potential, 265 V; and neubulizer gas, 20 psi.

Figure 5.5A shows the extracted ion current (EIC) for the $[M+H]^+$ ion of CoQ6 and CoQ10. Each standard was analyzed in separate LC-MS runs, and the spectra were overlaid. The CoQ6 elutes from the reverse phase column at about 11.1 min, and the CoQ10 elutes at about 13.7 min, consistent with the difference in the polyprenol chain length. Figure 5.5B and C show the positive ion MS and collision-induced dissociation MS (CID-MS) for the $[M+H]^+$ CoQ6 standard. The collision energy was set at 52 V (laboratory frame of reference), and nitrogen was used as the collision gas. The major fragment ion is m/z 197.079, which corresponds by exact mass to a proton adduct of the quinone ring and is formed by the elimination of the isoprenoid chain (Teshima and Kondo, 2005). The CID-MS of the CoQ10 standard produces a similar fragment ion pattern (data not shown). As shown in Fig. 5.5D, the peak area of the EICs of the CoQ6 and CoQ10 standards are linear for 1 ng to 50 ng injected onto the column. This also shows that, for CoQs, the number of isoprenoid repeats does not alter the ionization efficiency, as is sometimes seen with other lipids analyzed by MS (Callender et al., 2007). In addition, the peak area of the EIC of the CoQ6 standard increased linearly when increasing amounts of the standard were co-extracted with the RAW cells (data not shown).

Figure 5.5 Positive-ion liquid chromatography-mass spectrometry (LC-MS) analysis of CoQ6 and CoQ10 standards. (A) The extracted ion current (EIC) for the $[M+NH_4]^+$ ions of CoQ6 (608.464 m/z) and CoQ10 (880.734 m/z) standards. Under these liquid chromatography conditions, the CoQ6 elutes at about 10 minutes while the CoQ10 elutes at about 14 minutes. (B) The mass spectrum of the material eluting from minutes 10.6 to 11.1, representing the CoQ6 standard. The CoQ6 form $[M+H]^+$, $[M+NH_4]^+$, and $[M+Na]^+$ adduct ions. (C) The collision-induced mass spectrometry (CID-MS) of the $[M+H]^+$ ion of the CoQ6 standard. The major fragment ion (m/z 197.079) corresponds to the quinone ring and is formed by the elimination of the isoprenoid chain (Teshima and Kondo, 2005). (D) The plot of the peak area of the extracted ion current of the $[M+H]^+$ ion of CoQ6 and CoQ10 versus the nanogram (ng) of standard injected onto the Zorbax C-8 reverse phase column. The peak areas of the CoQ6 and the CoQ10 were similar when equivalent amounts of standard were analyzed. (See color insert.)

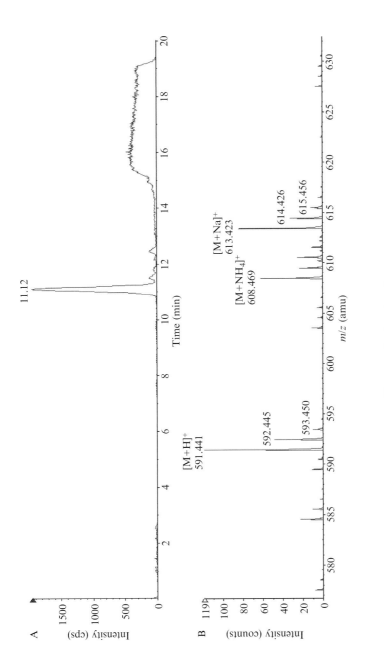

Figure 5.6 *(continued)*

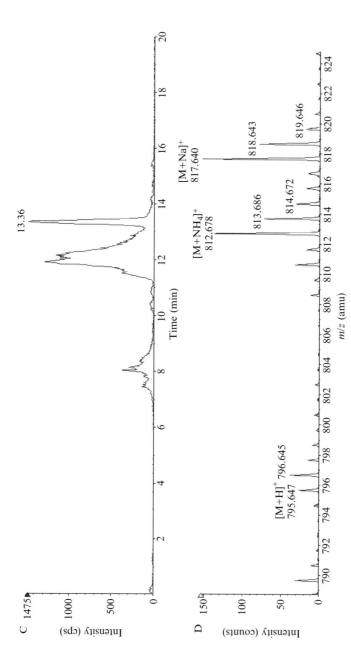

Figure 5.6 Positive ion liquid chromatography–mass spectrometry (LC-MS) analysis of CoQs in RAW lipid extracts. (A) The extracted ion current (EIC) for the $[M+NH_4]^+$ ion of CoQ6 standard after 1 μg of standard is co-extracted in the presence of RAW cells from one 150-mm plate. The retention time (11.12 min) is consistent with the retention time seen when the standard is injected alone (Fig. 5.5A). (B) The mass spectrum of the material eluting between 10.9 and 11.3 min. The $[M+H]^+$, $[M+NH_4]^+$, and $[M+Na]^+$ ions of the CoQ6 standard are easily resolved. (C) The extracted ion current of the $[M+NH_4]^+$ ion of the major CoQ of RAW cells, CoQ9. The CoQ9 elutes at approximately 13.3 min under these liquid chromatography conditions. The peaks at approximately 8 and 12 min are due to isobaric lipids found in the lipid extract and are not due to CoQ9. (D) The mass spectrum of the material eluting between 13.3 and 14.3 min. The $[M+H]^+$, $[M+NH_4]^+$, and $[M+Na]^+$ ions of the CoQ9 standard are detected. (See color insert.)

Therefore, the linear range established using the standard also applies to the analysis of CoQs from total lipid extracts.

Because the major CoQs found in mouse tissue are CoQ9 and CoQ10, we chose to use CoQ6 as the internal standard for our quantitative analysis (Olgun et al., 2003). CoQ6 is not found in RAW cells and is not isobaric with other ions eluting at a similar retention time.

In addition to establishing the linear range of the MS response, we needed to determine the amount of CoQ6 standard to add to the extraction mixture in order for the peak area of the EIC for the standard to be about equal to the peak area of the EIC for the CoQs being quantified in the sample. By varying the amount of CoQ6 added during the co-extraction, we determined that, for a 150-mm culture dish grown to ~90% confluence, 5 μg of CoQ6 standard gave an EIC peak area about equal to the EIC peak of the major coenzyme species of the cell, CoQ9.

Overall, our quantification procedure for CoQs in RAW cells is as follows: 5 μg of CoQ6 (dissolved in $CHCl_3$ at 1 mg/ml) standard is added to the cell suspension (1 ml) prior to the addition of chloroform (1.25 ml) and methanol (2.5 ml) to form the single-phase extraction mixture; the lipids are extracted as previously described, dried under a stream of nitrogen, and stored at $-20°$ until analysis; the dried lipids are redissolved in 0.2 ml of chloroform:methanol (1:1, v:v) and diluted twofold into chromatography solvent A; 10 μl is injected onto the reverse-phase column and analyzed using the LC-MS procedure previously shown. The extracted ion current and mass spectra for the CoQ6 standard and endogenous CoQ9 are shown in Figs. 5.6A–D. For CoQ9, the major ion species are observed as $[M+H]^+$ at m/z 795.647, $[M+NH_4]^+$ at m/z 812.678, and $[M+Na]^+$ at m/z 817.640 (Fig. 7D). For CoQ10, the major ion species are observed as $[M+H]^+$ at m/z 863.713, $[M+NH_4]^+$ at m/z 880.734, and $[M+Na]^+$ at m/z 885.673 (data not shown). The peak area of the EIC for the $[M+NH_4]^+$ is used

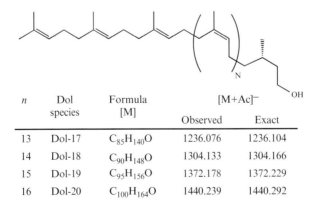

n	Dol species	Formula [M]	$[M+Ac]^-$ Observed	Exact
13	Dol-17	$C_{85}H_{140}O$	1236.076	1236.104
14	Dol-18	$C_{90}H_{148}O$	1304.133	1304.166
15	Dol-19	$C_{95}H_{156}O$	1372.178	1372.229
16	Dol-20	$C_{100}H_{164}O$	1440.239	1440.292

Figure 5.7 Structures and masses of the acetate adducts of dolichols-17, -18, -19, and -20.

to quantify the total content of CoQ9 and CoQ10 in the RAW cells. For example, the quantification of the CoQ9 is accomplished using the following equation:

$$[(CoQ9 \text{ peak area}/CoQ6 \text{ standard peak area}) \times 5\ \mu g]/MW \text{ of CoQ9} = \mu mol\ CoQ9$$

in the sample. Using this method, we have been able to detect CoQs in all cells and tissues we have analyzed.

6. LC-MS Detection and Quantification of Dolichol

Dolichols are linear polymers of five carbon isoprenoid units ranging in length from mostly 17 to 20 units in animals and up to 40 units in some plants (Figs. 5.1 and 5.7) (Burda and Aebi, 1999; Krag, 1998). Dolichol is found predominantly as the free alcohol in animal tissues, but it may also be esterified to fatty acids or derivatized with sugar phosphates (Chojnacki and Dallner, 1988; Elmberger *et al.*, 1989). The form and distribution of dolichol are tissues and species dependent (Elmberger *et al.*, 1989; Krag, 1998). Whether or not dolichols have functions other than their role in membrane-associated glycosylation reactions is unclear (Lai and Schutzbach, 1986; Rip *et al.*, 1981; Rupar *et al.*, 1982; Valtersson *et al.*, 1985).

Quantitative, chemically defined standards of dolichol are not commercially available. In order to quantify the dolichol content of the RAW cells, a nor-dolichol standard with 17 to 20 isoprene units, which differs from dolichol (Figs. 5.1, 5.7, and 5.8) by the absence of one CH_2 unit at the hydroxyl end of the molecule, was prepared by modification of natural dolichol at Avanti Polar Lipids.

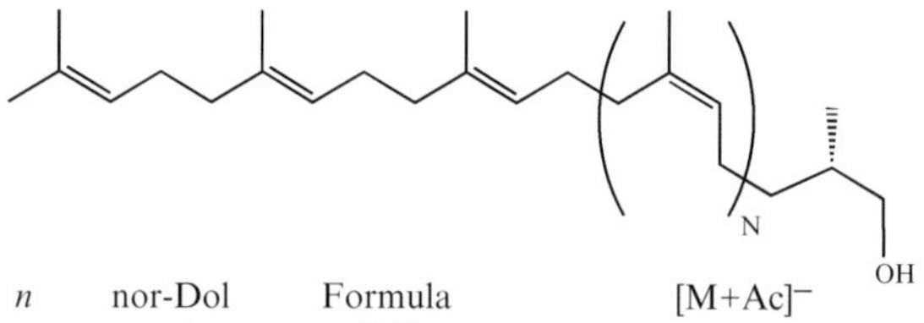

n	nor-Dol species	Formula [M]	[M+Ac]⁻ Observed	Exact
13	nor-Dol-17	$C_{84}H_{138}O$	1222.137	1222.086
14	nor-Dol-18	$C_{89}H_{146}O$	1290.196	1290.151
15	nor-Dol-19	$C_{94}H_{154}O$	1358.255	1358.214
16	nor-Dol-20	$C_{99}H_{162}O$	1426.316	1426.276

Figure 5.8 Structures and masses of the acetate adducts of nor-dolichols-17, -18, -19, and -20.

For the LC-MS analysis, we established, as with the coenzyme Q, the linear range of the MS response with respect to the amount of nor-dolichol injected onto the column. Using the same LC-MS system as previously shown and detecting in the negative ion mode, 5, 20, 50, or 100 ng in 10 μl of chromatography solvent A were injected onto the Zorbax SB-C8 (2.1 × 50 mm, 5 μm) reverse-phase column and eluted using the same scheme as previously described. The mass spectrometer settings are as follows: ESI voltage, −4400 V; declustering potential, −55 V; focusing potential, −265 V; neubulizer gas, 18 psi. The semisynthetic nor-dolichol is a mixture of species with the number of isoprene units ranging from mainly 16 to 21 (Fig. 5.8). The smaller nor-dolichols, such as nor-dolichol with 17 isoprene units, eluted from the column first, followed by the larger nor-dolichol species. The nor-dolichols are detected as the acetate adduct [M+Ac]$^-$ ions in the negative ion mode using the LC-MS method as previously described. Figure 5.9A shows the EIC for the predominant nor-dolichol-19. The MS of 16.3 to 17.3 min when the nor-dolichol-19 acetate adduct [M + Ac]$^-$ elutes from the column is shown (Fig. 5.9B). Small amounts of contaminants at m/z 1334.213 and 1344.241 may correspond to oxidative loss of additional carbon atoms from the dolichol during the preparation of the nor-dolichol. Similar to the CoQ standard, the peak area of the EIC for the dolichol standard was linear for 5 ng to 50 ng injected on the column (Fig. 5.9C).

We determined the amount of nor-dolichol standard to add to the extraction mixture in order to have the peak area of the EIC for the standard be about equal to the peak area of the EIC for the dolichols being quantified in the sample. By varying the amount of nor-dolichol added during the co-extraction, we determined that, for a 150-mm culture dish grown to ∼90% confluence, 1 μg of the nor-dolichol standard gave an EIC peak area of the nor-dolichol-18 about equal to the EIC peak of the major dolichol species of the cell, dolichol-18.

The ion intensities of dolichol species present in RAW cell lipid extracts are relatively low compared to those of CoQs. Dolichol cannot be detected by direct injection of total lipid extracts without prior chromatography. In addition, while CoQ ions can be readily detected when approximately 5% of the total lipid extract is analyzed using the above LC-MS system, endogenous dolichol ions are only effectively detected when 15% of the total lipid extract is injected onto the reverse-phase column.

Overall, our method for quantifying dolichol is as follows: 1 μg of nor-dolichol (dissolved in $CHCl_3$ at 1 mg/ml) is added to the cell suspension (1 ml) just prior to the addition of chloroform (1.25 ml) and methanol (2.5 ml) to form the single-phase extraction mixture; the lipids are extracted as previously described, dried under a stream of nitrogen, and stored at −20° until analysis; the dried lipids are redissolved in 0.2 ml of chloroform/methanol (2:1, v/v), and 30 μl is injected onto the reverse-phase column and analyzed using the LC-MS procedure previously described.

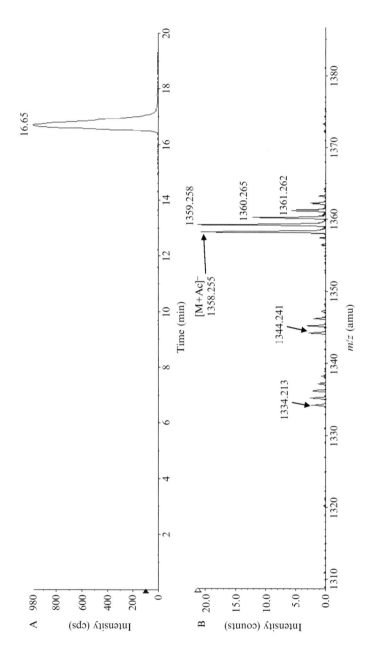

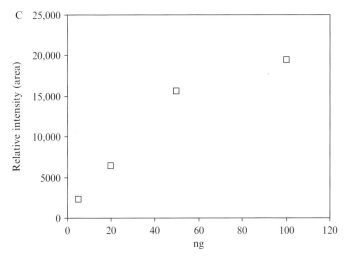

Figure 5.9 Negative-ion liquid chromatography-mass spectrometry (LC-MS) analysis of nor-dolichol standard. (A) The extracted ion current (EIC) for the [M+Ac]⁻ ions of the nor-dolichol-19 (1358.255 *m/z*) standard. Under these liquid chromatography conditions, the nor-dolichol-19 elutes at approximately 16.7 min. The shorter nor-dolichols, nor-dolichol-17 and 18, elute slightly earlier, while the nor-dolichol-20 elutes slightly later. (B) The mass spectrum of material eluting between 16.3 to 17.3 min, representing mainly nor-dolichol-19. The small peaks at *m/z* 1334.213 and 1344.241 are likely due to loss of additional carbons from the dolichol during the preparation of the nor-dolichol standard. (C) The plot of the peak area of the extracted ion current of the [M+Ac]⁻ ion of nor-dolichol-19 versus the nanogram (ng) of standard injected onto the Zorbax C-8 reverse-phase column. (See color insert.)

Figure 5.10A to D shows the EIC and MS for the endogenous dolichol with 17, 18, 19, and 20 isoprene units. The peak area of the EIC for the [M + Ac]⁻ ion of the nor-dolichol-17 was used to quantify dolichol-17; nor-dolichol-18 was used to quantify dolichol 18 and so forth. Because the nor-dolichol standard is a mixture, we needed to determine which fraction of the standard was nor-dolichol-17, nor-dolichol-18, etc. To do this, we summed the peak area of the EIC of nor-dolichol-16 to nor-dolichol-21 to obtain the total peak area when 1 µg of standard was injected onto the column. The total amount of nor-dolichol-18, for example, was then calculated by dividing the peak area of the EIC for nor-dolichol-18 by the summed peak areas. Using this method, we determined the total amount of dolichol in the sample using the following equation:

$$[(\text{Dol-18 peak area}/\text{nor-Dol-18 standard peak area}) \times 1 \text{ µg}]/\text{MW of Dol-18} = \text{µmol Dol-18}$$

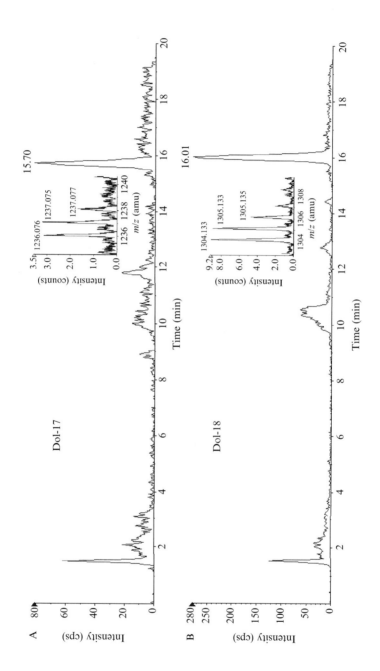
134

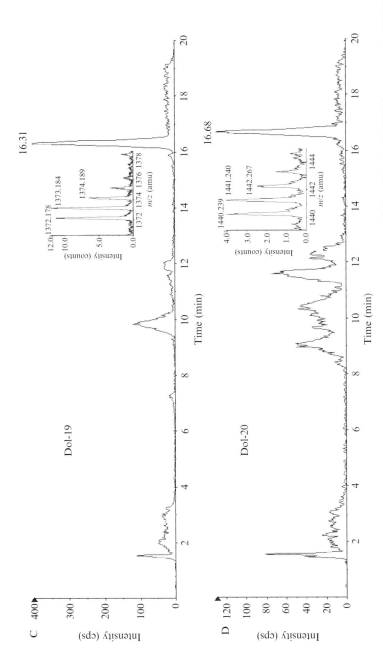

Figure 5.10 Negative ion liquid chromatography–mass spectrometry (LC–MS) analysis of dolichols in RAW lipid extracts. (A–D) The extracted ion current (EIC) of the acetate adducts of dolichol-17, -18, -19, and -20, respectively. For each, the peak labeled with a retention time corresponds to the relevant extracted ion current peak for a given dolichol. The inset on each panel shows the mass spectra of each of the dolichols detected in the RAW lipid extract. (See color insert.)

in the sample. Because large amounts of total lipid extract are being injected onto the column, a blank is run between each sample being analyzed to prevent carry-over between samples.

Note: Separate cell suspensions are not necessary for the quantification of the CoQs and dolichols. Routinely, a single cell suspension is co-extracted with both the nor-dolichol and CoQ6 standard. After extraction, the sample is divided in two and analyzed using positive-ion LC-MS for quantification of the CoQs and negative-ion LC-MS for quantification of the dolichols.

7. LC-MS AND LC-MS/MS CHARACTERIZATION OF DOLICHOL DIPHOSPHATE-LINKED OLIGOSACCHARIDES

Dolichols play a central role in the N-glycosylation and O-glycosylation of eucaryotic proteins (Imperiali and O'Connor, 1999; Schenk et al., 2001; Weerapana and Imperiali, 2006). Dol-PP-(GlcNAc)$_2$(Man)$_9$(Glc)$_3$ (Fig. 5.3) serves as the donor for en bloc transfer to select asparagine residues during the co-translational processing of secretory proteins in the lumen of the endoplasmic reticulum (Burda and Aebi, 1999; Helenius and Aebi, 2004). Removal and addition of sugars to this oligosaccharide occur as the protein moves through the secretory pathway and is involved, in part, in the targeting of proteins to their proper locations in cells (Burda and Aebi, 1999; Huet et al., 2003). Previously, the structure and biosynthesis of the 14-sugar oligosaccharide donor had been proposed based on the studies using a combination of radio-labeling, enzymatic digestion, chemical derivatization, thin layer chromatography, and gas-liquid chromatography (Kornfeld et al., 1978; Li and Kornfeld, 1979; Li et al., 1978). However, none of the structural characterization was performed on the intact donor molecule.

We wanted to develop a MS-based technique to detect and confirm the structure of the intact Dol-PP-(GlcNAc)$_2$(Man)$_9$(Glc)$_3$. A sample of Dol-PP-(GlcNAc)$_2$(Man)$_9$(Glc)$_3$, purified from porcine pancreas, was kindly provided by Dr. R. Gilmore. Due to the amphipathic nature of this molecule, we modified the LC-MS method previously described. Specifically, a new solvent mixture was used as chromatography solvent B (cholorform: methanol:1 mM ammonium acetate [2:3:1 v/v/v], supplemented with 0.1% piperidine). With this modification, we were able to detect the Dol-PP-(GlcNAc)$_2$(Man)$_9$(Glc)$_3$ eluting between 9 and 12 min, depending on the length of the dolichol to which the 14-sugar oligosaccharide was attached. Figures 5.11A–D shows the EIC and MS for the $[M-2H]^{2-}$ ion for the oligosaccharide attached to Dol-17, 18, 19, and 20. The exact mass predicted from the proposed structure matches the mass of the major $[M-2H]^{2-}$ (Fig. 5.3 and Table 5.1) and $[M-3H]^{3-}$ ions (not shown) within experimental error.

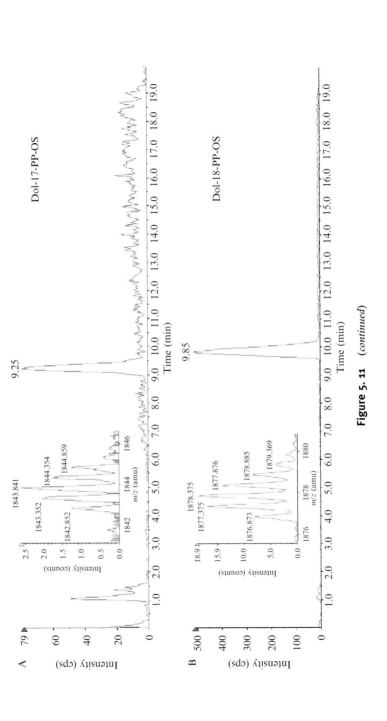

Figure 5. 11 (*continued*)

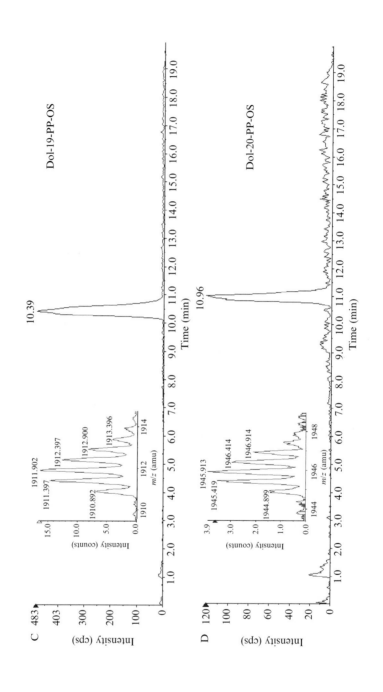

LC-MS of Ubiquinones, Dolichols, and Dolichol Diphosphate Oligosaccharides

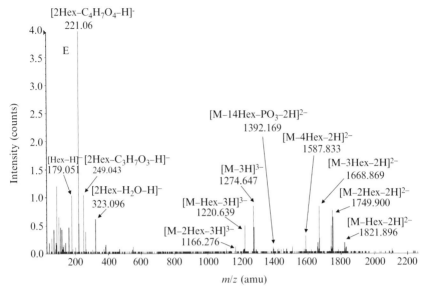

Figure 5.11 Negative-ion liquid chromatography-mass spectrometry (LC-MS) analysis of purified Dol-PP-(GlcNAc)$_2$(Man)$_9$(Glc)$_3$. (A–D) The extracted ion current (EIC) of Dol-PP-(GlcNAc)$_2$(Man)$_9$(Glc)$_3$ with 17, 18, 19, and 20 isoprenoid units on the dolichol, respectively. The inset shows the mass spectrum of the [M–2H]$^{2-}$ ion for the Dol-PP-(GlcNAc)$_2$(Man)$_9$(Glc)$_3$ detected in the purified sample. (E) The collision-induced mass spectrometry (CID-MS) of the most abundant isotope of the [M–3H]$^{3-}$ ion of the Dol-19-PP-(GlcNAc)$_2$(Man)$_9$(Glc)$_3$ (m/z 1274.647). Sequential loss of 162 u, which corresponds to the loss of one, two, three, or four hexoses, is observed. A fragment ion corresponding to Dol-19-P is also detected as a singly charged ion at m/z 1392.169. Fragment ions corresponding to a single hexose (m/z 179.051) and two hexoses (m/z 323.096) are detected in the low mass region. The remaining low mass fragment ions are most likely due to cross-ring cleavage as described by Sheeley and Reinhold (1998). (See color insert.)

Table 5.1 Masses of the [M-2H]$^{2-}$ Ions of Dol-PP-(GlcNAc)$_2$(Man)$_9$(Glc)$_3$

n	Dol-PP-OS species	Formula [M]	[M–2H]$^{2-}$ Observed (m/z)	Exact (m/z)
11	Dol-17-PP-OS	$C_{173}H_{288}N_2O_{77}P_2$	1842.852	1484.908
12	Dol-18-PP-OS	$C_{178}H_{296}N_2O_{77}P_2$	1876.873	1876.939
13	Dol-19-PP-OS	$C_{183}H_{304}N_2O_{77}P_2$	1910.892	1910.971
14	Dol-20-PP-OS	$C_{188}H_{312}N_2O_{77}P_2$	1944.899	1445.002

Using this technique, we also performed CID-MS on the most abundant isotope of the triply charged ion of the Dol-19-PP-(GlcNAc)$_2$ (Man)$_9$(Glc)$_3$ to verify the structure proposed by (Kornfeld *et al.*, 1978; Li and Kornfeld, 1979; Li *et al.*, 1978). This spectrum, shown in Fig. 5.11E, shows the sequential loss of 162 u, which corresponds to the loss of one, two, three, or four hexoses. A fragment ion corresponding to Dol-19-P is also detected as a singly charged ion at m/z 1392.169. In the lower mass region, fragment ions corresponding to a single hexose (m/z 179.051) and two hexoses (m/z 323.096) are also detected. The remaining low-mass fragment ions are most likely due to cross-ring cleavage as described by Sheeley and Reinhold (1998). Our data are consistent with the proposed structure shown in Fig. 5.2, but cannot be used to confirm the positions of the glycosidic linkages, the anomeric stereochemistry, or the actual sugar composition. These issues will have to be evaluated via NMR spectroscopy and other procedures.

We were unable to detect Dol-PP-(GlcNAc)$_2$(Man)$_9$(Glc)$_3$ in lipid extracts of RAW cells grown at the scale described above. The Bligh-Dyer extraction procedure that we use (Fig. 5.3) is not likely to extract this material, as it differs significantly from the procedure used by Kelleher *et al.* (2001). However, one would expect that dolichol phosphate, dolichol diphosphate, and dolichol phosphates derivatized with one or two sugars would be efficiently extracted by our procedure. Using a similar extraction, one does in fact readily detect undecaprenol phosphate in a total *Escherichia coli* lipid extract (Z. Guan and C. R. H. Raetz, unpublished), and both dodecaprenol phosphate and dodecaprenol phosphate-galacturonic acid in *Rhizobium leguminosarum* (Kanjilal-Kolar and Raetz, 2006). While we have been able to detect purified dolichol-phosphate and dolichol phosphate mannose (provided by Charles J. Waechter at University of Kentucky College of Medicine) using the LC-MS conditions described above (data not shown), our inability to detect dolichol phosphate, dolichol diphosphate, dolichol phosphate-hexose, or dolichol diphosphate-N-acetylglucosamine in the total lipids of RAW cells suggests that the endogenous levels of these compounds are extremely low.

ACKNOWLEDGMENT

We thank Dr. Robert Murphy for many stimulating discussions and advice with LC-MS conditions.

REFERENCES

Bligh, E. G., and Dyer, W. J. (1959). A rapid method of total lipid extraction and purification. *Can. J. Biochem. Phys.* **37,** 911–917.

Burda, P., and Aebi, M. (1999). The dolichol pathway of N-linked glycosylation. *Biochim. Biophys. Acta* **1426**, 239–257.

Callender, H. L., Forrester, J. S., Ivanova, P., Preininger, A., Milne, S., and Brown, H. A. (2007). Quantification of diacylglycerol species from cellular extracts by electrospray ionization mass spectrometry using a linear regression algorithm. *Anal. Chem.* **79**, 263–272.

Chojnacki, T., and Dallner, G. (1988). The biological role of dolichol. *Biochem. J.* **251**, 1–9.

Elmberger, P. G., Eggens, I., and Dallner, G. (1989). Conditions for quantitation of dolichyl phosphate, dolichol, ubiquinone and cholesterol by HPLC. *Biomed. Chromatogr.* **3**, 20–28.

Ericsson, J., Greene, J. M., Carter, K. C., Shell, B. K., Duan, D. R., Florence, C., and Edwards, P. A. (1998). Human geranylgeranyl diphosphate synthase: Isolation of the cDNA, chromosomal mapping and tissue expression. *J. Lipid Res.* **39**, 1731–1739.

Fahy, E., Subramaniam, S., Brown, H. A., Glass, C. K., Merrill, A. H. J., Murphy, R. C., Raetz, C. R., Russell, D. W., Seyama, Y., Shaw, W., Shimizu, T., Spener, F., et al. (2005). A comprehensive classification system for lipids. *J. Lipid Res.* **46**, 839–861.

Gelb, M. H., Brunsveld, L., Hrycyna, C. A., Michaelis, S., Tamanoi, F., Van Voorhis, W. C., and Waldmann, H. (2006). Therapeutic intervention based on protein prenylation and associated modifications. *Nat. Chem. Biol.* **2**, 518–528.

Gould, T. A., Herman, J., Krank, J., Murphy, R. C., and Churchill, M. E. (2006). Specificity of acyl-homoserine lactone synthases examined by mass spectrometry. *J. Bacteriol.* **188**, 773–783.

Haroldsen, P. E., Clay, K. L., and Murphy, R. C. (1987). Quantitation of lyso-platelet activiating factor molecular species from human neutrophils by mass spectrometry. *J. Lipid Res.* **28**, 42–49.

Hedl, M., Tabernero, L., Stauffacher, C. V., and Rodwell, V. W. (2004). Class II 3-hydroxy-3-methylglutaryl coenzyme A reductases. *J. Bacteriol.* **186**, 1927–1932.

Helenius, A., and Aebi, M. (2004). Roles of N-linked glycans in the endoplasmic reticulum. *Annu. Rev. Biochem.* **73**, 1019–1049.

Hubbard, S. C., and Ivatt, R. J. (1981). Synthesis and processing of asparagine-linked oligosaccharides. *Annu. Rev. Biochem.* **50**, 555–593.

Huet, G., Gouyer, V., Delacour, D., Richet, C., Zanetta, J. P., Delannoy, P., and Degand, P. (2003). Involvement of glycosylation in the intracellular trafficking of glycoproteins in polarized epithelial cells. *Biochimie* **85**, 323–330.

Imperiali, B., and O'Connor, S. E. (1999). Effect of N-linked glycosylation on glycopeptide and glycoprotein structure. *Curr. Opin. Chem. Biol.* **3**, 643–649.

Kanjilal-Kolar, S., and Raetz, C. R. (2006). Dodecaprenyl phosphate-galacturonic acid as a donor substrate for lipopolysaccharide core glycosylation in *Rhizobium leguminosarum*. *J. Biol. Chem.* **281**, 12879–12887.

Kean, E. L., Wei, Z., Anderson, V. E., Zhang, N., and Sayre, L. M. (1999). Regulation of the biosynthesis of N-acetylglucosaminylpyrophosphoryldolichol, feedback and product inhibition. *J. Biol. Chem.* **274**, 34072–34082.

Kelleher, D. J., Karaoglu, D., and Gilmore, R. (2001). Large-scale isolation of dolichol-linked oligosaccharides with homogeneous oligosaccharide structures: Determination of steady-state dolichol-linked oligosaccharide compositions. *Glycobiology* **11**, 321–333.

Kellogg, B. A., and Poulter, C. D. (1997). Chain elongation in the isoprenoid biosynthetic pathway. *Curr. Opin. Chem. Biol.* **1**, 570–578.

Kornfeld, R., and Kornfeld, S. (1985). Assembly of asparagine-linked oligosaccharides. *Annu. Rev. Biochem.* **54**, 631–664.

Kornfeld, S., Li, E., and Tabas, I. (1978). The synthesis of complex-type oligosaccharides. II. Characterization of the processing intermediates in the synthesis of the complex oligosaccharide units of the vesicular stomatitis virus G protein. *J. Biol. Chem.* **253**, 7771–7778.

Krag, S. S. (1998). The importance of being dolichol. *Biochem. Biophys. Res. Comm.* **243**, 1–5.
Kuzuyama, T., and Seto, H. (2003). Diversity of the biosynthesis of the isoprene units. *Nat. Prod. Rep.* **20**, 171–183.
Lai, C. S., and Schutzbach, J. S. (1986). Localization of dolichols in phospholipid membranes. An ESR spin label study. *FEBS Lett.* **203**, 153–156.
Leyes, A. E., Baker, J. A., and Poulter, C. D. (1999). Biosynthesis of isoprenoids in *Escherichia coli*: Stereochemistry of the reaction catalyzed by farnesyl diphosphate synthase. *Org. Lett.* **1**, 1071–1073.
Li, E., and Kornfeld, S. (1979). Structural studies of the major high mannose oligosaccharide units from Chinese hamster ovary cell glycoproteins. *J. Biol. Chem.* **254**, 1600–1605.
Li, E., Tabas, I., and Kornfeld, S. (1978). The synthesis of complex-type oligosaccharides. I. Structure of the lipid-linked oligosaccharide precursor of the complex-type oligosaccharides of the vesicular stomatitis virus G protein. *J. Biol. Chem.* **253**, 7762–7770.
Meganathan, R. (2001a). Biosynthesis of menaquinone (vitamin K2) and ubiquinone (coenzyme Q): A perspective on enzymatic mechanisms. *Vitam. Horm.* **61**, 173–218.
Meganathan, R. (2001b). Ubiquinone biosynthesis in microorganisms. *FEMS Microbiol. Lett.* **203**, 131–139.
Olgun, A., Serif, A., Tezcan, S., and Kutluay, T. (2003). The effect of isoprenoid side chain lenght on ubiquinone on life span. *Med. Hypotheses* **60**, 325–327.
Olson, J. A. (1964). The biosynthesis and metabolism of carotenoids and retinol (vitamin A). *J. Lipid Res.* **5**, 281–299.
Raetz, C. R. H., Garrett, T. A., Reynolds, C. M., Shaw, W. A., Moore, J. D., Smith, D. C., Jr., Ribeiro, A. A., Murphy, R. C., Ulevitch, R. J., Fearns, C., Reichart, D., et al. (2006). Kdo$_2$-lipid A of *Escherichia coli*, a defined endotoxin that activates macrophages via TLR-4. *J. Lipid Res.* **47**, 1097–1111.
Rip, J. W., Rupar, C. A., Chaudhary, N., and Carroll, K. K. (1981). Localization of a dolichyl phosphate phosphatase in plasma membranes of rat liver. *J. Biol. Chem.* **256**, 1929–1934.
Rodriguez-Concepcion, M. (2004). The MEP pathway: A new target for the development of herbicides, antibiotics, and antimalarial drugs. *Curr. Pharm. Des.* **10**, 2391–2400.
Rohdich, F., Bacher, A., and Eisenreich, W. (2004). Perspectives in anti-infective drug design. The late steps in the biosynthesis of the universal terpenoid precursors, isopentenyl diphosphate and dimethylallyl diphosphate. *Bioorg. Chem.* **32**, 292–308.
Rosner, M. R., Hubbard, S. C., Ivatt, R. J., and Robbins, P. W. (1982). N-asparagine-linked oligosaccharides: Biosynthesis of the lipid-linked oligosaccharides. *Methods Enzymol.* **83**, 399–408.
Rupar, C. A., Rip, J. W., Chaudhary, N., and Carroll, K. K. (1982). The subcellular localization of enzymes of dolichol metabolism in rat liver. *J. Biol. Chem.* **257**, 3090–3094.
Saraste, M. (1999). Oxidative phosphorylation at the fin de siecle. *Science* **283**, 1488–1493.
Schenk, B., Fernandez, F., and Waechter, C. J. (2001). The ins(ide) and outs(ide) of dolichyl phosphate biosynthesis and recycling in the endoplasmic reticulum. *Glycobiology* **11**, 61R–70R.
Sheeley, D. M., and Reinhold, V. N. (1998). Structural characterization of carbohydrate sequence, linkage, and branching in a quadrupole ion trap mass spectrometer: Neutral oligosaccharides and N-linked glycans. *Anal. Chem.* **70**, 3053–3059.
Swiezewskaa, E., and Danikiewiczb, W. (2005). Polyisoprenoids: Structure, biosynthesis and function. *Prog. Lipid Res.* **44**, 235–258.
Szkopinska, A. (2000). Ubiquinone. Biosynthesis of quinone ring and its isoprenoid side chain. Intracellular localization. *Acta Biochim. Pol.* **47**, 469–480.

Teshima, K., and Kondo, T. (2005). Analytical method for ubiquinone-9 and ubiquinone-10 in rat tissues by liquid chromatography/turbo ion spray tandem mass spectrometry with 1-alkylamine as an additive to the mobile phase. *Anal. Biochem.* **338,** 12–19.

Turunen, M., Olsson, J., and Dallner, G. (2004). Metabolism and function of coenzyme Q. *Biochim. Biophys. Acta* **1660,** 171–199.

Valtersson, C., van Duyn, G., Verkleij, A. J., Chojnacki, T., de Kruijff, B., and Dallner, G. (1985). The influence of dolichol, dolichol esters, and dolichyl phosphate on phospholipid polymorphism and fluidity in model membranes. *J. Biol. Chem.* **260,** 2742–2751.

Weerapana, E., and Imperiali, B. (2006). Asparagine-linked protein glycosylation: From eukaryotic to prokaryotic systems. *Glycobiology* **16,** 91R–101R.

CHAPTER SIX

EXTRACTION AND ANALYSIS OF STEROLS IN BIOLOGICAL MATRICES BY HIGH PERFORMANCE LIQUID CHROMATOGRAPHY ELECTROSPRAY IONIZATION MASS SPECTROMETRY

Jeffrey G. McDonald, Bonne M. Thompson,
Erin C. McCrum, *and* David W. Russell

Contents

1. Introduction	146
2. Supplies and Reagents	148
3. Extraction of Lipids from Cultured Cells and Tissues	149
4. Saponification of Lipid Extracts	151
5. Solid-Phase Extraction	152
6. Analysis by HPLC-ESI-MS	153
6.1. High performance liquid chromatography	153
6.2. Electrospray ionization mass spectrometry	153
7. Quantitation	156
8. Data	157
9. Discussion, Nuances, Caveats, and Pitfalls	162
9.1. Quantitation	162
9.2. Availability and purity of primary and deuterated standards	163
9.3. Resolving related sterols by HPLC	164
9.4. Acetonitrile and signal intensity	165
9.5. Auto-oxidation and use of SPE columns	165
9.6. Sample clean-up	166
9.7. Residual insoluble material	167
9.8. Comparing relative peak areas	168
9.9. Electrospray using other instrumental platforms	168
Acknowledgments	168
References	169

Department of Molecular Genetics, University of Texas Southwestern Medical Center at Dallas, Dallas, Texas

Abstract

We describe the development of a high performance liquid chromatography mass spectrometry (HPLC-MS) method that allows the identification and quantitation of sterols in mammalian cells and tissues. Bulk lipids are extracted from biological samples by a modified Bligh/Dyer procedure in the presence of eight deuterated sterol standards to allow subsequent quantitation and determination of extraction efficiency. Sterols and other lipids are resolved by HPLC on a reverse-phase C_{18} column using a binary gradient of methanol and water, both containing 5 mM ammonium acetate. Sterol identification is performed using an Applied Biosystems (Foster City, CA) 4000 QTRAP® mass spectrometer equipped with a TurboV electrospray ionization source and operated in the positive (+) selected reaction monitoring (SRM) mode. The total run time of the analysis is 30 min. Sterols are quantitated by comparison of the areas under the elution curves derived from the detection of endogenous compounds and isotopically labeled standards. The sensitivity of the method for sterol detection ranges between 10 and 2000 fmol on-column. Cultured RAW 264.7 mouse macrophages contain many different sterols, including the liver X receptor (LXR) ligand 24,25-epoxycholesterol. Tissues such as mouse brain also contain large numbers of sterols, including 24(S)-hydroxycholesterol, which is involved in cholesterol turnover in the brain. The extraction procedure described is flexible and can be tailored to sample type or information sought. The instrumental analysis method is similarly adaptable and offers high selectivity and sensitivity.

1. INTRODUCTION

Sterols play fundamental roles in many physiological processes in virtually all living organisms. Sterols have regulatory functions, are involved in cellular signaling, and are an integral building block of cell membranes in which they modulate fluidity. The most abundant sterol in mammals is cholesterol, which is a precursor to a host of important end-products including steroid hormones, bile acids, vitamin D, and others (Myant, 1981; Russell, 2003). Cholesterol has gained a certain amount of notoriety due to its negative impact on health when present in high levels, while other sterols, such as the phytosterols derived from plants, offer health benefits by lowering plasma cholesterol levels (Moruisi et al., 2006; Ostlund et al., 2003). In most mammalian tissues, sterols are both synthesized and absorbed through the diet, while in the brain they result exclusively from *de novo* synthesis (Dietschy and Turley, 2004).

Few advances have been made in the area of instrumental analysis of sterols. Many sterols are not well suited for analysis in their native state by modern instrumental analytical techniques due to their chemical structures and functional groups. Sterols have historically been analyzed by gas chromatographic MS (GC-MS) (Lund and Diczfalusy, 2003; Wilund et al., 2004). GC offers high temporal resolution, allowing for the separation of sterols of similar composition such as positional isomers and diastereomers. Analysis by GC-MS relies upon vaporization of the sample, and the polarity imparted to a sterol by functional groups decreases its ability to transition to the gas phase. This decrease is largely due to hydrogen bonding and analyte interactions with the glass liner of the injection port, which can interfere with vaporization and cause thermal degradation. These issues can be overcome by reducing the polarity of the molecule by derivatization with trimethylsilanes or methyl groups to achieve reasonable signal intensity and good chromatographic separation (Abidi, 2001; Isobe et al., 2002); however, derivatization adds additional steps to sample preparation, increases instrument maintenance requirements, and can increase the complexity of the resulting chromatogram. GCs are typically coupled to electron impact mass spectrometers, which offer the ability to compare acquired mass spectra to reference libraries for possible identification of unknowns. The almost exclusive use of single quadrupole electron impact mass spectrometers with GC systems does not offer the ability to perform advanced mass spectral investigations like tandem MS (MS/MS) or precursor/product ion scans.

Sterols are more amenable in their native state to separation by high performance liquid chromatography (HPLC), although few suitable detectors are available. Most sterols lack a chromophore, which prevents their detection by ultraviolet (UV)/visible or fluorescence spectroscopy. Sterols can be modified to better absorb light by derivatization or by oxidation of alcohol substituents to oxo-groups; however, these modifications add complexity and additional steps to the analysis. Evaporative light-scattering detection (ELSD) was developed for use with HPLC and has good sensitivity. An ELSD device generates particles from the eluent of the HPLC column via heated nebulization and produces a signal based on light scattered from an analyte. The ELSD responds only to the number of particles present, and thus does not discriminate based on the physical characteristics of the molecule being measured. Note that ELSD produces a nonlinear response, which should be considered when interpreting data. These features allow for relative comparisons to be made between chromatographic peaks, especially when the compounds of interest are present in similar amounts; however, ELSD does not provide selectivity in terms of mass or other compound-specific information, which renders the analysis dependent on the chromatographic performance of the HPLC system.

Historically, MS has been challenging to couple with HPLC, in part because taking microliter to milliliter quantities of HPLC eluent from atmospheric pressure to the high-vacuum environment of MS is difficult. Until recently, the eluent was thermally vaporized prior to MS analysis; however, this approach did not allow detection of thermally labile compounds.

Atmospheric Pressure Chemical Ionization (APCI) is a different ionization source that has been successfully employed for sterol analysis by HPLC-MS (Burkard et al., 2004; Palmgren et al., 2005; Tian et al., 2006).

The efficient generation of ions and their transport into the mass spectrometer was another technological hurdle. A variety of techniques were used to ionize analytes; however, no one technique proved to be broadly applicable. The advent of electrospray ionization (ESI) (Fenn et al., 1989) overcame these hurdles and made HPLC-ESI-MS a broadly applicable method for small molecule analysis. The electrospray interface can be used independently or to couple HPLC to most types of mass spectrometers, further diversifying the utility of this approach.

Although numerous biological functions of sterols have been described, many more are yet to be discovered. The elucidation of new functions will require not only advances in biology but also more sophisticated and sensitive analytical techniques to identify quantitative behaviors of sterols in situ. An increasing number of methods in the literature report on the extraction and analysis of sterol subclasses, such as oxysterols (Pulfer et al., 2005; Shan et al., 2003) and plant sterols (Abidi, 2001), using HPLC-MS; however, no comprehensive method exists for HPLC-MS of all sterols. We report here a method for the extraction and analysis of sterols from tissue and other biological matrices using HPLC-ESI-MS. The extraction procedure is flexible and can be tailored to sample type and information sought. The instrumental analysis offers excellent selectivity and sensitivity and is adaptable to virtually all sterols with a run time of 30 min per sample.

2. Supplies and Reagents

Phosphate Buffered Saline (PBS)
Cell lifters
Glass culture tubes with Teflon^{TM}-lined caps (7-, 11-, 14-, 18-ml)
Wheaton Science Products
   7-ml 358640
   11-ml 358646
   14-ml 358647
   18-ml 358648
10-ml glass pipettes
Pasteur pipettes
Mechanical pipette pump
Glass vials with Teflon^{TM}-lined caps (2-, 4-, 8-ml)
Wheaton Science Products
   2-ml W224581
   4-ml W224582
   8-ml W224584

Solid-phase extraction (SPE) cartridges (100-mg silica, Isolute)
Vacuum apparatus for SPE
0.5% (v/v) isopropanol in hexane
30% (v/v) isopropanol in hexane
HPLC-grade methanol, ethanol, chloroform, hexane, and H_2O
Potassium hydroxide (10 N)
Ammonium acetate
Saponification solution
Hyrolysis solution (6 ml of 10 N KOH diluted to 100 ml with ethanol)
Bligh/Dyer solvent (1:2 [v/v] chloroform:methanol)
Primary and deuterated sterol standards
Luna C_{18} reverse-phase HPLC column (250 × 2 mm; 3 μm-particle size)

3. Extraction of Lipids from Cultured Cells and Tissues

Extraction of cultured cells begins by placing the culture dishes (6-cm dishes seeded with $2e^6$ cells) on ice. The medium is removed with a glass pipette to a screw-cap glass culture tube and stored for extraction later. Cells are gently washed twice with 3 ml of cold PBS, which is removed and discarded. An additional 1.6 ml of cold PBS are added to the cells, which are then gently scraped from the surface of the dish using a cell lifter. Cells are transferred to a 14-ml screw-cap glass culture tube. Depending on the desired information, aliquots of the medium or cells may be taken at this stage for assays to determine levels of cytokines, DNA, or protein. Lipids are extracted from the cells using a Bligh/Dyer procedure (Bligh and Dyer, 1959) in which 6 ml of chloroform:methanol (1:2 v/v) is added to the resuspended cells in 14-ml screw-cap glass culture tubes. At this point, deuterated surrogate sterol standards are added for quantitative analysis, as indicated in Table 6.1 (see "Quantitation" section for definition of surrogate standards). Samples are vortexed at high speed for 10 s, then centrifuged at 2600 rpm (1360 rcf) for 5 min to pellet insoluble material. The supernatant is decanted to a fresh 14-ml glass culture tubes and 2 ml each of chloroform and PBS is added. Samples are again vortexed for 10 s and centrifuged at room temperature for 5 min at 2600 rpm. Two liquid phases should now be observed. The organic lower phase is removed using a Pasteur pipette and transferred to a 4-ml glass vial with a Teflon-lined cap. The upper aqueous phase is discarded. We recommend the use of a mechanical pipette pump (e.g., Scienceware Pipette Pump) versus a pipette bulb for better control when removing liquid near the aqueous–organic interface.

The organic phase is evaporated under a gentle stream of dry nitrogen while heating at approximately 35° to counter the evaporative

Table 6.1 Deuterated and primary sterols and the masses and volumes used for quantitative analysis

Compound[a]	Parts per million (PPM)
Surrogate mix 1 (10 μl)	
25-Hydroxycholesterol (D3)	4
27-Hydroxycholesterol-(25R) (D7)	4
24,25-Epoxycholesterol (D6) (R+S)	3
7α-Hydroxycholesterol (D7)	4
7-Ketocholesterol (D7)	4
4β-Hydroxycholesterol (D7)	2
Surrogate mix 2 (10 μl)	
Cholesterol (D7)	150
Desmosterol (D6)	80
Internal standard (10 μl)	
6α-Hydroxycholestanol (D7)	4
Sterol mix (10 μl)	
22R-Hydroxycholesterol	4
25-Hydroxycholesterol	4
27-Hydroxycholesterol-(25R)	4
24,25-Epoxycholesterol	3
7α-Hydroxycholesterol	4
7-Ketocholesterol	4
4β-Hydroxycholesterol	2
Desmosterol	80
7-Dehydrocholesterol	100
Cholestenone	15
Cholesterol	150
Lanosterol	100

[a] See Table 6.3 for systematic sterol names and Table 6.4 for deuterium positions.

cooling effect. The dried extracts are reconstituted in approximately 200 μl of 95% (v/v) methanol, shaken gently, and transferred to an auto-sampler vial containing the appropriate deuterated internal standard based on the amount given in Table 6.1 (see "Quantitation" section for definition of internal standard). An additional 200 μl of 95% methanol is added, the sample is gently shaken, and the solution is transferred to the same auto-sampler vial. This additional rinse ensures maximum transfer of sterols from the sample vial to the auto-sampler vial. Vortexing or sonicating the sample should not be done, as these treatments can disperse any residual insoluble material into the solution, which in turn adversely affects subsequent HPLC or MS analyses of the extracts. Occasionally, even gently shaking the extract can cause resuspension of insoluble material. If insoluble material is inadvertently resuspended, centrifugation and decanting the supernatant are usually sufficient to clarify the extract. If the final extract is not colorless and free of solid material, then the Bligh/Dyer extraction should be repeated. The auto-sampler vial should be equipped with an appropriate volume-reducing insert if the final volume is less than 500 μl.

This extraction method is also suitable for animal tissues (brain, liver, etc.), plasma, and medium from tissue culture. Adjustments to the volumes of extraction solvents must be made to maintain the water:chloroform: methanol ratio of 1:2:0.8 for the azeotrope and a 2:2:1.8 ratio for formation of the two-phase system (Bligh and Dyer, 1959). Exact volumes must be determined empirically for each sample type; however, we have found the following information useful as a guide: when extracting lipids from liquid solutions, the entire volume should be treated as the aqueous component; for tissues, the endogenous water content will contribute to the aqueous component in the azeotrope, and different tissues have different water contents. For example, an adult mouse brain ($\sim$400 mg) will typically contribute 0.4 ml of water, meaning that 1.2 ml of PBS would be added instead of the 1.6 ml suggested above. Final volumes must also be adjusted depending on initial mass of the sample and lipid content. Tissues often require homogenization in PBS using a Potter-Elvehjem device or a polytron prior to addition of chloroform and methanol, and again after addition of these organic solvents. The deuterated sterol standards surrogates should be added after homogenization to avoid nonspecific adsorption (Lund and Diczfalusy, 2003).

4. SAPONIFICATION OF LIPID EXTRACTS

Sterols can exist in biological matrices in free and esterified forms (Myant, 1981). The amount of total sterols or the ratio of free to esterified sterols may be of interest, in which case steryl conjugates, typically esters in most mammalian cells, must be saponified to remove fatty acyl groups and free the sterols. Cholesteryl esters require different chromatographic and mass spectral methods to measure their abundance compared to those used to measure free sterols. Saponifying steryl esters to free sterols requires additional steps in the extraction and purification scheme, but is a more judicious use of time and resources compared to employing two HPLC-ESI-MS methods, which can more than double instrument time.

Our saponification procedure is derived from those previously described by Lund and Diczfalusy (2003) and Yang et al. (2006). Hydrolysis reagents should be prepared immediately prior to use according to the prescribed method already listed. The lipid extract is dried under nitrogen at approximately 35° in a glass vial with a Teflon-lined cap, and 1 ml of hydrolysis solution is added. The vial is capped, vortexed for 10 s at maximum speed, heated to 90° for 2 h, and then allowed to cool to room temperature. Lipids are extracted from the solution using a modified Bligh/Dyer extraction in which ethanol is substituted for methanol as follows. The hydrolyzed lipid solution is transferred to a 14-ml screw-cap glass culture tube using a Pasteur pipette, and the glass vial is rinsed once with 1 ml of ethanol. To the 2-ml volume of hydrolyzed extract in ethanol, 2 ml of chloroform and 1.8 ml of PBS are added. Samples are vortexed for 10 s at maximum speed and centrifuged at 2600 rpm for 5 min. Two phases should be observed, and the organic phase (lower) is removed using a Pasteur pipette and placed in a fresh 4-ml glass vial with a Teflon-lined cap, dried down, and reconstituted in 95% methanol.

5. Solid-Phase Extraction

Bulk-lipid extracts made with the Bligh/Dyer procedure contain many nonpolar species such as steryl esters, monoacylglycerols, diacylglycerols, and triacylglycerols, which are all strongly retained on reverse phase HPLC columns. Accumulation of these compounds on the column can lead to changes in retention time, decreased resolution, and increased back pressure. For these reasons, it is often advisable to resolve the major lipid classes by SPE prior to HPLC-MS analysis. Sterols can be isolated from other classes of lipids by sequential development of a single SPE column and, with the appropriate solvents, oxysterols, or sterols containing additional hydroxy-, oxo-, or epoxy-groups, can be separated from cholesterol and other sterols that contain a single hydroxyl functional group.

For routine SPE (Lund and Diczfalusy, 2003), a 100-mg Isolute silica cartridge (Biotage, Charlottesville, VA) is prewashed by passing 2 ml of hexane through the column. A dried lipid extract prepared as previously described (with or without saponification) is dissolved in 1 ml of toluene and passed through the cartridge. Nonpolar compounds such as cholesteryl esters are eluted first with 1 ml of hexane. Cholesterol and other related sterols are eluted next with 8 ml of 30% isopropanol in hexane. The eluted sterols are dried down and then resuspended in 95% methanol prior to analysis by HPLC-MS.

To resolve sterols containing a single hydroxyl group from those containing more than one hydroxyl group (e.g., cholesterol from oxysterols), the SPE column is pre-washed and loaded with lipids from the extraction procedure as described above. Steryl esters are again eluted from the column with 1 ml of hexane. Cholesterol and other mono-hydroxysterols are then eluted with 8 ml of 0.5% isopropanol in hexane. Oxysterols are eluted next with 5 ml of 30% isopropanol in hexane. Eluates from the individual chromatographic steps are dried under nitrogen and resuspended in 95% (v/v) methanol/water.

Our laboratory has determined that, in general, a 400-μl final volume for 5 to 10 × 10^6 cells is suitable for the instrumental analysis described in the "Analysis by HPLC-ESI-MS" section. The final volume-to-cell number ratio can be scaled to accommodate varying amounts of cells.

6. ANALYSIS BY HPLC-ESI-MS

6.1. High performance liquid chromatography

Sterols are resolved using reverse-phase HPLC (RP-HPLC). A 10-μl aliquot of lipid extract (in 95% methanol) is loaded onto a RP-HPLC column (Luna C_{18} 2 × 250 mm, 3-μm particle; Phenomenex, Torrance, CA) equipped with a guard column (C_{18}, 4 × 2 mm). From this point on, lipids are resolved by gradient elution. The gradient program begins with 100% Solvent B (85% methanol, 5 mM ammonium acetate) for 2 min, and is thereafter ramped to Solvent A (100% methanol, 5 mM ammonium acetate) over a 13-min period and held for 10 min. Solvent B is then passed through the column for 5 min to re-equilibrate the column prior to the next run. The flow rate is 0.25 ml/min, and the column is maintained at 30° using a column oven.

6.2. Electrospray ionization mass spectrometry

The HPLC is coupled to a 4000 QTrap triple quadrupole mass spectrometer (Applied Biosystems, Foster City, CA) through a Turbo V^{TM} ESI source. The source is operated in the positive (+) mode with a spray voltage of 5500 V, a curtain gas of 15 psi (nitrogen), and ion Source Gas 1 and 2 set to 60 and 20 psi, respectively (both nitrogen). Gas 2 is maintained at 50°. The mass spectrometer is operated in standard reaction monitoring (SRM) mode to achieve maximum selectivity. To increase signal intensity, the first quadrupole is set to low resolution and the third quadrupole is set to unit resolution. These settings allow ions of wider mass distribution to pass through quadrupole 1 into the collision cell where ions are fragmented and resolved by quadrupole 3. Individual transitions, optimal declustering potentials, collision energies, limits of detection, and linear ranges are given in Table 6.2. SRM pairs are chosen by infusing individual sterols into the MS and generating product ion MS/MS from which these values are selected and optimized. Examples of product-ion spectra from several different types of sterols are shown in Fig. 6.1. A more comprehensive list of sterol product-ion MS can be found at http://www.lipidmaps.org. Common ions formed for most sterols by electrospray include $[M+NH_4]^+$, $[M+H]^+$, $[M+H-H_2O]^+$, and $[M+H-2H_2O]^+$ (Pulfer et al., 2005). Note that the $[M+H]^+$ ion does not result from gaining a proton during electrospray, but is likely the result of losing gaseous ammonia (NH_3) from the NH_4 adduct (Murphy, 2006).

Table 6.2 Molecular weight, instrumental parameters, and figures of merit for representative sterols

Common name	Peak # (Fig. 6.3)	MW[a]	SRM ions	DP (v)	CE (v)	Dwell time (msec)	RT (min)	LOD (fmol on-column)	Linear range (fmol on-column)
22R-Hydroxycholesterol	1	402.7	420/385	50	15	110	10.5	50	5.00e1–1.25e5
24-Hydroxycholesterol (R+S)	2/4	402.7	420/385	50	15	110	12.2	60	6.50e1–1.50e5
25-Hydroxycholesterol	3	402.7	420/367	50	15	110	13.3	25	2.50e1–2.50e5
27-Hydroxycholesterol-(25R)	5	402.7	420/385	50	15	110	13.9	55	5.50e1–1.25e5
24,25-Epoxycholesterol	6	400.4	418/383	45	12	80	14.5	20	2.00e1–1.25e5
7α-Hydroxycholesterol	7	402.7	385/367	50	15	110	16.5	60	6.00e1–1.25e5
7-Ketocholesterol	8	400.6	401/383	80	35	70	17.3	85	9.00e1–6.30e4
5,6β-Epoxycholesterol	9	402.7	420/385	50	15	110	18.7	5	5.00e0–8.70e4
5,6α-Epoxycholesterol	10	402.7	420/385	50	15	110	19.5	5	6.00e0–8.70e4
4β-Hydroxycholesterol	11	402.7	420/385	50	15	110	20.2	25	2.50e1–6.20e4
Zymosterol	a	384.6	385/367	85	20	70	N/A	N/A	N/A
Desmosterol	12	384.6	402/367	55	20	110	21.9	2000	6.50e4–3.20e6
7-Dehydrocholesterol	13	384.6	385/367	85	20	70	22.6	520	5.20e2–3.30e6
Cholest-4-en-3-one	14	384.6	385/367	85	20	70	23.4	390	3.90e2–1.60e6
Lathosterol	15	386.7	404/369	110	13	70	24	1300	9.00e3–2.50e6
Cholesterol	16	386.7	404/369	110	13	70	24.3	1000	2.60e4–3.20e6
Lanosterol	17	426.7	444/409	80	15	100	24.9	175	2.30e3–3.00e6
Cholestanol	18	388.7	404/369	110	13	70		b	b
24-Dihydrolanosterol	19	428.7	429/411	100	17	70		b	b

[a] Compound no longer commercially available. Existing material is limited and impure; thus, LOD and linear ranges are not measured.
[b] Standards are of questionable purity; thus, LOD and linear range were not measured.
Key to abbreviations: CE, collision energy; DP, declustering potential; LOD, limit of detection; MW, molecular weight; N/A, not applicable; RT, retention time; SRM, selected reaction monitoring.

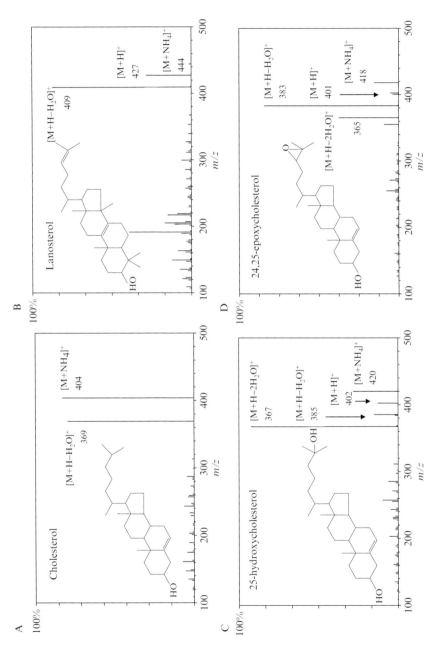

Figure 6.1 Product-ion mass spectra for selected sterols. See http://www.lipidmaps.org for additional spectra.

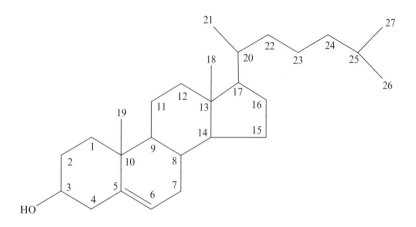

Cholest-5-en-3β-ol (cholesterol)

Figure 6.2 Structure of cholesterol with positional numbering system.

7. Quantitation

Deuterated analogs of sterols are used for quantitation because they have similar physical properties as non–isotopically labeled (i.e., endogenous) sterols, but can be resolved from the latter by MS due to their increased mass. Quantitation relies on the principle of isotope dilution MS in which known amounts of deuterated sterol standards are added to the biological sample at the beginning of the extraction and are thereafter carried through the extraction procedure with endogenous sterols. The amount of endogenous sterol in the cell or tissue extract is then calculated using a predetermined relative response factor (RRF) between peak areas and masses of the deuterated versus endogenous sterols. To obtain RRFs, a standard mixture of deuterated surrogates and internal standard is prepared as indicated in Table 6.1 prior to instrumental analysis. An autosampler vial is fitted with a 500-μl insert, and ~400 μl of 95% methanol is added. Deuterated surrogates and internal standards are then added to the solution in the exact manner in which they were added to the sample as previously described (same volume, pipette, stock mixtures, etc.). Corresponding primary sterols are added to the mixture as well. The masses and volumes for all compounds are given in Table 6.1.

The quantitation mixture is analyzed at the beginning and end of the instrument sequence, and after every 8 to 10 samples. Using the masses of standard added to the quantitation mixture and the areas under the elution curves generated by HPLC-ESI-MS, a response factor can be generated using Equation 1.

$$RRF = \frac{(Area_{Analyte} \times Mass_{std})}{(Area_{std} \times Mass_{analyte})} \qquad (1)$$

The response factors are averaged over the course of the sequence to account for any instrumental drift. Should the response factor drift excessively between two standard analyses, specific response factors should be used only for the samples they bracket. This approach establishes the relationship between mass and peak area, which can be used to quantitate the amount of each sterol present in individual samples. Equation 2 takes into account the area under the elution curve for the sterol of interest ($Area_{analyte}$), the deuterated analog ($Area_{std}$), the mass of the deuterated analog ($Mass_{std}$) added to the sample, and the RRF generated with Equation 1.

$$Mass_{analyte} = \frac{(Area_{Analyte} \times Mass_{std})}{(Area_{std} \times RRF)} \qquad (2)$$

Equation 2 can be seen as a rearrangement of Equation 1 in which the RRF established for each sterol and deuterated analog in Table 6.1 is incorporated.

Deuterated analogs are not available for all sterols. To overcome this problem, an alternate deuterated analog of similar chemical composition to that of the analyte to be measured can be substituted. For example, quantitative analysis of 22-hydroxycholesterol can be performed using (d3) 25-hydroxycholesterol as an alternative surrogate. Common primary and deuterated sterol standards and their sources are given in Tables 6.3 and 6.4.

8. DATA

An HPLC-ESI-MS chromatogram of a mixture of sterols is shown in Fig. 6.3A. Using our standard gradient elution program, sterols elute from the column between 10 and 28 min, and, in general, more hydrophilic oxysterols elute before cholesterol and other more hydrophobic sterols. Experience has shown that the earliest eluting sterols (i.e., the most hydrophilic) extracted from cultured mammalian cells are trihydroxy compounds, which elute several minutes before a 22-hydroxycholesterol standard. The latest eluting sterol (i.e., most hydrophobic) is 24-dihydrolanosterol. All other sterols detected in this laboratory elute between these two extremes. ak widths are generally less than 20 s at the full-width, half maximum setting,

Table 6.3 Sources of representative sterol standards

Systematic name[a]	Common name[d]	MW	Source
Cholest-5-en-3β,22-diol (22R)	22R-Hydroxycholesterol	402.7	Sigma Aldrich[b]
Cholest-5-en-3β,24-diol	24-Hydroxycholesterol(R+S)	402.7	Avanti Polar Lipids[c]
Cholest-5-en-3β,25-diol	25-Hydroxycholesterol	402.7	Avanti Polar Lipids
Cholest-(25R)-5-en-3β,27-diol	27-Hydroxycholesterol-(25R)	402.7	Avanti Polar Lipids
24,25-Epoxycholest-5-en-3β-ol	24,25-Epoxycholesterol	400.6	Avanti Polar Lipids
Cholest-5-en-3β,7α-diol	7α-Hydroxycholesterol	402.7	Avanti Polar Lipids
7-Keto-5-cholesten-3β-ol	7-Ketocholesterol	400.6	Avanti Polar Lipids
5,6β-Epoxy-5β-cholestan-3β-ol	5,6β-Epoxycholesterol	402.7	Avanti Polar Lipids
5,6α-Epoxy-5α-cholestan-3β-ol	5,6α-Epoxycholesterol	402.7	Avanti Polar Lipids
Cholest-5-en-3β,4β-diol	4β-Hydroxycholesterol	402.7	Avanti Polar Lipids
8,24(5α)-Cholestadien-3β-ol	Zymosterol	384.7	[a]
Cholest-5,24-dien-3β-ol	Desmosterol	384.7	Avanti Polar Lipids
Cholesta-5,7-dien-3β-ol	7-Dehydrocholesterol	384.7	Sigma Aldrich
Cholest-4-en-3-one	Cholestenone	384.7	Sigma Aldrich
Cholest-7-en-3β-ol	Lathosterol	386.7	Steraloids[e]
Cholest-5-en-3β-ol	Cholesterol	386.7	Avanti Polar Lipids
Lanosta-8,24-dien-3β-ol	Lanosterol	426.7	Steraloids
Cholestan-3β-ol	Cholestanol	388.7	Sigma Aldrich
5α-Lanost-8-en-3β-ol	24-Dihydrolanosterol	428.7	Steraloids

[a] Compound no longer commercially available.
[b] St. Louis, MO.
[c] Alabaster, AL.
[d] See Fahy et al. (2005) for systematic nomenclature of sterols.
[e] Newport, RI.

Table 6.4 Sources of deuterated sterol standards

Common name (deuterium positions)[a]	MW	Source
24-Hydroxycholesterol(D6) (26,26,26,27,27,27-D6)	408.7	Avanti Polar Lipids[b]
25-Hydroxycholesterol (D3) (26,26,26-D3)	405.7	Avanti Polar Lipids
27-Hydroxycholesterol-(25R)(D5) (26,26,26,27,27-D5)	407.7	Medical Isotopes, Inc.[c]
24,25-Epoxycholesterol (D6) (R+S) (26,26,26,27,27,27-D6)	406.6	Avanti Polar Lipids
6α-Hydroxycholestanol(D7) (25,26,26,26,27,27,27-D7)	411.7	Avanti Polar Lipids
7α-Hydroxycholesterol(D7) (25,26,26,26,27,27,27-D7)	409.7	Avanti Polar Lipids
7β-Hydroxycholesterol(D7) (25,26,26,26,27,27,27-D7)	409.7	Avanti Polar Lipids
7-ketocholesterol(D7) (25,26,26,26,27,27,27-D7)	407.6	Avanti Polar Lipids
5,6α-Epoxycholesterol (D7) (25,26,26,26,27,27,27-D7)	409.7	Avanti Polar Lipids
4β-Hydroxycholesterol(D7) (25,26,26,26,27,27,27-D7)	409.7	Avanti Polar Lipids
Desmosterol (D6) (26,26,26,27,27,27-D6)	390.7	Avanti Polar Lipids
Cholesterol (D7) (25,26,26,26,27,27,27-D7)	393.7	Avanti Polar Lipids

[a] See Table 6.3 for systematic sterol names.
[b] Alabaster, AL.
[c] Pelham, NH.

which gives excellent resolution. Note that some sterols show multiple SRM pairs for a single compound. For example, 4β-hydroxycholesterol (peak 11, Fig. 6.3B) shows an SRM signal of 420/385, 420/367, and 385/367 u. These additional SRM pairs add confidence to the identification of a specific sterol and can provide further information regarding compound identification based on their relative intensities.

An HPLC-ESI-MS chromatogram derived from analysis of a Bligh/Dyer extract from wild-type mouse brain is shown in Fig. 6.4. Early in the chromatogram, a group of compounds was observed that was not retained by the stationary phase and eluted with the solvent front. This observation is common in biological extracts, especially when no additional pre-purification

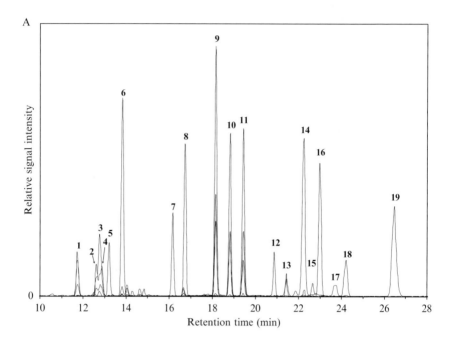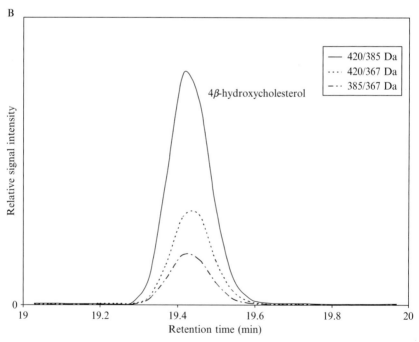

Extraction and Analysis of Sterols in Biological Matrices by HPLC-ESI-MS   161

or separation is performed. Multiple compounds are observed eluting in the sterol region, some of which can be identified based on SRM pairs and retention times. For example, 24-hydroxycholesterol eluting at 12.8 min is abundant in the brain and readily observed in the chromatogram. Other sterols identified in the brain extract include 24,25-epoxycholesterol (13.5 min), 4β-hydroxycholesterol (19.7 min), desmosterol (21.3 min), cholesterol (23.7 min), and lanosterol (24.3 min). Note that the region around the cholesterol peak is shown at one-tenth scale. The chromatographic peak resulting from cholesterol is typically an order of magnitude or larger than any other sterol, which reflects the large abundance of this sterol in most mammalian tissues and cells. In this example, the cholesterol peak has a split top indicative of poor chromatographic performance due to column overloading. Injecting a smaller volume or a more dilute sample will improve peak shape. Additional sterols are present in the extract as shown by

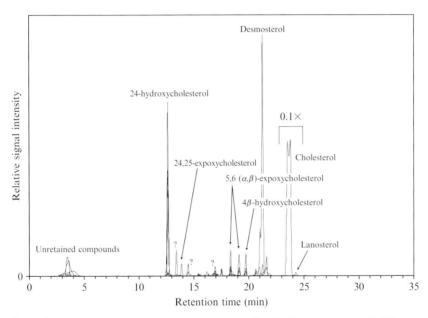

Figure 6.4 Reconstructed SRM chromatogram of a Bligh/Dyer extract of wild-type mouse brain. Known sterols are identified by name, and unknown compounds with sterol-like signatures are indicated with a "?." Note the reduced scale (×0.1 between 23 and 25 min).

Figure 6.3 (A) Reconstructed SRM ion chromatogram derived from the analysis of a mixture of common sterols. Corresponding peak numbers are given in Table 6.2. The baseline is truncated for clarity to show only 10 to 28 minutes. (B) Expanded view of multiple SRM pairs for peak 11 (4β-hydroxycholesterol).

signals from other SRM pairs. These represent known or unidentified sterols, or non-sterol compounds that happen to produce a signal identical to one or more of the SRM pairs but do not match the retention time of any of the sterol standards.

9. Discussion, Nuances, Caveats, and Pitfalls

9.1. Quantitation

A practical description and discussion of quantitative MS by isotope dilution are provided here. For additional information, see Duncan et al. (2006) or USEPA (2003). Surrogate standards are deuterated analogs added to the sample prior to extraction and carried through the extraction process with the endogenous compounds. Theoretically, deuterated analogs will behave identically to their non-isotopically labeled and endogenous counterparts during the extraction process, including any losses due to incomplete transfer, oxidation, degradation, or others. When surrogates are used for quantitation, losses are accounted for and corrected automatically. In contrast, internal standards are those added to the sample after extraction but immediately prior to HPLC-ESI-MS analysis. As internal standards are added after the extraction, they cannot account for losses during extraction.

A combination of surrogate and internal standards provides additional information, allowing for the measure of extraction efficiency. The amount of surrogate recovered at the end of an extraction can be calculated providing extraction efficiency using the mass and peak areas of a surrogate and an internal standard. While there is no specified level of recovery necessary for a suitable extraction, a minimum of 50% recovery is our internal target. Values that are consistently lower than 50% recovery indicate the need for evaluation and modification of the extraction procedure. Occasional recovery values below 50% are typically not of concern, but should be evaluated on a case-by-case basis. Extraction efficiency is especially important in sterol analysis as many are present in trace quantities within cells and tissues. The extraction procedure previously described for cells typically yields between 70 and 80% average recovery using (d6) 27-hydroxycholesterol as an internal standard.

When measuring the response factor of a sterol and its deuterated analog, the ratio should ideally be unity; however, due to impurities in the primary standards and to potential deuterium exchange during the ESI process, this ratio may be higher or lower than one. The ratio should be determined empirically for each sterol and deuterated analog. The RRF between a sterol and a related, but non-ideal deuterated surrogate may be very different from unity and must be determined empirically. The linear working ranges of response factors (i.e., calibration curves) must also be determined empirically, but generally reflect the linear ranges of the individual sterols themselves.

Deuterated analogs are not available for all sterols. In this situation, we use (d3) 25-hydroxycholesterol as an alternate deuterated analog of 22-hydroxycholesterol and (d7) cholesterol as the deuterated surrogate for 7-dehydrocholesterol, cholestenone, and lathosterol. Other common primary and deuterated sterol standards and their sources are given in Tables 6.3 and 6.4.

Isotope dilution has the benefit of alleviating the analyst from having to know the volumes of the extract, including the final volume. Because the calculations are based on ratios between peak area and mass, no knowledge of extract volume is necessary. The calculated value will have the units (mass/extract) that allow the end user to normalize the mass of sterol to a variety of standard measures, including DNA, protein, cell number, or tissue weight.

9.2. Availability and purity of primary and deuterated standards

Primary sterol standards are available from several commercial sources. Some standards are prepared synthetically while others are extracted from biological matrices and purified. It has been our experience that the purity of individual sterols varies greatly depending on the specific sterol and the supplier. For example, cholesterol is available from multiple sources as a highly pure compound (>99%), but 24-dihydrolanosterol has limited availability and typically contains many impurities. Some stereochemically pure compounds are available, such as 22R-hydroxycholesterol, 27-hydroxycholesterol-(25R), and 24(S)-hydroxycholesterol. Others are currently only available as the racemic mixture. Suppliers generally offer limited analytical data to demonstrate purity, which ranges from a basic thin layer chromatographic (TLC) analysis, to an HPLC-UV/vis analysis and nuclear magnetic resonance (NMR) analysis.

Evaluating the purity of primary and deuterated sterol standards can be complex and time-consuming and may be beyond the scope of many research laboratories. As described in the introductory section, sterols are not well suited for analysis in their native state by most modern analytical techniques. Even when detected, the response between two similar sterols can differ by more than an order of magnitude, which makes it difficult to ascertain relative percent purity without having standards for each impurity detected. HPLC-ELSD is currently the most widely used and nondiscriminatory technique for evaluation of sterol purity. HPLC-ELSD is optimal for evaluation of standards but, because it does not provide mass information, has limited applicability with complex extracts. Finally, NMR is often considered a definitive technique for identifying purity and structure of organic compounds; however, access to NMR is not always readily available to those studying sterols.

A collection of deuterated sterol standards and an increasing number of primary sterol standards has become available from Avanti Polar Lipids (Alabaster, AL) as part of the LIPID MAPS consortium. The deuterated sterols available at the time of this writing are listed in Table 6.4, and many of the primary sterol standards available are listed in Table 6.3. Each standard from Avanti has been prepared with a full complement of available analytical data, including HPLC-ESI-MS, ESI-MS, HPLC-ELSD, and NMR. This characterization far exceeds the QA/QC data offered by other commercial vendors of sterol standards at this time. Additionally, Avanti performs routine stability testing of stocked compounds and will notify users should the purity of a given standard fail their specified compliance of ±5%.

9.3. Resolving related sterols by HPLC

Many sterols are fully resolved by the chromatographic method previously described; however, some, such as 24-hydroxycholesterol $(R+S)$ and 25-hydroxycholesterol, co-elute. Mass spectrometry can offer an additional level of resolution by differentiating co-eluting compounds by mass but, because many sterols are isobaric, this dimension is of variable worth. 24-Hydroxycholesterol $(R+S)$ and 25-hydroxycholesterol share three SRM pairs, but they vary in relative intensities between 24- and 25-hydroxycholesterol. 24-Hydroxycholesterol shows 420/385 u as the most abundant transition, whereas 25-hydroxycholesterol shows 420/367 u as the most abundant transition. In some cases, a difference in the tissue-specific distribution of two closely related sterols obviates the need for HPLC separation. In this example, a variety of tissues contain 25-hydroxycholesterol, but 24-hydroxycholesterol is selectively present in the brain.

We have made an extensive effort to develop a single gradient elution method that resolves all closely related sterols present in a typical mammalian cell, and this effort has largely succeeded. To date, our most favorable results allow for a qualitative identification of the known co-eluting sterols; however, resolution is still inadequate for quantitation. Should resolution of two co-eluting sterols be necessary, various isocratic elution methods can be employed. This approach, however, usually produces broader peaks and lower signal intensity (Uomori et al., 1987). Also, more hydrophobic sterols, such as cholesterol and lanosterol, will have long isocratic elution times that fall outside of optimal chromatographic conditions. GC-MS remains a viable alternative for the quantitation of 24- and 25-hydroxycholesterol (Lund and Diczfalusy, 2003).

9.4. Acetonitrile and signal intensity

Some methods describing the analysis of sterols by HPLC-MS utilize acetonitrile as a component of the mobile phase (Burkard et al., 2004; Pulfer et al., 2005; Shan et al., 2003). When we incorporated acetonitrile into the gradient mobile phases, a modest to significant decrease in signal intensity was observed for all sterols analyzed by LC-ESI-MS. Oxysterols suffered the largest reduction in signal intensity in the presence of acetonitrile, whereas the reduction observed for other mono-hydroxylated sterols was not as severe. Chromatographic parameters were similar in terms of peak-width and retention time. Thus, reductions in signal intensity are not likely explained by changes in chromatographic performance. We have no clear explanation for this phenomenon at this time, but it appears that acetonitrile has a detrimental effect on the electrospray process and adduct formation, which in turn causes decreased signal intensity for oxysterols. Recent research by Lawrence and Brenna (2006) and Van Pelt et al. (1999) has described acetonitrile as a reactive gas-phase reagent with atmospheric pressure ionization methods that may be related to the decreased signal intensity observed here. For this reason, and because no increase is observed in chromatographic resolution when using acetonitrile, the methanol/water gradient previously described was implemented.

9.5. Auto-oxidation and use of SPE columns

It has been shown that cholesterol is subject to autooxidation, forming oxidized end-products such as 22-, 24-, 25-, and 27-hydroxycholesterol, and $5/6(\alpha/\beta)$-epoxycholesterol (Breuer and Björkhem, 1995; Kudo et al., 1989). In our experience, the most readily formed oxidation product is $5/6(\alpha/\beta)$-epoxycholesterol, which has been shown to be physiologically irrelevant (Breuer and Björkhem, 1995; Kudo et al., 1989). Other oxidation end-products are formed in negligible amounts. We have found that extraction of biological samples using the Bligh/Dyer approach above described typically results in minimal oxidation. When further purification steps are conducted on SPE silica columns, the abundance of $5/6(\alpha/\beta)$-epoxycholesterol increases. A pilot study using deuterated compounds to allow quantitation and the full extraction/SPE procedure previously described showed that as much as 1 to 3% of cholesterol can be oxidized to $5/6(\alpha/\beta)$-epoxycholesterol. No other autooxidation products of cholesterol were detected. Addition of the antioxidant butylhydroxy toluene (BHT) to the sample prior to extraction produced only a modest decrease in oxidation. In most experiments, we do not consider these oxidation products problematic. Minimizing oxidation by using good lab practices is important, but to eliminate oxidation would require processing the samples in an inert environment, which is not logistically possible for most laboratories. It should be noted that 5/6α- and 5/6β-epoxycholesterol are among the most sensitive sterols analyzed by this instrumental method (Table 6.2), which can result in large signals being observed in the SRM chromatograms relative to other signals. This imbalance in relative signal strength adds complexity and, possibly, confusion to the interpretation of the data.

9.6. Sample clean-up

Depending on sample type and desired information, additional sample processing may be desirable; however, in addition to the increased formation of oxidation products already described, we find that typical clean-up steps such as saponification and SPE require additional time and resources and produce only limited benefit. While we have described here a comprehensive extraction procedure involving extraction, saponification, and solid-phase purification, the last two steps are not always required for successful analysis of sterols. In specific cases, such as plasma, oxysterols are found as free and esterified compounds (Bodin et al., 2001; Dzeletovic et al., 1995; Smith et al., 1981). Saponifying the extracts will increase the amount of free oxysterols present in the extract, resulting in stronger signal intensities for these compounds. Otherwise, unless knowledge of free-to-esterified sterol ratios is required or the final extracts are not suitable for analysis after multiple Bligh/Dyer extractions, we do not recommend performing additional separation and purification steps beyond the Bligh/Dyer extraction. Use of a guard column with the primary HPLC column will trap many highly retained compounds and can be replaced with minimal expense relative to the HPLC column. Should diminished chromatographic performance be observed, back-flushing the column with a strong reverse-phase solvent such as chloroform or isooctane will often remove most retained compounds and extend the life of the column. Alternatively, HPLC columns are consumable items, and the cost of replacing a column with greater frequency is usually smaller than that associated with the increased equipment and time required for additional separation and purification steps. The chromatographic method outlined above adequately separates a majority of sterols in mammalian cells, and SPE cannot itself resolve problems associated with the analysis of co-eluting sterols or diastereomers. Lastly, with every new step in processing comes the possibility of additional sample loss and, although surrogate standards can account for these losses, if the sterol of interest exists at trace levels, excess processing may reduce its signal below the limits of detection. Figure 6.5 shows various work-flows for the extraction of sterols that can be tailored to individual user's requirements.

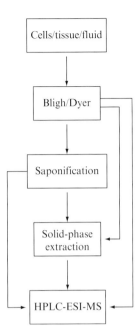

Figure 6.5 Flow chart of various procedures for the extraction and preparation of sterols from biological extracts.

9.7. Residual insoluble material

Regardless of how extensively an extract is separated and purified, residual insoluble material is often present in the final dried extract. When this situation is encountered, we redissolve lipids in 95% methanol and make every effort to minimize resuspension of insoluble material. We investigated the composition of this insoluble material by performing additional extractions with both methanol and chloroform followed by ESI-MS analysis. Methanol extraction of particulates showed only small amounts of cholesterol, which may represent residual compound remaining after the extract was transferred to the auto-sampler vial. Chloroform extraction showed no additional sterols, but did dissolve some of the residual material. It would appear that the minor amounts of insoluble material that sometimes form using the above methods is of little concern for subsequent sterol analyses.

Recent experiments (unpublished work) demonstrate that the majority of residual insoluble material previously described is due to the use of plastics in conjunction with organic solvents. We recommend using only glass and Teflon when contact with any organic solvents will occur. The only steps where plastics and organic solvents should come into contact are during the addition of the surrogate and internal standards (when using positive displacement pipettes with disposable tips) or during pre-purification of extracts with SPE columns (the columns are made of plastic). Completely eliminating plastics in these steps would add considerable effort and cost.

9.8. Comparing relative peak areas

Individual sterols have varying responses to ESI. The more polar oxysterols (e.g., 24,25-epoxycholesterol) are generally more amenable to ESI ionization compared to less polar compounds (e.g., lanosterol). This difference is reflected in Table 6.2, which shows that limits of detection span several orders of magnitude. Also, later eluting sterols are subject to longitudinal diffusion, which causes peak-broadening and results in a less-intense signal compared to the narrower peaks eluting early in the elution profile. Both of these phenomena prevent interpretation of the relative amounts of sterols in an extract solely by evaluation of their signal intensities. Equivalent signals for two sterols may be the result of very different amounts. Interpretation of relative amounts can only be evaluated by assessing the instrumental response of primary standards for the compounds of interest.

9.9. Electrospray using other instrumental platforms

Electrospray ionization and adduct formation of sterols is dependent on the mass spectrometer and, more importantly, the ionization source used. Two laboratories analyzing cholesterol by ESI-MS using the previously mentioned parameters but with instruments and/or ionization sources from other manufacturers or other models from the same manufacturer, may obtain different mass spectra or have varying optimal SRM pairs. The method described here is developed and optimized for an Applied Biosystems 4000 QTrap® equipped with a Turbo V^{TM} ionization source. Through discussions with other mass spectrometrists, it is clear that, with some instruments, sterol-ammonium adducts are not observed, but in-source decay products (e.g., loss of NH_3, H_2O, etc.) are. Furthermore, some instruments ionize and transfer sterols to the MS so inefficiently that atmospheric pressure chemical ionization (APCI) is employed in place of ESI (Burkard et al., 2004; Palmgren et al., 2005; Tian et al., 2006). The parameters and results described here are specific to the 4000 QTrap triple quadrupole system, although, on other platforms, they can serve as a foundation from which a suitable HPLC-MS method can be developed, including those for single quadrupole systems.

ACKNOWLEDGMENTS

This work was supported by the National Institutes of Health (NIH) Glue Grant 1 U54 GM69338. We thank Carolyn Cummins for reviewing this manuscript.

REFERENCES

Abidi, S. L. (2001). Chromatographic analysis of plant sterols in foods and vegetable oils. *J. Chrom. A.* **935,** 173–201.
Bligh, E. G., and Dyer, W. J. (1959). A rapid method of total lipide extraction and purification. *Can. J. Biochem. Physiol.* **37,** 911–917.
Bodin, K., Bretillon, L., Aden, Y., Bertilsson, L., Broome, U., Einarsson, C., and Diczfalusy, U. (2001). Antiepileptic drugs increase plasma levels of 4β-hydroxycholesterolin humans. Evidence for involvement of cytochrome P450 3A4. *J. Biol. Chem.* **276,** 38685–38689.
Breuer, O., and Björkhem, I. (1995). Use of an $^{18}O_2$ inhalation technique and mass isotopomer distribution analysis to study oxygenation of cholesterol in rat. *J. Biol. Chem.* **270,** 20278–20284.
Burkard, I., Rentsch, K. M., and von Eckardstein, A. (2004). Determination of 24S- and 27-hydroxycholesterol in plasma by high-performance liquid chromatography-mass spectrometry. *J. Lipid Res.* **45,** 776–781.
Dietschy, J. M., and Turley, S. D. (2004). Thematic review series: Brain lipids. Cholesterol metabolism in the central nervous system during early development and in the mature animal. *J. Lipid Res.* **45,** 1375–1397.
Duncan, M. W., Gale, P. J., and Yergey, A. L. (2006). "The Principles of Quantitative Mass Spectrometry." Rockpool Productions, LLC., Denver, CO.
Dzeletovic, S., Breuer, O., Lund, E., and Diczfalusy, U. (1995). Determination of cholesterol oxidation products in human plasma by isotope dilution-mass spectrometry. *Anal. Biochem.* **225,** 73–80.
Fahy, E., Subramaniam, S., Brown, H. A., Glass, C. K., Merrill, A. H., Jr., Murphy, R. C., Raetz, C. R. H., Russell, D. W., Seyama, Y., Shaw, W., Shimizu, T., Spener, F., et al. (2005). A comprehensive classification system for lipids. *J. Lipid Res.* **46,** 839–862.
Fenn, J. B., Mann, M., Meng, C. K., Wong, S. F., and Whitehouse, C. M. (1989). Electrospray ionization for mass spectrometry of large biomolecules. *Science* **246,** 64–71.
Isobe, K. O., Tarao, M., Zakaria, M. P., Chiem, N. H., Minh, L. Y., and Takada, H. (2002). Quantitative application of fecal sterols using gas chromatography-mass spectrometry to investigate fecal pollution in tropical waters: Western Malaysia and Mekong Delta, Vietnam. *Environ. Sci. Technol.* **36,** 4497–4507.
Kudo, K., Emmons, G. T., Casserly, E. W., Via, D. P., Smith, L. C., St. Pyrek, J., and Schroepfer, G. J., Jr. (1989). Inhibitors of sterol synthesis. Chromatography of acetate derivatives of oxygenated sterols. *J. Lipid Res.* **30,** 1097–1111.
Lawrence, P., and Brenna, J. T. (2006). Acetonitrile covalent adduct chemical ionization mass spectrometry for double bond localization in non-methylene-interrupted polyene fatty acid methyl esters. *Anal. Chem.* **78,** 1312–1317.
Lund, E. G., and Diczfalusy, U. Quantitation of receptor ligands by mass spectrometry. *Methods Enzymol.* **364,** 24–37.
Moruisi, K. G., Oosthuizen, W., and Opperman, A. M. (2006). Phytosterols/stanols lower cholesterol concentrations in familial hypercholesterolemic subjects: A systematic review with meta-analysis. *J. Am. Coll. Nutr.* **25,** 41–48.
Murphy, R. C. (2006). Personal Communication.
Myant, N. B. (1981). "The Biology of Cholesterol and Related Steroids." Heinemann Medical Books, London, England.
Ostlund, R. E., Jr., Racette, S. B., and Stenson, W. F. (2003). Inhibition of cholesterol absorption by phytosterol-replete wheat germ compared with phytosterol-depleted wheat germ. *Am. J. Clin. Nutr.* **77,** 1385–1389.
Palmgren, J. J., Toyras, A., Mauriala, T., Monkkonen, J., and Auriola, S. (2005). Quantitative determination of cholesterol, sitosterol, and sitostanol in cultured Caco-2 cells

by liquid chromatography-atmospheric pressure chemical ionization mass spectrometry. *J. Chrom. B.* **821,** 144–152.

Pulfer, M. K., Taube, C., Gelfand, E., and Murphy, R. C. (2005). Ozone exposure *in vivo* and formation of biologically active oxysterols in the lung. *J. Pharmacol. Exp Ther.* **312,** 256–264.

Russell, D. W. (2003). The enzymes, regulation, and genetics of bile acid synthesis. *Annu. Rev. Biochem.* **72,** 137–174.

Shan, H., Pang, J., Li, S., Chiang, T. B., Wilson, W. K., and Schroepfer, J. G. J. (2003). Chromatographic behavior of oxygenated derivatives of cholesterol. *Steroids* **68,** 221–233.

Smith, L. L., Teng, J. I., Yong Yeng, L., Seitz, P. K., and McGehee, M. F. (1981). Sterol metabolism–XLVII. Oxidized cholesterol esters in human tissues. *J. Steroid Biochem.* **14,** 889–900.

Tian, Q., Failla, M. L., Bohn, T., and Schwartz, S. J. (2006). High-performance liquid chromatography/ atmospheric pressure chemical inonization tandem mass spectrometry determination of cholesterol uptake by Caco-2 cells. *Rapid Commun. Mass Spectrom.* **20,** 3056–3060.

United States Environmental Protection Agency (USEPA) (2003). EPA, SWA-846 Method 8000C. Determinative chromatographic separations. Available at: http://nlquery.epa.gov.

Uomori, A., Seo, S., Sato, T., Youshimura, Y., and Takeda, K. J. (1987). Synthesis of (25R)-[26-2H1]cholesterol and 1H n.m.r. and h.p.l.c. resolution of (25R)- and (25S)-26-hydroxycholesterol. *J. Chem. Soc. Perkin. Trans. I.* **33,** 1713–1718.

Van Pelt, C. K., Carpenter, B. K., and Brenna, J. T. (1999). Studies of structure and mechanism in acetonitrile chemical ionization tandem mass spectrometry of polyunsaturated fatty acid methyl esters. *J. Am. Soc. Mass Spectrom.* **10,** 1253–1262.

Wilund, K. R., Yu, L., Xu, F., Vega, G. L., Grundy, S. M., Cohen, J. C., and Hobbs, H. H. (2004). No association between plasma levels of plant sterols and atherosclerosis in mice and men. *Arterioscler. Thromb. Vasc. Biol.* **24,** 2326–2332.

Yang, C., McDonald, J. G., Patel, A., Zhang, Y., Umetani, M., Xu, F., Westover, E. J., Covey, D. F., Mangelsdorf, D. J., Cohen, J. C., and Hobbs, H. H. (2006). Sterol intermediates from cholesterol biosynthetic pathway as liver X receptor ligands. *J. Biol. Chem.* **281,** 27816–27826.

CHAPTER SEVEN

THE LIPID MAPS INITIATIVE IN LIPIDOMICS

Kara Schmelzer,[*,†] Eoin Fahy,[‡] Shankar Subramaniam,[*,‡,§] and Edward A. Dennis[*,†]

Contents

1. Introduction 172
2. Building Infrastructure in Lipidomics 173
3. Classification, Nomenclature, and Structural Representation of Lipids 174
4. Mass Spectrometry as a Platform for Lipid Molecular Species 177
5. Future Plans 180
Acknowledgments 182
References 182

Abstract

The Lipid Metabolites and Pathways Strategy (LIPID MAPS) initiative constitutes the first broad scale national exploration of lipidomics and is supported by a U.S. National Institute of General Medical Sciences Large Scale Collaborative "Glue" Grant. The emerging field of lipidomics faces many obstacles to become a true systems biology approach on par with the other "omics" disciplines. With a goal to overcome these hurdles, LIPID MAPS has been developing the necessary infrastructure and techniques to ensure success. This review introduces a few of the challenges and solutions implemented by LIPID MAPS. Among these solutions is the new comprehensive classification system for lipids, along with a recommended nomenclature and structural drawing representation. This classification system was developed by the International Lipids Classification and Nomenclature Committee (ILCNC) in collaboration with LIPID MAPS and representatives from Europe and Asia. The latest changes implemented by the committee are summarized. In addition, we discuss the adoption of mass spectrometry (MS) as the instrumental platform to investigate lipidomics. This platform has the versatility to quantify known individual lipid molecular species and search for novel lipids affecting biological systems.

---

[*] Department of Chemistry and Biochemistry, University of California, San Diego, La Jolla, California
[†] Department of Pharmacology, University of California, San Diego, La Jolla, California
[‡] San Diego Supercomputer Center, University of California, San Diego, La Jolla, California
[§] Department of Bioengineering, University of California, San Diego, La Jolla, California

1. INTRODUCTION

Remarkable technological advances in the biological sciences have forged a new era of research in the field of systems biology. These comprehensive investigations of living organisms at the molecular system level can be classified into different fields: genomics, transcriptomics, proteomics, and metabolomics, otherwise known as the "omics cascade." The integrative analysis of an organism's response to a perturbation on the "omics" levels will lead to a more complete understanding of the biochemical and biological mechanisms in complex systems.

An explosion of information has occurred in the fields of genomics, transcriptomics, and proteomics because each field was able to adopt a single instrumental methodology to analyze all of their respective components in a given sample. However, whereas genomics, transcriptomics, and proteomics have made significant strides in technological development, the tools for the comprehensive examination of the metabolome are still emerging. Although metabolomics is the endpoint of the "omics cascade" and the closest to phenotype, no single instrument can currently analyze all metabolites (Bino et al., 2004).

Comprehensive investigation of the metabolome is hindered by its enormous complexity and dynamics. The metabolome represents a vast number of components that belong to a wide variety of compound classes, including nucleic acids, amino acids, sugars, and lipids. These compounds are diverse in their physical and chemical properties and occur in vastly different concentration ranges. Additionally, metabolite concentrations can vary spatially and temporally. Furthermore, diet- and/or nutrient-dependent biological variability confounds such analysis (Vigneau-Callahan et al., 2001). For example, within the lipid classes, highly abundant compounds exist, such as fatty acids, triglycerides, and phospholipids, and trace level components, such as eicosanoids, which have important regulatory effects on homeostasis and disease states.

In addition to analyzing a vast number of metabolites, a comprehensive lipidomic investigation includes studying the metabolizing enzymes and lipid transporters. Together, these components play a role in cell signaling, metabolism, physiology, and disease, making them excellent targets for systematic measurements. The LIPID MAPS initiative is a result of our quest to characterize the mammalian lipidome to an extent similar to that of the genome and proteome (Dennis et al., 2005). To accomplish this goal, we have formed a consortium of twelve independent laboratories from seven academic institutions and one company. We have six lipidomics cores that are charged with identifying, quantifying, and becoming experts in six major categories of lipids: fatty acyls, glycerolipids, glycerophospholipids, sphingolipids, sterol lipids, and prenol lipids. This division of labor is essential, as there are hundreds of thousands of individual molecular species of metabolites in each category of lipids, presenting a challenge for routine quantification. Additionally, the Lipid Synthesis/Characterization Core is charged with generating quantitative lipid standards and the synthesis and characterization of novel lipids. We also have a Bioinformatics Core, a Cell Biology Core, and a MS Development Core. Finally, five independent sections explore more specific aspects of lipidomics as related to future core methodologies and the medical field.

The following text will discuss each of the LIPID MAPS contributions to lipidomics. The "Building Infrastructure in Lipidomics" section discusses the development of critical infrastructure for lipidomics to become a true systems biology approach. "Classification, Nomenclature, and Structural Representation of Lipids" describes the standardization of the lipid classification scheme, lipid nomenclature, and structural representation (Fahy et al., 2005). "Mass Spectrometry as a Platform for Lipid Molecular Species" describes the adoption of MS as the major platform to investigate lipidomics and the challenges in quantifying the complex lipid metabolites in a biological system. This section addresses the complications associated with integrating the molecular species lipidomic data and the need for analyte standards. After the first 3 years of the project, we have implemented most of the infrastructure and have evolved our future plans to expand lipidomics, as described in the "Future Plans" section.

2. Building Infrastructure in Lipidomics

In 2003, the consortium identified several important challenges that any lipidomics effort would face. These obstacles included developing a robust and versatile platform to quantify the lipidome, specific reagents and experimental protocols, high throughput methods, data export standardization, computer algorithms for automated data analysis, visualization tools, libraries, and databases for integration of lipidomics with the other omic cascades.

Because no single quantization platform is capable of measuring the entire lipidome, the first goal of the initiative was to develop the requisite quantitation technology. In order for different laboratories to minimize instrument variability and to ensure interoperability and method sharing capabilities, LIPID MAPS chose for all cores to use the same ABI-4000 Q-trap mass spectrometers (Foster City, CA) for quantitative comparisons. We have and will continue to design stringent sets of isotopic-labeled reference compounds that will allow the accurate quantification of a wide range of lipid metabolites. The reference standards are intended to be commercially available to all researchers worldwide, both within and beyond the lipid field, which will facilitate inter-laboratory comparison.

Once a requisite quantitation technology was established, we realized that, to successfully conduct bioinformatic analysis of data collected from six different laboratories, a significant level of reproducibility between experiments and between laboratories would be required. To achieve this reproducibility, we have employed a rigorously maintained set of common biological, biochemical, and analytical technologies in each of the consortium laboratories. To this end, we have tried to standardize all reagents. We purchased a large lot of fetal calf serum, which is being employed by all consortium laboratories for the duration of this project. We also prepared a large batch of RAW 264.7 cells obtained from ATCC (Manassas, VA) and froze enough stabs to supply fresh cells from this one batch to all laboratories for the duration of the project. Thus, all cells in these experiments were similarly passaged and treated. For example, to investigate changes associated with the binding of the Toll-like receptor 4 (TLR-4), we developed the methodology to prepare highly purified 3-deoxy-D-manno-octulosonic acid Kdo_2-Lipid A, and then evaluated it by electrospray ionization (ESI)-MS (ESI-MS), liquid chromatography (LC)-MS (LC-MS), and ^{1}H-NMR (nuclear magnetic resonance). Finally, we compared its bioactivity with laser plasma spectrometry (LPS) in RAW 264.7 cells and bone marrow macrophages from wild-type and TLR-4–deficient mice. Now, standardized Kdo_2-Lipid A ensures a crucial intra-laboratory quality control agonist for TLR-4 (Raetz et al., 2006).

No universal classification system for lipids was suitable for modern informatics and experimental investigations. Unlike existing lipid legacy databases, such as Lipid Bank and Lipidat, our new system plans to integrate all the omic cascades associated with lipidomics. Integrating the MS data on lipidomics with the genomic data will lead to a more complete understanding of how complex lipidomic networks function, from biosynthesis to the removal of cellular lipids and the important roles of metabolites as second messengers. With thorough and extensive experimental planning, we are able to integrate and analyze large amounts of data that will be developed into "road maps."

3. Classification, Nomenclature, and Structural Representation of Lipids

With the emergence of lipidomics as a rapidly expanding field came an urgent need for an internationally accepted method of describing and classifying lipid molecules. The first step toward classification of lipids was the establishment of an ontology that was extensible, flexible, and scalable. One must be able to classify, name, and represent these molecules in a logical manner that is amenable to databasing and computational manipulation. Lipids have been loosely defined as biological substances that are generally hydrophobic in nature and, in many cases, soluble in organic solvents (Smith, 2000). These chemical features are present in a broad range of molecules, such as fatty acids, phospholipids, sterols, sphingolipids, and terpenes (Christie, 2003). The LIPID MAPS consortium has taken a chemistry-based approach by defining lipids as hydrophobic or amphipathic small molecules that may originate entirely or in part by carbanion-based condensations of thioesters such as fatty acids and polyketides (Fig. 7.1A) and/or by carbocation-based condensations of isoprene units such as prenols and sterols (Fig. 7.1B) (Fahy et al., 2005).

The classification scheme shown in Table 7.1 organizes lipids into well-defined categories that cover eukaryotic and prokaryotic lipid origins, and is equally applicable to archaea and synthetic (manufactured) lipids (Fahy et al., 2005). Biosynthetically related compounds that are not technically lipids due to their water solubility are included for completeness in this classification scheme.

Lipids are divided into eight categories (fatty acyls, glycerolipids, glycerophospholipids, sphingolipids, sterol lipids, prenol lipids, saccharolipids, and polyketides) containing distinct classes and subclasses of molecules, and a 12-digit unique identifier is associated with each distinct lipid molecule.

For each lipid, the unique 12-character identifier based on this classification scheme is an important database field in the Lipids Database (Sud et al., 2007). The format of the LIPID ID, outlined in Table 7.2, provides a systematic means of assigning a unique identification to each lipid molecule and allows for the addition of large numbers of new categories, classes, and subclasses in the future. The last four characters of the ID comprise a unique identifier within a particular subclass and are randomly assigned. Initially using numeric characters allows 9999 unique IDs per subclass; however, with the additional use of 26 uppercase alphabetic characters, a total of 1.68 million possible combinations can be generated, providing ample scalability within each subclass.

The classification system is under the guidance of the ILCNC, which meets periodically to propose changes and updates to classification, nomenclature, and structural representation. The ILCNC currently consists of Dr. Edward A. Dennis (chair), Dr. Robert C. Murphy, Dr. Masahiro Nishijima, Dr. Christian R.H. Raetz, Dr. Takao Shimizu, Dr. Friedrich Spener, Dr. Gerrit van Meer, and Dr. Michael Wakelam. The most recent meeting was on July 7, 2006 in La Jolla, CA, during which a number of recommendations were implemented to extend the original system (Fahy et al., 2005). Some key changes can be seen in Table 7.3.

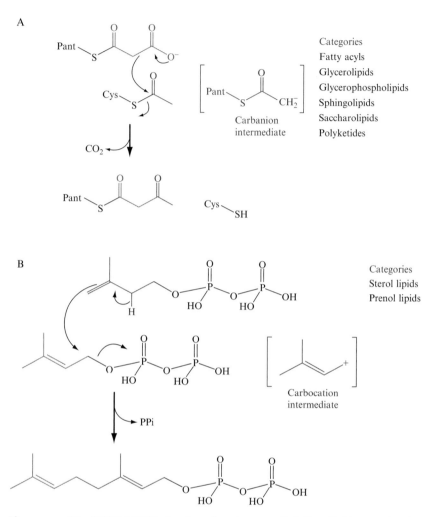

Figure 7.1 The LIPID MAPS chemistry-based approach defines lipids as molecules that may originate entirely or in part by carbanion-based condensations of (A) thioesters and/or by (B) carbocation-based condensations of isoprene units. (See color insert.)

The ILCNC has been greatly assisted by many experts throughout the world who have given helpful advice and comments on enhancing the scope and utility of the lipid classification scheme. The scheme can be conveniently browsed on the LIPID MAPS website (http://www.lipidmaps.org), in which the various classes and subclasses are linked to the LIPID MAPS structure database (Fahy and Subramaniam, 2007).

The LIPID MAPS classification scheme has now gained widespread international acceptance and has recently been adopted by KEGG (Kyoto

Table 7.1 LIPID MAPS lipid categories and examples

Category	Abbreviation	Example
Fatty acyls	FA	Dodecanoic acid
Glycerolipids	GL	1-Hexadecanoyl-2-(9Z-octadecenoyl)-*sn*-glycerol
Glycerophospholipids	GP	1-Hexadecanoyl-2-(9Z-octadecenoyl)-*sn*-glycero-3-phosphocholine
Sphingolipids	SP	N-(tetradecanoyl)-sphing-4-enine
Sterol lipids	ST	Cholest-5-en-3β-ol
Prenol lipids	PR	2E,6E-farnesol
Saccharolipids	SL	UDP-3-O-(3R-hydroxy-tetradecanoyl)- βD-N-acetylglucosamine
Polyketides	PK	Aflatoxin B1

Table 7.2 Format of 12-character LIPID ID

Characters	Description	Example
1–2	Fixed database designation	LM
3–4	Two-letter category code	FA
5–6	Two-digit class code	03
7–8	Two-digit subclass code	02
9–12	Unique four-character identifier within subclass	AG12

Encyclopedia of Genes and Genomes), where functional hierarchies involving lipids, reactions, and pathways have been constructed (http://www.genome.ad.jp/brite/). In addition, LIPID MAPS lipid structures are now available on NCBI's PubChem website (http://pubchem.ncbi.nlm.nih.gov/), where entries are hyperlinked to the LIPID MAPS classification system.

4. MASS SPECTROMETRY AS A PLATFORM FOR LIPID MOLECULAR SPECIES

Mass spectrometry is an established and invaluable tool for the characterization of changes in lipidomics and lipid-mediated signaling processes resulting from disease, toxicant exposure, genetic modifications, or drug therapy

Table 7.3 Updates to LIPID MAPS classification system

1. Under the fatty acyls category, the hydroxyeicosatrienoic acids, hydroxyeicosatetraenoic acids, and hydroxyeicosapentaenoic acids subclasses have been expanded to include the corresponding hydroperoxy and keto analogues. The N-acyl amide and N-acyl ethanolamide subclasses have been renamed to N-acyl amines and N-acyl ethanolamines.
2. In order to improve representation of plant lipids in the classification scheme, the glycosylmonoradyl and glycosyldiradylglycerol classes were created within the glycerolipid category.
3. Several new bile acids (C22, C23, C25, and C29 bile acids, alcohols, and derivatives) and secosteroids (vitamin D4, D5, D6, and D7 and derivatives) have been added to the sterol lipids category.
4. It was decided not to rely on phylum- or species-based references in the classification scheme, but rather to use structural core units where possible. Accordingly, the phytosterol, marine sterol, and fungal sterol subclasses were removed from the sterol lipids category and replaced with a more extensive, structurally based list (ergosterols, stigmasterols, C24-propyl sterols, gorgosterols, furostanols, spirostanols, furospirostanols, calysterols, cardenolides, bufanolides, brassinolides, and solanidines/alkaloids) of sterols.
5. The retinoids subclass was added to the prenol lipids category. The hopanoids class was also moved to this category (from sterol lipids).
6. In the case of lipids with multiple functional groups (especially fatty acyls) where it may be difficult to objectively classify a structure, the Cahn-Ingold-Prelog rules are applied to place the lipid in the subclass with the highest Cahn-Ingold-Prelog "score." However, to ensure that lipids containing a particular functional group (e.g., a hydroxyl-fatty acid) can be located in a database that uses this classification system, an ontology-based system has been implemented, where a user may locate all lipids with the specified functionality, regardless of their subclass designation.
7. The LIPID MAPS glycerophospholipid abbreviations (GPCho, GPEtn, etc.) are used to refer to species with one or two radyl side-chains, where the structures of the side chains are indicated within parentheses, such as GPCho(16:0/18:1(9Z)).

(Watkins and German, 2002). MS provides a wide dynamic range to quantify lipids. In addition, it offers excellent selectivity for examining compounds using parameters such as ionization mode (positive/negative), mass selection, MS/MS characterization, and numerous MS^n techniques for structural elucidation. MS-based lipidomics can deliver detailed, quantitative information about the cellular lipidome constitution and provide insights into biochemical mechanisms of lipid metabolism, lipid–lipid, and lipid–protein interactions.

Traditionally, lipids have been analyzed using gas chromatographic (GC) separation, electron impact ionization MS (EI-MS), and flame ionization detection (FID). GC is limiting because compounds must be thermally stable with high enough vapor pressure to volatilize during injection. Therefore, extensive sample manipulation is required on complex lipid samples to produce accurate data. Sample preparation can include pre-separation of lipid classes, hydrolysis, derivatization, or pyrolysis. Although highly quantitative, these labor-intensive preparations destroy structural information regarding lipid composition. One of the LIPID MAPS goals is to improve the ability to quantify discrete molecular species, utilizing MS.

Detection of the specific molecular species was made possible by the development of solid state ion sources (e.g., fast atom bombardment [FAB] or matrix-assisted laser desorption/ionization [MALDI]) and liquid-phase ion sources (e.g., electrospray ionization [ESI], atmospheric pressure chemical ionization [APCI], or atmospheric pressure photo ionization [APPI]) (Griffiths, 2003). Although MALDI and other laser-based ionization methods have great potential for nonpolar species, our initial work has focused on ESI ion sources, as they are efficient for a wide variety of metabolites and are easily mated to high performance liquid chromatography (HPLC). ESI in positive mode allows for the detection of phosphatidylcholine, lysophosphatidylcholine, phosphatidylethanolamine, and sphingomyelins. ESI in negative mode allows for the detection of phosphatidic acid, phosphatidylserine, phosphatidylinositol, phosphatidylglycerol, phosphotidylethanolamine, free fatty acids, and eicosanoids. Finally, compounds that are not readily ionized by ESI, such as neutral lipids (e.g., triacylglycerols), can be detected by examining ammonium, lithium, or sodium adducts with ESI in positive mode (Duffin *et al.*, 1991; Han and Gross, 2005; Hsu and Turk, 1999).

Ionized lipids are further studied by tandem MS (MS/MS) to identify lipid species within specific classes. For example, the specific headgroups of phospholipids produce characteristic fragmentation patterns that are used for their identification (Griffiths, 2003; Pulfer and Murphy, 2003; Merrill *et al.* 2005) used both neutral loss and precursor-ion methods to identify sphingolipids. Precursor-ion scans for m/z 184 in positive mode are used to identify sphingomyelin while scans for m/z 264 are used to identify ceramide. More complex lipids, such as gangliosides, require advanced MS^n experiments for elucidation.

Certain lipid classes lend themselves to automated computational lipidomic approaches because of structural similarities and differentiation only in their headgroups and acyl groups. Forrester *et al.* (2004) developed methods to automate spectral interpretation and quantification of multiple phospholipid species utilizing direct injection ESI-MS/MS analysis. Han and Gross (2003, 2005) have also attempted to eliminate the need for HPLC by developing a "shotgun lipidomic" approach. They utilized the traditional Bligh/Dyer extraction to produce tissue extracts, and then directly injected the sample onto the mass spectrometer in both positive and negative mode ESI. Hermansson *et al.* (2005), using LC-ESI-MS/MS, quantified more than 100 polar lipids using a fully automated method from sample handling to computer algorithms analysis of the spectra produced. In most cases, only semiquantitative data were reported.

The current goal of LIPID MAPS focuses on the complete quantitative analysis of the exact molecular species. Accurate quantitative data require a known standard for each molecular species, as well as internal standards to mimic analytes in these biological samples. Internal standards are most commonly deuterated analogues or lipids containing odd chain fatty acids. LIPID MAPS has collaborated with suppliers who provide analytical, chemically pure standards for precise quantization. In some categories, such identification of discrete molecular species is extremely labor-intensive. While it would be labor- and cost-prohibitive to develop the standards for quantifying all of the known molecular species, we continue to develop as many standards as possible. Moreover, when unknown species are detected and characterized, their final identification requires chemically synthesized standards (Harkewicz *et al.*, 2007) Progress to date in developing standards is shown in Table 7.4.

5. Future Plans

Signaling through lipid metabolism constitutes key cellular events that participate in innate immunity, inflammation, and a host of other physiological phenomena. The complexity of lipid metabolites in biological

Table 7.4 Mass spectrometry standards

Core	Library MS/MS SPECTRA[a]	Internal standard[b]	Primary standard[c]	LIPID MAPS standard[d]
Fatty acyls eicosanoids	74	16	16	0
Glycerolipids	21	15		15
Glycerophospholipids	182	24	40	24
Sphingolipids	5	15		15
Sterols	23	13	13	11
Prenols/other	13	1		7
Total	**318**	**84**	**69**	**72**

[a] MS/MS spectra of chemically pure analytes, which are located in the LIPID MAPS library on www.lipidmaps.org.
[b] Analyte analog (e.g., stable isotope, odd carbon fatty acid).
[c] Chemically pure, accurately quantitated analyte.
[d] Internal or primary standard that is a certified, chemically pure analyte, quantitated, and stability-tested for shelf life.

samples has presented a great challenge. As stated previously, the different subclasses of lipids are affected by various stimuli and produce both temporal and spatial metabolites in a biological system. Therefore, the true goal of "lipidomics," namely the quantitative analysis of all lipid metabolites in a biological system, requires different tools to probe the lipidome for dynamic changes. Two complementary approaches are lipid fingerprinting and lipid profiling.

Lipid fingerprinting identifies patterns, or "fingerprints," of lipid metabolites that change in response to various stimuli. These methods, which are not intended to quantify compounds, cast a wide net and generate and test hypotheses. Specifically, LIPID MAPS–collected lipid extracts will be subjected to a general lipid fingerprinting analysis using LC–time-of-flight (TOF)–MS. This technique is used to screen for differences in lipid expression among treatment groups and identify important lipid metabolites and potential biomarkers. The bioinformatics team will process the data and, when significant changes are detected, further efforts will be taken to identify and quantify these compounds. This approach allows for the discovery of novel lipids that change in response to a given stimulus as well as interpretation of the role in biochemical processes.

Lipid profiling uses analytical methods for quantifying metabolites in a pathway or for a class of compounds. LIPID MAPS has developed infrastructure for quantifying analytes in each lipid class using this approach. The quantitative lipid analysis information has allowed us to produce information that can be applied to known biochemical pathways and physiological interactions. Profiling methods are used to test specific hypotheses and investigate mechanisms of action within the biological systems. Individual analyte data comprise a platform-independent legacy database, while the collection of lipid profiles provide temporal snapshots that lead to identification of different response stages among the different various groups. In combination with fingerprinting, various biomarkers can potentially be discerned from each stage.

In future work, we will use both approaches to comprehensively investigate lipid changes resulting from a genetic modification, toxicant exposure, diet, disease, or drug therapy. LIPID MAPS plans to carry out systematic and quantitative measurements of lipid changes in an effort to reconstruct normal and pathological networks associated with inflammation. Our vehicle for these experiments will be primary macrophages and macrophage cell lines as well as mouse plasma and various tissues. We will continue to develop novel MS techniques, as well as methods for integrating lipidomic and genomic technologies. We anticipate that our efforts will provide the foundation for building a bridge to translational medicine and serve as a paradigm for interdisciplinary systems medicine projects.

ACKNOWLEDGMENTS

We wish to thank our LIPID MAPS colleagues, including H. Alex Brown, Christopher Glass, Alfred Merrill, Jr., Robert C. Murphy, Christian R. H. Raetz, David Russell, Walter Shaw, Michael VanNieuwenhze, Stephen H. White, and Joseph Witztum, and their laboratories for their contributions to this work. Financial support of the National Institute of General Medical Science U54 GM069338-3 is gratefully acknowledged, as is the advice of NIGMS Program Officer Jean Chin.

REFERENCES

Bino, R. J., Hall, R. D., Fiehn, O., Kopka, J., Saito, K., Draper, J., Nikolau, B. J., Mendes, P., Roessner-Tunali, U., Beale, M. H., Trethewey, R. N., Lange, B. M., et al. (2004). Potential of metabolomics as a functional genomics tool. Trends Plant. Sci. **9**, 418–425.

Christie, W. (2003). "Lipid analysis, 3rd ed." Oily Press, Bridgewater, UK.

Dennis, E. A., Brown, H. A., Deems, R., Glass, C. K., Merrill, A. H., Murphy, R. C., Raetz, C. R. H., Shaw, W., Subramaniam, S., and Russell, D. W. (2005). The LIPID MAPS approach to lipidomics. In "Functional Lipidomics" (L. Feng and G. D. Prestwich, eds.), pp. 1–15. CRC Press/Taylor and Francis Group, Boca Raton, FL.

Duffin, K. L., Henion, J. D., and Shieh, J. J. (1991). Electrospray and tandem mass spectrometric characterization of acylglycerol mixtures that are dissolved in nonpolar solvents. Anal. Chem. **63**, 1781–1788.

Fahy, E., Subramaniam, S., Brown, H. A., Glass, C. K., Merrill, A. H., Jr., Murphy, R. C., Raetz, C. R. H., Russell, D. W., Seyama, Y., Shaw, W., Shimizu, T., Spener, F., et al. (2005). A comprehensive classification system for lipids. J. Lipid Res. **46**, 839–862.

Fahy, E., and Subramaniam, S. (2007). New resources in lipid classification and databases. Meth. Enzymol. In press.

Forrester, J. S., Milne, S. B., Ivanova, P. T., and Brown, H. A. (2004). Computational lipidomics: A multiplexed analysis of dynamic changes in membrane lipid composition during signal transduction. Mol. Pharmacol. **65**, 813–821.

Griffiths, W. J. (2003). Tandem mass spectrometry in the study of fatty acids, bile acids, and steroids. Mass Spectrom. Rev. **22**, 81–152.

Han, X., and Gross, R. W. (2003). Global analyses of cellular lipidomes directly from crude extracts of biological samples by ESI mass spectrometry: A bridge to lipidomics. J. Lipid Res. **44**, 1071–1079.

Han, X., and Gross, R. W. (2005). Shotgun lipidomics: Electrospray ionization mass spectrometric analysis and quantitation of cellular lipidomes directly from crude extracts of biological samples. Mass Spectrom. Rev. **24**, 367–412.

Harkewicz, R., Fahy, E., Andreyev, A., and Dennis, E. A. (2007). Arachidonate-derived dihomoprostaglandin production observed in endotoxin-stimulated macrophage-like cells. J. Biol. Chem. **282**, 2899–2910.

Hermansson, M., Uphoff, A., Käkelä, R., and Somerharju, P. (2005). Automated quantitative analysis of complex lipidomes by liquid chromatography/mass spectrometry. Anal. Chem. **77**, 2166–2175.

Hsu, F. F., and Turk, J. (1999). Structural characterization of triacylglycerols as lithiated adducts by electrospray ionization mass spectrometry using low-energy collisionally activated dissociation on a triple stage quadrupole instrument. J. Am. Soc. Mass Spectrom. **10**, 587–599.

Merrill, A. H., Jr., Sullards, M. C., Allegood, J. C., Kelly, S., and Wang, E. (2005). Sphingolipidomics: High-throughput, structure-specific, and quantitative analysis of sphingolipids by liquid chromatography tandem mass spectrometry. *Methods* **36,** 207–224.

Pulfer, M., and Murphy, R. C. (2003). Electrospray mass spectrometry of phospholipids. *Mass Spectrom. Rev.* **22,** 332–364.

Raetz, C. R., Garrett, T. A., Reynolds, C. M., Shaw, W. A., Moore, J. D., Smith, D. C., Jr., Ribeiro, A. A., Murphy, R. C., Ulevitch, R. J., Fearns, C., Reichart, D., Glass, C. K., et al. (2006). Kdo2-Lipid A of *Escherichia coli*, a defined endotoxin that activates macrophages via TLR-4. *J. Lipid Res.* **47,** 1097–1111.

Smith, A. (2000). "Oxford Dictionary of Biochemistry and Molecular Biology," 2nd ed. Oxford University Press, New York.

Sud, M., Fahy, E., Cotter, D., Brown, A., Dennis, E. A., Glass, C. K., Merril, A. H., Jr., Murphy, R. C., Raetz, C. R., Russell, D. W., and Subramaniam, S. (2007). LMSD: LIPID MAPS structure database. *Nucleic Acids Res.* **35**(Database issue), D527–D532.

Vigneau-Callahan, K. E., Shestopalov, A. I., Milbury, P. E., Matson, W. R., and Kristal, B. S. (2001). Characterization of diet-dependent metabolic serotypes: Analytical and biological variability issues in rats. *J. Nutr.* **131,** 924S–932S.

Watkins, S. M., and German, J. B. (2002). Toward the implementation of metabolomic assessments of human health and nutrition. *Curr. Opin. Biotechnol.* **13,** 512–516.

CHAPTER EIGHT

BASIC ANALYTICAL SYSTEMS FOR LIPIDOMICS BY MASS SPECTROMETRY IN JAPAN

Ryo Taguchi,[*,†] Mashahiro Nishijima,[‡] and Takao Shimizu[§]

Contents

1. Introduction	186
2. Lipid Bank and Related Databases	187
3. Strategies for Lipid Identification and Quantitative Analysis by Mass Spectrometry	188
3.1. Accurate mass value of molecular weight-related ions	188
3.2. Characteristic fragment ions	189
3.3. Good separation and reproducible retention time in liquid chromatography	189
4. Several Practical Lipidomics Methods by Mass Spectrometry	190
4.1. Global and untargeted	190
4.2. Focused	191
4.3. Targeted	192
5. Strategies for Identification of Individual Molecular Species in Glycerolipids and Glycerophospholipids by Lipid Search	192
5.1. Theoretical database and search engine for identification of phospholipid molecular species and their fragments	193
6. Quantitative Analysis or Profiling of Lipid Molecular Species	197
6.1. Compensation by a few selected internal standards for correcting relative ionization efficiency	198
6.2. Compensation with limited number of proper internal standards	198
6.3. Compensation with stable isotope-labeled standard	198
7. Application of Several Different Methods in Lipidomics	199
7.1. Global and comprehensive analyses in lipidomics under non-targeted strategies	199

---

[*] Department of Metabolome, Graduate School of Medicine, The University of Tokyo, Tokyo, Japan
[†] Core Research for Evolutional Science and Technology (CREST), Japan Science and Technology Agency, Tokyo, Japan
[‡] National Institute of Health Sciences, Tokyo, Japan
[§] Department of Molecular Biology, Graduate School of Medicine, The University of Tokyo, Tokyo, Japan

7.2. Application of focused methods by mass spectrometry
for lipidomics   200
7.3. Application of targeted method using expanded MRM for
oxidized phospholipids   207
8. Future Program for Lipidomics   208
   8.1. MassBank: Mass fragment database for metabolomics in
   Japan's BIRD project   208
   8.2. Flux analysis by stable isotope labeling   208
   8.3. Mass imaging of lipids by MALDI-MS and TOF-SIMS   208
Acknowledgments   209
References   209

Abstract

In recent analyses of phospholipids, the application of mass spectrometry (MS) has become increasingly popular. To elucidate the function of phospholipids, it is necessary to analyze not only their classes and subclasses, but also their molecular species. In choosing analytical methods for lipidomics, we selected several different approaches in the identification of phospholipid molecular species. The first approach, global and shotgun liquid chromatographic tandem MS (LC-MS/MS) analysis, uses data-dependent MS/MS scanning, whereas the second approach is structure related, focusing on methods using precursor ion scanning or neutral loss scanning. The third approach uses theoretically expanded multiple reaction monitoring for the analysis of targeted molecules in extremely small amounts.

Data from the first and second types of analyses can be obtained with the use of our search engine, Lipid Search (http://lipidsearch.jp), and most probable molecular species can be obtained with their compensated ion intensities. Identified individual molecular species can be automatically profiled according to their compensated ion intensities. Profiled data can be visualized in the state of relative increase or decrease. These data are available via the Lipid Bank (http://lipidbank.jp) and the MassBank (http://www.massbank.jp).

1. INTRODUCTION

Lipidomics has become a prominent research field in metabolomics through recent advances in MS (Han and Gross, 2005; Pulfer and Murphy, 2003). The aim of lipidomics is to identify lipid molecules from MS data and to obtain profiling patterns of the changes in these molecules under specific circumstances. New enzyme proteins can be investigated through the analytical processes of profiling to elucidate unknown pathways or to precisely characterize lipid substrates.

Electrospray ionization-tandem MS (ESI-MS/MS) is one such advance that has been applied to the analysis of lipids (Brugger et al., 1997; Han and

Gross, 1994; Kerwin et al., 1994; Kim et al., 1994; Lehmann et al., 1997; Weintraub et al., 1991). Since ESI is a soft ionization method, each molecule in a mixture can be detected without any fragmentation (Fenn et al., 1989). The analytical methods of lipidomics by MS, adding to the comprehensive and untargeted analysis, require focused analyses for categorical components or targeted analyses for individual molecular species. It is very difficult to obtain exact annotation for all molecules, even in the limited categories of metabolites such as lipids. This difficulty is due to differences in the extraction efficiency of individual metabolites, solubility in analytical solvents, and ionic efficiency. In addition, an extensive and dynamic range of lipid metabolites is found in biological samples. It is very difficult in proteomics to detect small amounts of peptides or proteins in mammalian plasma due to the presence of a very diverse range of protein contents.

Normally, only the major peaks will be detected if the sample is injected as a mixture without any separation. One of the solutions to this problem is to use a specific detection method, such as precursor ion scanning or neutral loss scanning (Ekroos et al., 2002, 2003; Han and Gross, 2003; Houjou et al., 2004; Taguchi et al., 2005). These scanning modes are often used for the measurement of particularly focused phospholipids (Houjou et al., 2004; Ishida et al., 2004, 2003). For the detection of minor but physiologically important lipid molecules, such as lipid mediators, several technical improvements are required. One such improvement involves selective detection methods, including the choice of effective high performance liquid chromatography (HPLC) columns and the most suitable MS systems and collision conditions in MS/MS.

We now move on to a presentation of our recent approaches by MS as well as the strategies for identification and other analytical tools in lipidomics, which focus mainly on phospholipids and their metabolites.

2. LIPID BANK AND RELATED DATABASES

The online project, Lipid Bank, was started in 1996 and ran until 2002 through joint research of the International Medical Center of Japan and the Japan Science and Technology Agency (JST). From 2003 to 2006, this database was supported by a Grant-in-Aid for Scientific Research from the Japan Society for the Promotion of Science and by the Department of Metabolome, Graduate School of Medicine, Tokyo University.

The database contains factual data, such as lipid names, chemical and physical properties, biological activities, metabolism, and genetic information. It also contains graphical data, such as structural formulae, UV (ultraviolet), IR (infrared spectrometry), NMR (nuclear magnetic resonance), MS, LC, and

TLC (thin-layered chromatography). Further, some specific reference data linked to PubMed (http://www.ncbi.nlm.nih.gov/entrez) are included.

In August 2006, Lipid Bank was transferred to the Japanese Conference on the Biochemistry of Lipids (JCBL; President, M. Nishijima). At the present time, a revision process by an expert committee of JCBL is underway. The new version of Lipid Bank, to be launched at the end of 2007, will contain practical names, structures, and so on. It will be actively connected to Mass-Bank (http//www.massbank.jp), which is supported by JST, and BIRD (Institute for Bioinformatics Research and Development) in Japan. MassBank contains a database of molecular weight-related ions of metabolites and their fragments obtained by various MS systems and collisional conditions. The new version of Lipid Bank will use an ID number system, which is constructed by consortium work with LIPID MAPS (LIPID Metabolites and Pathways Strategy, http://www.lipidmaps.org) in the United States and ELI (European Lipidomics Initiative, http://www.lipidomics.net) in Europe. The new database will contain more than 10,000 factual data items on lipid metabolites. A description of the new Lipid Bank will be available until the end of 2007 (http://lipidbank.jp). Data elements in Lipid Bank will also be linked directly to Lipid Search (http://lipidsearch.jp), ARM (Atomic Reconstruction of Metabolites, http://www/metabolome.jp), and KEEG (Kyoto Encyclopedia of Genes and Genomes, http://www.genome.jp/kegg).

3. Strategies for Lipid Identification and Quantitative Analysis by Mass Spectrometry

Several basic factors are required for the correct identification of lipid metabolites, as listed within this section.

3.1. Accurate mass value of molecular weight-related ions

The most important factor is to obtain accurate mass-to-charge ratio (m/z) values of molecular weight-related ions of individual molecules, which preferably have been obtained by high mass resolution. Fourier transform ion cyclotron resonance (ICR) is presently the best instrument for obtaining accurate mass data, followed by time-of-flight MS (TOF-MS), and then quadrupole MS. Less than 100 ppm mass accuracy can be obtained by using the quadrupole machine. Furthermore, less than 20 ppm and less than 3 ppm can be obtained by TOF-MS and FT-MS, respectively, even with external calibration. It is very difficult to separate the mass peak of target molecule from the close mass peaks of other molecules, from those with different numbers of carbon chains and double bonds, or from those with different numbers of oxygen molecules. To discriminate, for instance, nearly 0.03 mass units observed as differences in those of alkyl-acyl or alkenyl-acyl species and odd-number diacyl species, resolution of more than 60,000 is required (Ishida et al., 2004b). Even for discrimination of mass differences, less than 20 ppm mass accuracy is probably required. Thus, to obtain fully reliable identification of these peaks both for resolution and mass accuracy, FT-MS is typically required; TOF-MS is required only for discriminating the differences in m/z values. Effective separation of the peaks would be required for the correct identification by TOF-MS.

3.2. Characteristic fragment ions

The second requirement in correct identification is to obtain characteristic fragment ions from molecular weight-related ions of individual metabolites. In this case, even unit mass accuracy of fragment ions is significantly effective for identification of the characteristic structure of precursor ions. In proteomics, a common and effective method for identifying proteins from fragment data of each trypsin hydrolyzed peptide uses the fragment ions, such as b- and y-type series. Similarly, fragments of fatty acyl anions from the *sn*-1 and *sn*-2 positions in the negative ion mode and neutral loss of ketenes or fatty acids observed in the negative or positive ion mode are very important for identification of individual molecular species of glycerolipids and glycerophospholipids (Hsu and Turk, 2000, 2003; Ramanadham et al., 1998). However, high mass accuracy in fragment data obtained from TOF-MS or FT-MS is very helpful, especially where target metabolites are unknown, new, or not yet identified. Fragment ions of polar headgroups or specific neutral loss from each class of phospholipids and glycerolipids are also very important for identification of each different category of polar lipids. In some cases, fragments obtained in MS^3 or MS^4 spectra would be required for more reliable identification. High coverage of detected fragments results in more accurate identification.

3.3. Good separation and reproducible retention time in liquid chromatography

Liquid chromatographic separation is also important and necessary for accurately identifying and quantifying isobaric molecules with identical molecular elements. However, to identify molecular species with close m/z values, which are different molecular compositions within 0.1 mass units, effective separation of these molecules by LC is also required. In particular, very small amounts of lipid molecules, or those with low ionization efficiency, are difficult to detect and identify without separating them from major metabolites or from those with high ionization efficiency. Identification of a number of lipid species can be greatly improved by good separation of lipid molecules in LC (Houjou et al., 2005; Kakela et al., 2003; Merrill et al., 2005;

Reis et al., 2005; Taguchi et al., 2003; Wang et al., 2005). Recent column technology using less than 2-μm particles with high pressure and a monolithic column makes it possible to obtain more than 100 peaks within a single LC run. Reproducible and accurate retention time is important for identifying each individual molecule. Once profiles concerning the elution order of isobaric molecules or those of different molecules with close m/z values by LC-MS/MS are established, we can easily use them to predict the molecular species, even without MS/MS data.

The methods introduced here were created by using a combination of these strategies to acheive the objectives of the specific researchers.

4. SEVERAL PRACTICAL LIPIDOMICS METHODS BY MASS SPECTROMETRY

There are several different approaches for lipid analysis. We have essentially classified the comprehensive analysis of MS in three different categories: untargeted, focused, and targeted (Fig. 8.1).

4.1. Global and untargeted

The global (i.e., comprehensive) and untargeted method is used to detect all metabolites contained within extracted lipid samples without preliminary information molecular on weight-related ions or their fragments (Houjou et al., 2005; Ishida et al., 2004b; Taguchi et al., 2003).

The strategy of this method is to subject all detected peaks of molecular weight-related ions to further analysis (with or without exact identification methods such as principal component analysis). The untargeted method has been applied to lipidomics by using MS with high resolution, such as FT-MS, without LC separation (Fridriksson et al., 1999; Ishida et al., 2004b; Ivanova et al., 2001; Marto et al., 1995); however, in most cases, this method is applied by using quadrupole or TOF-MS combined with LC-MS or LC-MS/MS (Houjou et al., 2005; Taguchi et al., 2003). A combination of appropriate HPLC separation and MS with high resolution, such as FT-MS or TOF-MS, is preferred, both for the correct identification of known metabolites and the annotation of newly found metabolites. By using normal-phase LC, different classes of phospholipids can be effectively separated, and rough identification is possible from both positive and negative molecular weight-related ions (Taguchi et al., 2003). However, information of retention time is important for obtaining separate identification of isobaric molecular ions, even in the cases of identification by only mass values from molecular weight-related ions.

Another important factor is information of fragment ions obtained by MS/MS data. Thus, the application of LC-MS/MS with data-dependent

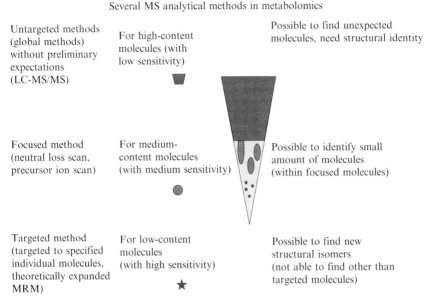

Figure 8.1 Several mass spectrometric analytical methods in metabolomics. High-content metabolites exist in relatively few molecular species, while low content metabolites exist in a large number of species. These three methods (i.e., untargeted, focused, and targeted) can be available in a more or less comprehensive manner. (See color insert.)

scanning mode is a useful and practical method in lipidomics, as it is in proteomics. Data-dependent MS/MS can be performed as a shotgun strategy from molecular weight-related ion peaks with high intensity (Houjou et al., 2005). The use of fragmentation is applicable to several different situations or methods. Methods of obtaining information about fragment ions from targeted molecular weight-related ions are commonly applied where structural confirmation of suspected molecules or structural identification of unknown molecules is required.

4.2. Focused

The focused method is used for detecting molecules in some categories while comprehensively utilizing specific fragments or neutral loss caused by a specific feature of their partial structures (Ekroos et al., 2002, 2003; Han and Gross, 2003; Houjou et al., 2004; Taguchi et al., 2005). In this case, peaks of molecular weight-related ions that are lower than the detection limit of the signal-to-noise (S/N) value in the mass spectrum can be identified effectively and separately from noise peaks. This method is most frequently used in lipidomics, especially for glycerophospholipids and glycerolipids

(Ekroos et al., 2002, 2003; Han and Gross, 2003; Houjou et al., 2004; Taguchi et al., 2005).

The focused method can also be effectively combined with LC (Taguchi et al., 2005). In this case, detection limits of minor components or molecules with low ionic efficiency can be greatly improved by lowering the ion suppression through separate elution from other major ions with high ion efficiency. Data on target fragments or target neutral losses for these surveys are used in the focused method, but m/z of molecular weight-related ions are comprehensively surveyed.

4.3. Targeted

The targeted method can be applied to targeted molecules by using information both for molecular weight-related ions and their specific fragment ions. Thus, each combination of ionic m/z values should be selected for detection of each individual molecule. This popular method is known as selected reaction monitoring (SRM) or multiple reaction monitoring (MRM) and is commonly used by pharmaceutical companies for quantification of target drugs or drug metabolites. Individual m/z data pairs of molecular weight-related ions and characteristic fragment ions are used in the targeted method. Furthermore, scanning for pairs of many theoretically expected m/z values can be used comprehensively as a survey. Comprehensively collected data for hypothetical pairs of characteristic fragment ions and molecular weight-related ions are used for the application of this method in lipidomics. Recent triple-stage quadrupole MS can detect up to 100 pairs of MRM (SRM) in only one LC run. Therefore, we selected this method as a comprehensive analytical technique to detect minor individual, oxidized phospholipid molecular species.

5. STRATEGIES FOR IDENTIFICATION OF INDIVIDUAL MOLECULAR SPECIES IN GLYCEROLIPIDS AND GLYCEROPHOSPHOLIPIDS BY LIPID SEARCH

Once we had developed strategies for analysis, we created our search engine, Lipid Search (http://lipidsearch.jp), which is now available on the web. We constructed this search engine for the identification of individual lipids from law mass data through collaboration with Mitsui Knowledge Industry. Our automated search engine can indicate the most probable candidates for each MS data. We began using this search tool in our laboratory in 2000. In 2002, the search tool known as Lipid Search was opened to the public. The database was constructed with theoretical m/z data of molecular weight-related ions and their fragment ions for all molecular species of phospholipids, fatty acids, glycerolipids, and their metabolites. These databases are theoretically constructed using the fragment data obtained from commercially available standards. In the first version opened in 2002, only text-based Excel files of MS data could be handled.

5.1. Theoretical database and search engine for identification of phospholipid molecular species and their fragments

The new version of Lipid Search, combined with Lipid Navigator, was opened to the public in November 2005 (Figs. 8.2, 8.3, and 8.4). This tool can directly handle law MS data obtained from Q-TOF (Micromass, Waters, Milford, MA), LTQ, LTQ-FT, Quantum-Ultra (Thermo Fisher Scientific, Waltham, MA), 4000 Q Trap (Applied Biosystems, Foster City, CA), and LC-MS-IT-TOF (Shimadzu, Kyoto, Japan). It will be updated in mid-2007 with an expanded database and refined search efficiency. The identification efficiency of Lipid Search highly depends on the qualities of MS data.

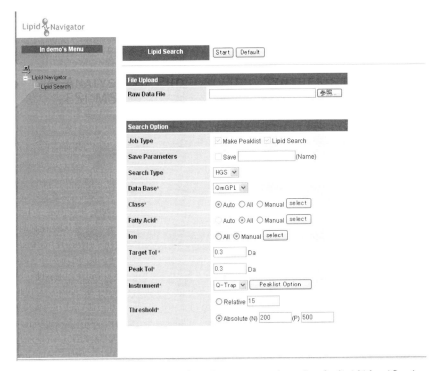

Figure 8.2 Lipid Search with Lipid Navigator, a search engine for lipid identification (http://lipidsearch.jp). The user selects the proper description in each box of the indicated search condition. (See color insert.)

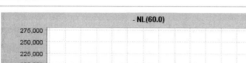
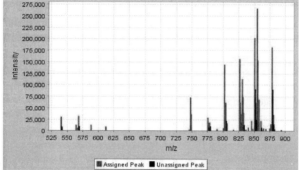

Moleular Species	Ion	Obs m/z(Da)	Calc m/z(Da)	Delta(Da)	Intensity	Intensity(mono)	Fragment	Delta(Da)
LPC,acyl,16:0	+HCOO	540.2	540.07	-0.129	30255.5	39886.9	-	-
PAF,alk,14:0	+HCOO	540.2	540.07	-0.129	30255.5	39886.9	-	-
LPC,acyl,18:2	+HCOO	564.3	564.27	-0.029	13202.4	17793.5	-	-
LPC,acyl,18:1	+HCOO	566.5	566.654	0.154	6601.2	8124.15	-	-
PAF,alk,16:1	+HCOO	566.5	566.654	0.154	6601.2	8124.15	-	-
LPC,acyl,18:0	+HCOO	568.3	568.239	-0.06	31905.8	42648.2	-	-
PAF,alk,16:0	+HCOO	568.3	568.239	-0.06	31905.8	42648.2	-	-
LPC,acyl,20:4	+HCOO	588.2	588.07	-0.129	13202.4	18190.3	-	-
LPC,acyl,22:6	+HCOO	612.4	612.47	0.07	9351.7	13171.9	-	-
SM(SP),amido,16:1	+HCOO	745.5	745.45	-0.049	3300.6	5132.36	-	-
SM(SP),amido,16:0	+HCOO	747.7	747.835	0.135	73163.3	113272.0	-	-
PC,acyl,30:1	+HCOO	748.6	748.687	0.087	36856.7	4403.62	-	-
SM(SP),amido,18:0	+HCOO	775.6	775.603	0.0030	29155.3	46340.8	-	-
PC,acyl,32:1	+HCOO	776.6	776.656	0.056	18153.3	4533.74	-	-
PC,alk,34:7	+HCOO	778.6	778.698	0.098	18703.4	18242.9	-	-
PC,acyl,32:0	+HCOO	778.6	778.64	0.04	18703.4	18242.9	-	-
PC,alk,34:1	+HCOO	790.5	790.404	-0.095	4400.8	4400.8	-	-
PC,acyl,34:3	+HCOO	800.6	800.656	0.056	6601.2	6601.2	-	-
PC,acyl,34:2	+HCOO	802.5	802.44	-0.059	144676.0	144676.0	-	-
SM(SP),amido,20:0	+HCOO	803.3	802.972	-0.327	37956.9	61666.0	-	-
PC,36:8	+HCOO	804.5	804.482	-0.017	62161.3	43585.8	-	-
PC,acyl,34:1	+HCOO	804.5	804.425	-0.074	62161.3	43585.8	-	-
PC,alk,36:7	+HCOO	806.6	806.666	0.066	17053.1	16361.4	-	-
PC,acyl,34:0	+HCOO	806.6	806.609	0.0090	17053.1	16361.4	-	-
PC,alk,36:2	+HCOO	816.7	816.788	0.088	3300.6	3300.6	-	-
PC,acyl,36:4	+HCOO	826.5	826.44	-0.059	156779.0	156779.0	-	-
SM(SP),amido,22:2	+HCOO	827.5	827.372	-0.127	63261.5	105049.0	-	-
PC,acyl,36:3	+HCOO	828.6	828.625	0.025	37406.9	5040.45	-	-

Figure 8.3 Results of Lipid Search with mass spectrometry law data of headgroup survey. Molecular species with summed number of carbon chain and double bond both on *sn*-1 and *sn*-2 are indicated as search results for headgroup survey law data. (See color insert.)

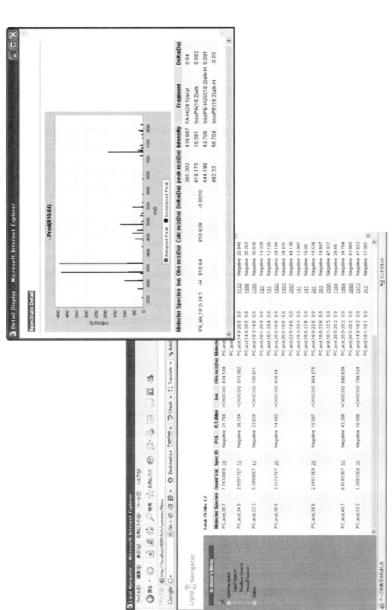

Figure 8.4 Results of Lipid Search for liquid chromatography mass spectrometry law data of molecular weight-related ion survey. Search results of molecular weight-related ion survey law data from liquid chromatography mass spectrometry are also indicated in the same manner as in Fig. 8.3. The user can also obtain results of molecular species with individual pairs of fatty acyl chains on sn-1 and sn-2 positions from law data of fatty acyl survey or product ion survey. (See color insert.)

Several important factors can be selected by the user during the search process concerning the efficiency of identification. These selections are described in this section.

5.1.1. Select the database to which the search should be applied

Our database contains more than 200,000 theoretical m/z values of molecular weight-related ions of individual lipid metabolites and their theoretical fragment ions. Patterns of ion fragments have been theoretically constructed and improved by using experimental data obtained from synthesized standards and various natural samples.

We constructed three differently sized databases for different experiments. The first database contains only molecules with an even number of fatty acyl chains. The second database also contains an odd number of fatty acyl chains, while the third database contains lipid metabolites with oxidized fatty acyl chains. Each user can select one of these three databases according to rationale; however, if the second or third database is selected, a higher number of accidentally fitted results may occur. Therefore, in these cases, a higher mass of accurate data should be used.

5.1.2. Select positive or negative ion state and probable adduct ions

The negative ion mode is normally selected for the identification of the molecular species of glycerophospholipids. Conversely, the positive ion mode is used to identify glycerolipids, such as diacylglycerols and triglycerides.

The type of adduct ions are normally dependent on the eluting conditions and on the solvent used in mass analysis. In our experiments, protons and ammonium are selected as positive adduct ions, and formate or acetate are selected as negative adduct ions, depending on the solvent used. In some cases, Na^+ and K^+ can be selected as positive adducts, and Cl^- can be selected as a negative adduct.

5.1.3. Select mass tolerance or mass error in experimental data

Tolerance of mass data is mainly dependent on the MS equipment and the assay conditions. If a high tolerance value is selected, a large number of probable candidate molecules will be obtained, including those accidentally fitted to the search algorithm. On the other hand, a genuine candidate, which exceeds the selected tolerance value from the MS data, may not be found if too low a tolerance value is selected. Thus, appropriate tolerance values should be selected for each data set. Furthermore, to obtain the correct results, it is very important to compensate in calibration by using the proper standard in the spectrum data. Good results will not be obtained from poor data. Tolerances from 0.2 to 0.4 are normally recommended for the data obtained from quadrupole MS, and 0.02 to 0.1 are recommended for the data from TOF-MS. Tolerances from 0.01 to 0.001 are recommended for the data from FT-MS.

5.1.4. Select mass survey method used

In this version, we set four different survey methods: molecular ion survey (MIS), headgroup survey (HGS), fatty acyl survey (FAS), and product ion survey (PIS). The second and third survey systems were initially constructed for the identification of molecular species of glycerophospholipids; however, these methods are also available for all metabolites with structural similarities. Thus, the second and third systems will be reconstructed as precursor ion and neutral loss surveys in the version available in mid-2007 (http://lipidsearch.jp).

5.1.5. Others

The user must select the name of the mass spectrometer used in the experiments, the mass range for molecular weight-related ions, and the level of mass intensity to be analyzed.

6. Quantitative Analysis or Profiling of Lipid Molecular Species

It has been said that it is difficult to obtain quantitative data from mass analysis. However, we and many researchers engaged in this field consider that quantitation of significant changes in metabolites can be effectively obtained by simple compensation of profile data. The problem of quantifying mass data can often arouse stern criticism from two sides—those who are accustomed to MS and those who work outside this field. In the field of biological sciences, errors of less than 30% absolute and less than 15% relative are sufficient to recognize some specific changes in metabolic profiling. Thus, even simple profiling of lipid molecular species seems to be sufficiently useful to discover new candidates relating to the physiological phenomenon of interest.

Most problems in quantitative analysis exist due to several differences, such as the variations between individual animals, the sampling conditions such as time-dependent alterations, and circumstances such as food supply.

Several factors can affect differences in ionization efficiency of phospholipids and glycerolipids. Ionic efficiencies mainly depend on each class of lipids and the experimental conditions. Ionic efficiency largely depends on the structure of polar headgroups within the molecules. Various headgroups, such as choline phosphate, ethanolamine phosphate, serine phosphate, and inositol phosphate, greatly affect ionic efficiency. The balance of polar groups and nonpolar groups is also an issue, especially at higher concentrations.

In the new version of Lipid Search, quantitative data will be mainly obtained from peak areas of molecular weight-related ions found in positive or negative ions. These data will be combined with the identification data obtained from data-dependent MS/MS data.

Essentially, for exact compensation of relative ionization efficiency in individual molecular species, use of individual standards containing stable isotopes is most accurate; however, these approaches are difficult both technically and economically because more than several hundred molecules exist even in single LC-MS/MS analysis for glycerophospholipids or glycerolipids.

We normally use multiple compensation methods to obtain quantitative profiles of lipids for each research strategy, all of which are described in this section.

6.1. Compensation by a few selected internal standards for correcting relative ionization efficiency

We typically use peak area data of molecular weight-related ions for quantitative profiling in lipid molecular species of glycerophospholipids, glycerolipids, sphingomyelins, and their metabolites. In this case, peak area data were subjected to compensation of the peak ratio of isotope patterns calculated by molecular elements obtained after automatic identification by Lipid Navigator. This compensation can also be applied automatically to the peak area obtained from MS analysis by precursor ion scanning and neutral loss scanning.

6.2. Compensation with limited number of proper internal standards

Peak area of molecular weight-related ions obtained by polar group–related precursor ion scanning or neutral loss scanning is also useful. Ionization efficiency is largely influenced by the nature of the polar headgroups. Thus, ionization efficiency is almost identical where molecules have the same polar headgroups, such as those in the same class of phospholipids. We normally add diacyl 12:0–12:0 PE as an internal standard, and separately compensate ionic efficiency in both negative and positive ion modes by using the same amount of diacyl 16:0/16:0 or diacyl 16:0/18:1 species for each class of phospholipids (PC, PE, PS, PI, PG).

6.3. Compensation with stable isotope-labeled standard

After obtaining the target molecules or locating where an exact target is to be quantified, the most accurate and commonly selected method of compensation uses a stable isotope-labeled standard. Cold multireaction monitoring, or selected ion monitoring, is the most popular quantitative analysis method by MS used by pharmaceutical companies. Appropriately selected combinations of m/z values of molecular weight-related ions and their fragments are important for obtaining accurate quantitative data. Compensation with a stable isotope-labeled standard can also be useful for compensation of the peak areas of molecular weight-related ions obtained by precursor ion scanning or neutral loss scanning of polar head–related mass. Stable-labeled positions of the element should be carefully selected for this purpose, such as those within targeted fragments or escaping from specific fragments.

7. Application of Several Different Methods in Lipidomics

7.1. Global and comprehensive analyses in lipidomics under non-targeted strategies

7.1.1. Identification by FT-MS without liquid chromatographic separation

When using FT-MS, an accurate mass of less than 2 ppm is used as effective annotation. High mass accuracy and high separation ability obtained from FT-MS give very limited combinations of molecular elements. Accurate identification of metabolites can be difficult using only m/z values of the molecular weight-related ions, even with FT-MS, and it is more effective in combination with separation by LC or capillary electrophoresis (CE). Yet, data-dependent MS/MS in combination with LC is difficult when using FT-MS because of the relatively longer acquisition time in MS/MS; however, many isobaric molecular species in natural metabolites and, in particular, various pairs of fatty acyl chains on sn-1 and sn-2 in glycerolipids cannot be determined without MS/MS data. Hence, obtaining data of fragment ions by MS/MS for each individual molecule is important for identification.

Several different molecular ions exist that have close m/z values within 0.1 mass units or mass difference data normally obtained by quadrupole MS. The separation of these ions can only be obtained after separation by HPLC in the case of quadrupole TOF. Mass resolution of more than 100,000 and mass accuracy within 2 ppm obtained by FT-MS can effectively discriminate molecular species with 0.02 mass units of difference. However, isobaric ions with identical atomic compositions but different structures cannot be effectively identified, even by FT-MS.

7.1.2. Application results of untargeted and comprehensive analysis by LC-MS/MS

Experiment A global lipidomics approach for the comprehensive and precise identification of molecular species in a crude lipid mixture was performed using LTQ Orbitrap and C30 reverse-phase LC (RPLC) (Ishikawa *et al.*, in preparation). LTQ Orbitrap provides high mass accuracy MS spectra via the FT-MS mode and can perform rapid MS^n by ion trap MS (IT-MS) mode. The negative ion mode was selected to detect fragment ions (fatty acids, etc.) from phospholipids by MS/MS or MS^3. The data-dependent MS/MS or MS^3 automatically performed analyses of molecular peaks in order of intensity. We established the specific detection approach by MS^3 for the specific detection of phosphatidyl choline (PC) and phosphatidylserine (PS) using IT spectrometry. The identification of molecular species was performed by using both the high mass accuracy of the MS data obtained from FT mode and structural data obtained from fragments in IT mode.

Results Some alkyl-acyl and alkenyl-acyl species have the same m/z value of molecular weight-related ions and fragment ions (e.g., alkyl-acyl 36:2, with m/z 830.6275; alkenyl-acyl 36:1, with m/z 830.6275). After direct acid hydrolysis analysis was performed, the RPLC-LTQ Orbitrap method was applied to identify alkyl-acyl and alkenyl-acyl species. As a result, 276 species from six different classes of phospholipids in mouse brain and 310 species in mouse liver were identified (Ishikawa *et al.*, in preparation). Furthermore, some novel molecular species were identified using this RPLC-LTQ Orbitrap method. The ion intensities of each molecular weight-related ion were used for quantitative profiling of molecular species of phospholipids after appropriate compensation by standard phospholipids of identical classes through Lipid Search.

7.2. Application of focused methods by mass spectrometry for lipidomics

The methods for focused lipidomics by MS were examined. Precursor ion scanning and neutral loss scanning of the polar headgroups of phospholipids were valuable techniques for detecting most molecular species within the same class of phospholipids (Ekroos *et al.*, 2002, 2003; Han and Gross, 2003; Houjou *et al.*, 2004; Taguchi *et al.*, 2005). Furthermore, the data were useful for profiling these lipids in biological samples. Precursor ion scanning of carboxylic anions was also very useful for identifying molecular species of phospholipids with specified fatty acyl chains (Ekroos *et al.*, 2002, 2003; Han and Gross, 2003; Houjou *et al.*, 2004; Taguchi *et al.*, 2005).

The automatic, programmed analysis systems for the scanning modes, along with the automatic identification Lipid Search tool, allowed for identification of more than 200 molecular species of phospholipids within 2 h of extraction from total phospholipids (Taguchi et al., 2005).

7.2.1. Application of focused lipidomics with flow injection

Experiment Precursor ion scanning and neutral loss scanning were performed on a 2000 Qtrap with flow injection at 4 μl/min. Optimal conditions for collision-induced decay (CID) were selected by individual fragments or neutral loss (Houjou et al., 2004; Taguchi et al., 2005). Optimal conditions for the detection of appropriate precursor ions and neutral losses were obtained by MS/MS analyses of each phospholipid preliminary class. Automatic and programmed scanning were performed for each class of phospholipids. Precursor ion scanning at m/z 184 was used for choline-containing phospholipids in the positive ion mode. Neutral scanning of 141, 185, 189, and 277 u were used for PE, PS, phosphatidylglycerol (PG), and phosphatidylinositol (PI), respectively. In the negative ion mode, neutral loss scanning of 60 (loss of $HCOO+CH_3$) and 87 u (loss of serine $-H_2O$) was used for choline-containing phospholipids and serine-containing phospholipids, respectively. Finally, precursor ion scanning at m/z 153 and 241 in the negative ion mode was used for glycerol-containing phospholipids and inositol-containing phospholipids, respectively. Quantitative profiling of identical phospholipid molecular species was also obtained by focused methods with specified precursor scanning or neutral loss scanning of headgroup-related mass (Taguchi et al., 2005).

Results Precursor ion scanning and neutral loss scanning for polar headgroups were performed on the phospholipid mixture extracted from THP-1 cells (Taguchi et al., 2005). The major individual molecular species of each class of phospholipids was effectively detected by the m/z values of the specific fragment and neutral loss for each polar headgroup (Fig. 8.5) (Taguchi et al., 2005). Precursor ion scanning of carbonic anions in the negative ion mode was used for the identification of the molecular species having specified fatty acyl moieties (Fig. 8.6) (Taguchi et al., 2005). The data obtained from these two different methods were compared in both qualitative and quantitative aspects. Neutral loss scanning and precursor ion scanning are more effective for the identification and quantitation of the minor class of phospholipids, such as PS. Furthermore, they are also useful for the comprehensive analysis of categorical metabolites with structural similarities.

7.2.2. Combination analysis by reverse-phase LC-MS/MS surveyed with precursor ion scanning or neutral loss scanning

We recently constructed combination methods for group-specific surveys and LC separation (Ishikawa et al., in preparation). This method is more sensitive than normal headgroup scanning without LC. A combination of

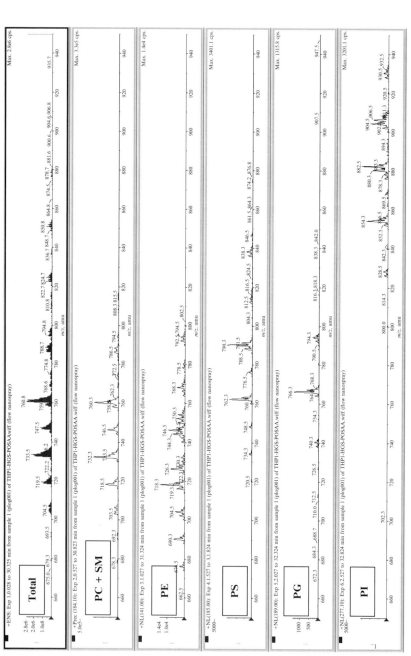

Figure 8.5 Mass spectra of phospholipid molecular species obtained by precursor ion scanning and neutral loss scanning of each headgroup of phospholipids in the positive ion mode (Taguchi et al., 2005). These are application results of a headgroup-specific survey (HGS) for phospholipids extracted from THP-1 cells. Choline-containing lipids such as PC and SM, and PE, PS, PG, and PI were separately detected without liquid chromatographic separation by specific precursor or neutral loss caused by class-specific structure. (See color insert.)

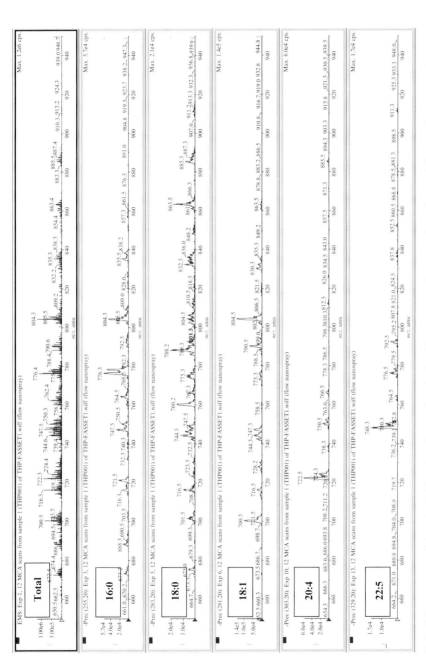

Figure 8.6 Mass spectra of phospholipid molecular species obtained by precursor ion scanning of several fatty acyl anions in the negative ion mode (Taguchi et al., 2005). These results were obtained by fatty acid–specific survey (FAS) for same samples. These data were automatically obtained by applying mass spectrometry law data to our search engine, as described in the text. (See color insert.)

headgroup scanning with reverse-phase separation is particularly effective. We identified very small components. A combination of RPLC-MS/MS and neutral loss scanning is effective for the detection of individual molecular species with low amounts of a focused class of phospholipids at the level of exact fatty acyl pairs. This method is applicable without preliminary separation of phospholipid classes. Major molecular species can be analyzed without LC separation; however, in this case, precursor ions sometimes contained other classes of phospholipids with close molecular weight-related m/z. Thus, the fragments obtained were contaminated and identification of exact molecular species was difficult. Therefore, we combined RPLC-MS/MS and neutral loss scanning for exact identification of minor molecular species.

Experiment The ESI-MS analyses were performed using a TSQ Quantum triple quadrupole mass spectrometer (Thermo Electron, Waltham, MA) equipped with an ESI source. The online chromatographic separations were performed isocratically at room temperature using the HPLC system (Gilson, Middleton, WI) and a Deverosil C30 (Nomura Chemicals, Aichi, Japan) reverse-phase column (150 × 0.3 mm i.d., 3-μm particles). The solvent consisting of acetonitrile/methanol 1:1 (0.1% acetic acid, 0.1% ammonium, 0.2% H_2O) was eluted at 5 μl/min. The ESI-MS and ESI-MS/MS analyses of column eluent were performed using a TSQ Quantum triple quadrupole mass spectrometer.

Two types of ESI-MS/MS surveys—precursor ion scan and neutral loss scan for HGS or FAS—were performed in the positive or negative ion mode. Optimum conditions for collision-induced dissociation (CID) were selected for each individual ion fragment and neutral loss (NL) to detect the appropriate precursor ions. NL was obtained by preliminary MS/MS analysis of examples of each phospholipid class. In the positive ion mode, precursors of m/z 184.1 were used to detect choline-containing phospholipids (PC, lysoPC, sphingomyelin [SM]), and the NL of 141.0 u was used to detect ethanolamine-containing phospholipids (PE, lysoPE). Protonated ions $[M + H]^+$ are the main type of ionized PC, PE, and SM. Collision energy of 25 to 35 eV was applied, and nitrogen was used as the collision gas at 0.8 mTorr. In the negative ion mode, precursor ion scanning for carboxylate anions was applied for the identification of the molecular species containing specified fatty acyl moieties (16:0, 16:1, 18:0, 18:1, 18:2, 20:4, 22:5, and 22:6). Acetate adduct ions, $[M + CH_3COO]^-$, are the main type of ionized choline-containing phospholipids. Deprotonated ions, $[M - H]^-$, are the main type of ionized PE and other phospholipids.

Results We applied the RPLC-ESI-MS/MS analysis with HGS in the positive ion mode for focused detection of choline- and ethanolamine-containing phospholipids in crude lipid extracts from rat liver. Ionized phospholipids were

Analytical Systems for Lipidomics in Japan 205

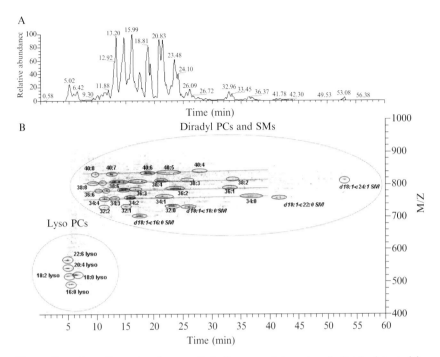

Figure 8.7 Two-dimensional map of choline-containing phospholipids detected by reverse-phase liquid chromatography electrospray ionization mass spectrometry with headgroup survey in the positive ion mode. (A) Total ion chromatogram of precursor for m/z 184. (B) Two-dimensional map. By reverse-phase liquid chromatography/electrospray ionization tandem mass spectrometry analysis, approximately 80 peaks were detected. (See color insert.)

detected as protonated ions ($[M + H]^+$) under used mobile phase. The results of choline-containing phospholipid separations in reverse-phase LC are shown in Fig. 8.7. The separation was observed by creating a two-dimensional (2D) map, where the m/z value was set at the vertical axis and retention time was set at the horizontal axis. Approximately 110 molecules were detected by this RPLC-ESI-MS/MS method. Among choline-containing phospholipids, lysoPCs were eluted first at 4 to 7 min, and then diradyl PCs at 10 to 40 min, and finally SMs. PCs and SMs were detected by precursor ion scanning of m/z 184.1. However, all PCs had even m/z values and all SMs had odd values. Therefore, PCs and SMs could be identified separately by confirming retention times of isotopic peak PCs. Thus, this RPLC-ESI-MS/MS analysis method enabled focused detection of different classes of phospholipids, such as PCs and PEs, in a single LC run.

The RPLC-ESI-MS/MS analysis with focused FAS (16:0, 16:1, 18:0, 18:1, 18:2, 20:4, 22:5, and 22:6) was performed in the negative ion mode to identify fatty acyl chains in the peak detected by HGS analysis. Ionized PCs and SMs were detected as acetate adduct ions $[M + CH_3COO]^-$, and ionized PEs were detected as depronated ions [M − H]⁻ under used mobile phase in the negative ion mode. Therefore, the difference in m/z values between positive ions and negative ions is 58 u ([M + H]⁺/[M + CH$_3$COO]⁻) for choline-containing phospholipids, and 2 u ([M + H]⁺/[M − H]⁻) for ethanolamine-phospholipids. As a result, approximately 80 molecular species were identified in eight major fatty acyl fragments.

Molecular species that differed in their pairs of fatty acyl chains were identified by the RPLC-ESI-MS/MS analysis with FAS. The total numbers of acyl carbons and unsaturated bonds contained in these different molecules were the same. Furthermore, an elution property was observed, which depended on unsaturated bonds of molecular species (Fig. 8.8). Retention time of each of the following peaks was plotted on the horizontal axis: saturated (u0), 14:0–16:0/16:0–16:0/16:0–18:0 diacyl PCs; one unsaturated bond (u1), 16:0–16:1/16:0–18:1/18:0–18:1 diacyl PCs; two unsaturated bonds (u2), 14:0–18:2/16:0–18:2/18:0–18:2 diacyl PCs; four unsaturated bonds (u4), 14:0–20:4/16:0–20:4/18:0–20:4 diacyl PCs; and six unsaturated bonds (u6), 16:0–22:6/18:0–22:6/20:0–22:6. The m/z values were plotted on the vertical axis. As a result, the secondary approximation curves were obtained for each unsaturated bond plot. Molecular species of PCs

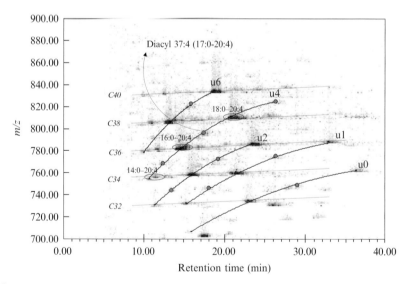

Figure 8.8 Elution characteristics depending on the number of fatty acyl chains and unsaturated bonds. u0, diacyl 14:0–16:0/16:0–16:0/16:0–18:0; u1, diacyl 16:0–16:1/16:0–18:1/18:0–18:1; u2, diacyl 14:0–18:2/ 16:0–18:2/18:0–18:2; u4, diacyl 14:0–20:4/16:0–20:4/18:0–20:4; u6, diacyl 14:0–22:6/16:0–22:6/18:0–22:6. The molecules, which are marked with a closed circle, were supposed to be diacyl molecular species with odd number acyl chains, not even number alkyl- or alkenyl-acyl species, from their retention time. (See color insert.)

with odd numbered fatty acyl chains were effectively identified in the same unsaturated categories with this plot.

7.3. Application of targeted method using expanded MRM for oxidized phospholipids

7.3.1. Experiment

By using data-dependent enhanced product ion scanning (EPI) after the MRM survey, precise data of fragment ions from the detected precursor can be obtained by the high sensitive linear ion trap efficiency of 4000 Q TRAP (Iida *et al.*, in preparation). Mass chromatograms of characteristic product ions can be obtained by continuous product ion scanning. This allows separation of structural isomers that could not be separated by MRM with HPLC. For this reason, we directly obtained the fragment data from interesting peaks by continuous EPI experiments for listed analytes.

In this experiment, non-oxidized and oxidized (ox) PCs were detected as several different pairs of precursors and fragments (Harrison *et al.*, 2000; Murphy *et al.*, 1999; Sun *et al.*, 2006). Normally, detection was operated in negative ion mode for acquiring effective fragments of fatty acyl anions. Precursor ions of PCs were mainly detected as formate adducts, except for carboxylic derivatives of oxPCs. The first product ion for MRM pairs used was m/z 255.2, fatty acyl anion of 16:0 at the *sn*-1 position. In the case of 16:0–18:2 (+nO) PCs, the fatty acyl anion of 16:0 (m/z 255.2) at the *sn*-1 position was used as the first product ion for MRM pairs. These MRM pairs are common to all non-oxidized and oxidized derivatives of 16:0–18:2 (+nO). Thus, the anion seems to be useful for quantitative analyses of oxidized PCs. Fragment ions from the oxidized form of 18:2 (+nO), such as m/z 295.2 (+O), 311.2 (+2O), 327.2 (+3O), and 343.2 (+4O), were used as second fragments for MRM pairs. Those with a loss of H_2O from *sn*-2 oxidized form were used third.

7.3.2. Results

In this investigation, targeted molecular species are those containing two hydroxyl, ketene, or epoxy groups, or one hydroperoxyl group. In MRM survey, the formate adduct of 16:0–18:2 (+2O) as precursor ions was selected, and *sn*-1 acyl anion m/z 255.2 and *sn*-2 oxidized acyl anion m/z 311.2 and 293.3 for product ions were selected. The ion at m/z 293.2 was thought from preliminary experiments to occur from the loss of one H_2O from the *sn*-2 fatty acyl anion. These molecular species were also detected in multiple peaks on the chromatogram.

By EPI analyses, fragmentation patterns obtained from oxFA were applied. For dihydroxyl derivative oxPCs, fragmentation patterns of 9-hydroxyoctadecadienoic acid (HODE) and 13-HODE were applied. In consideration of the presence of another oxygen group, fragment ions such as m/z

171.1 and 211.1 are expected. For hydroperoxyl derivative oxPCs, the fragment ions of 9-hydroperoxyoctadecadienoic acid (HPODE) (m/z 185.1) and 13-HPODE (m/z 195.1) were applied. By reconstructing the mass chromatograms of these four fragment ions, the real existence of these molecular species was assumed. Multiple peaks are still detected for each ion fragment, and might be E and Z isomers. Furthermore, we can use these ion fragments for future MRM experiments.

8. FUTURE PROGRAM FOR LIPIDOMICS

8.1. MassBank: Mass fragment database for metabolomics in Japan's BIRD project

In December 2006, mass data of metabolomics, including molecular weight-related ions and isotope patterns and their fragments, were made publicly available by the Consortium for Mass Data of Metabolomics via MassBank (http://massbank.jp/index-e.html). Data concerning lipid metabolites will be also open to the public in early summer 2007 on the same website. Mammalian lipid real MS data will be supplied and maintained by Taguchi's Lab through the same database software. This database will link to MassBank and include several thousand real data concerning mammalian lipids and lipid metabolites. Those in plants, such as terpenoids, carotenoid, and polyphenols, will also be accessible in the already public data of soluble metabolites from MassBank. All individual MS data included in MassBank will link to Lipid Bank, Lipid Search, ARM, KEEG, LIPID MAPS, and PubChem databases.

8.2. Flux analysis by stable isotope labeling

Using stable isotope labeling and following the half-life after dilution seem to be useful for clarifying fast turnover rates of lipid metabolites (Bleijerveld et al., 2006; Enjalbal et al., 2004). Stable isotope labeling of fatty acyl or polar groups is mapped using special software constructed in ARM to follow the metabolites. Several groups (Keio University and Tokyo University) have started this kind of work in Japan.

8.3. Mass imaging of lipids by MALDI-MS and TOF-SIMS

As is already well known, phospholipids seem to be good targets for mass imaging because of their hydrophobic nature and special localization in the membrane. Recent data on the localization of lipid molecular species in mouse brain have revealed that specific localized functions or events should exist on special domains of mammalian tissue (Garret, et al., 2006; Sjovall *et al.*, 2004). Discrimination between co-localization of lipid metabolites and related proteins, such as lipid metabolizing enzymes or receptor of lipid ligand, is very important for elucidating valid localized positions for specific cell events. MALDI (matrix-assisted laser desorption/ionization)-MS and SIMS (secondary ion MS) are commonly used for this purpose. We will collaborate with two major groups, Dr. Seto's group at the National Institute for Physiological Sciences, Okazaki and Dr. Naitou's group at the Graduate School for the Creation of New Photonics Industries, Hamamatsu. These two groups started their projects for Imaging MS in Japan.

ACKNOWLEDGMENTS

We thank all the members of the Department of Metabolome in Tokyo University, especially for the experimental supports by Dr. Toshiaki Houjou in global analytical system by LCMS/MS, Mr. Masaki Ishikawa in focused analysis combination with LC, and Mr. Yasuhiro Iida in targeted analysis with expanded MRM methods. We also thank all other collaborators, especially Mr. Yasuhito Yokoi, Mitsui Knowledge Industry, for the construction of our search engine Lipid Search in combination with Lipid Navigator.

This study was performed with the help of Special Coordination Funds of JST, CREST from the Japanese government.

REFERENCES

Bleijerveld, O. B., Houweling, M., Thomas, M. J., and Cui, Z. (2006). Metabolipidomics: Profiling metabolism of glycerophospholipid species by stable isotopic precursors and tandem mass spectrometry. *Anal. Biochem.* **352,** 1–14.

Brugger, B., Erben, G., Sandhoff, R., Wieland, F. T., and Lehmann, W. D. (1997). Quantitative analysis of biological membrane lipids at the low picomole level by nanoelectrospray ionization tandem mass spectrometry. *Proc. Natl. Acad. Sci. USA* **94,** 2339–2344.

Ekroos, K., Chernushevich, I. V., Simons, K., and Shevchenko, A. (2002). Quantitative profiling of phospholipids by multiple precursor ion scanning on a hybrid quadrupole time-of-flight mass spectrometer. *Anal. Chem.* **74,** 941–949.

Ekroos, K., Ejsing, C. S., Bahr, U., Karas, M., Simons, K., and Shevchenko, A. (2003). Charting molecular composition of phosphatidylcholines by fatty acid scanning and ion trap MS^3 fragmentation. *J. Lipid Res.* **44,** 2181–2192.

Enjalbal, C., Roggero, R., Cerdan, R., Martinez, J., Vial, H., and Aubagnac, J. L. (2004). Automated monitoring of phosphatidylcholine biosyntheses in *Plasmodium falciparum* by electrospray ionization mass spectrometry through stable isotope labeling experiments. *Anal. Chem.* **76,** 4515–4521.

Fenn, J. B., Mann, M., Meng, C. K., Wong, S. F., and Whitehouse, C. M. (1989). Electrospray ionization for mass spectrometry of large biomolecules. *Science* **246,** 64–71.

Fridriksson, E. K., Shipkova, P. A., Sheets, E. D., Holowka, D., Baird, B., and McLafferty, F. A. (1999). Quantitative analysis of phospholipids in functionally important membrane domains from RBL-2H3 mast cells using tandem high-resolution mass spectrometry. *Biochemistry* **38,** 8056–8063.

Garret, J., Prieto-Conaway, M. C., Kovtoun, V., Bui, H., Izgarian, N., Stafford, G., and Yost, R. A. (2006). Mass spectrometric imaging of lipids in brain tissue. *Int. J. Mass Spectrom.* doi:10.1016/j.ijims.2006.09.019.

Han, X., and Gross, R. W. (1994). Electrospray ionization mass spectroscopic analysis of human erythrocyte plasma membrane phospholipids. *Proc. Natl. Acad. Sci. USA* **91,** 10635–10639.

Han, X., and Gross, R. W. (2003). Global analyses of cellular lipidomes directly from crude extracts of biological samples by ESI mass spectrometry: A bridge to lipidomics. *J. Lipid Res.* **44,** 1071–1079.

Han, X., and Gross, R. W. (2005). Shotgun lipidomics: Electrospray ionization mass spectrometric analysis and quantitation of cellular lipidomes directly from crude extracts of biological samples. *Mass Spectrom. Rev.* **24,** 367–412.

Harrison, K. A., Davies, S. S., Marathe, G. K., McIntyre, T., Prescott, S., Reddy, K. M., Falck, J. R., and Murphy, R. C. (2000). Analysis of oxidized glycerophosphocholine lipids using electrospray ionization mass spectrometry and microderivatization techniques. *J. Mass Spectrom.* **35,** 224–236.

Houjou, T., Yamatani, K., Nakanishi, H., Imagawa, M., Shimizu, T., and Taguchi, R. (2004). Rapid and selective identification of molecular species in phosphatidylcholine and sphingomyelin by conditional neutral loss scanning and MS^3. *Rapid Commun. Mass Spectrom.* **18,** 3123–3130.

Houjou, T., Yamatani, K., Nakanishi, H., Imagawa, M., Shimizu, T., and Taguchi, R. (2005). A shotgun tandem mass spectrometric analysis of phospholipids with normal-phase and/or reverse-phase liquid chromatography/electrospray ionization mass spectrometry. *Rapid Commun. Mass Spectrom.* **19,** 654–666.

Hsu, F. F., and Turk, J. (2000). Characterization of phosphatidylinositol, phosphatidylinositol-4-phosphate, and phosphatidylinositol-4,5-bisphosphate by electrospray ionization tandem mass spectrometry: A mechanistic study. *J. Am. Soc. Mass Spectrom.* **11,** 986–999.

Hsu, F. F., and Turk, J. (2003). Electrospray ionization/tandem quadrupole mass spectrometric studies on phosphatidylcholines: The fragmentation processes. *J. Am. Soc. Mass Spectrom.* **14,** 352–363.

Ishida, M., Imagawa, M., Shimizu, T., and Taguchi, R. (2004a). Specific detection of lysophosphatidic acids in serum extracts by tandem mass spectrometry. *J. Mass Spectrom. Soc. Jpn.* **53,** 25–32.

Ishida, M., Yamazaki, T., Houjou, T., Imagawa, M., Harada, A., Inoue, K., and Taguchi, R. (2004b). High-resolution analysis by nano-electrospray ionization Fourier transform ion cyclotron resonance mass spectrometry for the identification of molecular species of phospholipids and their oxidized metabolites. *Rapid Commun. Mass Spectrom.* **18,** 2486–2494.

Ivanova, P. T., Cerda, B. A., Horn, D. M., Cohen, J. S., McLafferty, F. W., and Brown, H. A. (2001). Electrospray ionization mass spectrometry analysis of changes in phospholipids in RBL-2H3 mastocytoma cells during degranulation. *Proc. Natl. Acad. Sci. USA* **98,** 7152–7157.

Kakela, R., Somerharju, P., and Tyynela, J. (2003). Analysis of phospholipid molecular species in brains from patients with infantile and juvenile neuronal-ceroid lipofuscinosis using liquid chromatography-electrospray ionization mass spectrometry. *J. Neurochem.* **84,** 1051–1065.

Kerwin, J. L., Tuininga, A. R., and Ericsson, L. H. (1994). Identification of molecular species of glycerophospholipids and sphingomyelin using electrospray mass spectrometry. *J. Lipid Res.* **35,** 1102–1114.

Kim, H. Y., Wang, T. C., and Ma, Y. C. (1994). Liquid chromatography/mass spectrometry of phospholipids using electrospray ionization. *Anal. Chem.* **66,** 3977–3982.

Lehmann, W. D., Koester, M., Erben, G., and Keppler, D. (1997). Characterization and quantification of rat bile phosphatidylcholine by electrospray-tandem mass spectrometry. *Anal. Biochem.* **246,** 102–110.

Marto, J. A., White, F. M., Seldomridge, S., and Marshall, A. G. (1995). Structural characterization of phospholipids by matrix-assisted laser desorption/ionization Fourier transform ion cyclotron resonance mass spectrometry. *Anal. Chem.* **67**, 3979–3984.

Merrill, A. H., Jr., Sullards, M. C., Allegood, J. C., Kelly, S., and Wang, E. (2005). Sphingolipidomics: High-throughput, structure-specific, and quantitative analysis of sphingolipids by liquid chromatography tandem mass spectrometry. *Methods* **36**, 207–224.

Murphy, R. C., Khaselev, N., Nakamura, T., and Hall, L. M. (1999). Oxidation of glycerophospholipids from biological membranes by reactive oxygen species: Liquid chromatographic-mass spectrometric analysis of eicosanoid products. *J. Chromatogr. B Biomed. Sci. Appl.* **731**, 59–71.

Pulfer, M., and Murphy, R. C. (2003). Electrospray mass spectrometry of phospholipids. *Mass Spectrom. Rev.* **22**, 332–364.

Ramanadham, S., Hsu, F. F., Bohrer, A., Nowatzke, W., Ma, Z., and Turk, J. (1998). Electrospray ionization mass spectrometric analyses of phospholipids from rat and human pancreatic islets and subcellular membranes: Comparison to other tissues and implications for membrane fusion in insulin exocytosis. *Biochemistry* **37**, 4553–4567.

Reis, A., Domingues, M. R., Amado, F. M., Ferrer-Correia, A. J., and Domingues, P. (2005). Identification of linoleic acid free radicals and other breakdown products using spin trapping with liquid chromatography-electrospray tandem mass spectrometry. *Biomed. Chromatogr.* **19**, 129–137.

Sjovall, P., Lausmaa, J., and Johansson, B. (2004). Mass spectrometric imaging of lipids in brain tissue. *Anal. Chem.* **76**, 4271–4278.

Sun, M., Finnemann, S. C., Febbraio, M., Shan, L., Annangudi, S. P., Podrez, E. A., Hoppe, G., Darrow, R., Organisciak, D. T., Salomon, R. G., Silverstein, R. L., and Hazen, S. L. (2006). Light-induced oxidation of photoreceptor outer segment phospholipids generates ligands for CD36-mediated phagocytosis by retinal pigment epithelium: A potential mechanism for modulating outer segment phagocytosis under oxidant stress conditions. *J. Biol. Chem.* **281**, 4222–4230.

Taguchi, R., Hayakawa, J., Takeuchi, Y., and Ishida, M. (2003). Two-dimensional analysis of phospholipids by capillary liquid chromatography/electrospray ionization mass spectrometry. *J. Mass Spectrom.* **35**, 953–966.

Taguchi, R., Houjou, T., Nakanishi, H., Yamazaki, T., Ishida, M., Imagawa, M., and Shimizu, T. (2005). Focused lipidomics by tandem mass spectrometry. *J. Chromatogr. B* **823**, 26–36.

Wang, C., Kong, H., Guan, Y., Yang, J., Gu Yang, S., and Xu, G. (2005). Plasma phospholipid metabolic profiling and biomarkers of type 2 diabetes mellitus based on high-performance liquid chromatography/electrospray mass spectrometry and multivariate statistical analysis. *Anal. Chem.* **77**, 4108–4116.

Weintraub, S. T., Pinckard, R. N., and Hail, M. (1991). Electrospray ionization for analysis of platelet-activating factor. *Rapid Commun. Mass Spectrom.* **5**, 309–311.

Wenk, M. R., Lucast, L., Di Paolo, G., Romanelli, A. J., Suchy, S. F., Nussbaum, R. L., Cline, G. W., Shulman, G. I., McMurray, W., and De Camilli, P. (2003). Phosphoinositide profiling in complex lipid mixtures using electrospray ionization mass spectrometry. *Nat. Biotechnol.* **21**, 813–817.

CHAPTER NINE

THE EUROPEAN LIPIDOMICS INITIATIVE: ENABLING TECHNOLOGIES

Gerrit van Meer,[*,†] Bas R. Leeflang,[*] Gerhard Liebisch,[‡] Gerd Schmitz,[‡] and Felix M. Goñi[§]

Contents

1. Introduction	214
2. Methods of Lipidomics: Quantitative Analysis in Time	215
2.1. Ongoing technology development	215
2.2. Need for further technology development	217
3. Imaging Lipids	218
4. Methods to Study the Physical Properties of Lipids	219
4.1. Lipids as amphiphiles	220
4.2. Lipid–lipid interactions and mesomorphism	220
4.3. Lipid structural functions	220
4.4. Lipid signaling functions	223
5. Data Handling and Standardization	224
5.1. Software requirements	224
5.2. Existing lipid-related databases	224
5.3. Lipidomics expertise platform	226
6. Perspectives	226
Acknowledgments	227
References	227

Abstract

Lipidomics is a new term to describe a scientific field that is a lot broader than lipidology, the science of lipids. Besides lipidology, lipidomics covers the lipid-metabolizing enzymes and lipid transporters, their genes and regulation; the quantitative determination of lipids in space and time, and the study of lipid function. Because lipidomics is concerned with all lipids and their enzymes and

---

[*] Bijvoet Center, Utrecht University, Utrecht, The Netherlands
[†] Institute of Biomembranes, Utrecht University, Utrecht, The Netherlands
[‡] Institute of Clinical Chemistry and Laboratory Medicine, University Hospital Regensburg, Regensburg, Germany
[§] Unidad de Biofísica (CSIC-UPV/EHU), and Department of Biochemistry and Molecular Biology, University of the Basque Country, Leioa, Spain

genes, it faces the formidable challenge to develop enabling technologies to comprehensively measure the expression, location, and regulation of lipids, enzymes, and genes in time, including high-throughput applications. The second challenge is to devise information technology that allows the construction of metabolic maps by browsing through connected databases containing the subsets of data in lipid structure, lipid metabolomics, proteomics, and genomics. In addition, to understand lipid function, on the one hand we need a broad range of imaging techniques to define where exactly the relevant events happen in the body, cells, and subcellular organelles; on the other hand, we need a thorough understanding of how lipids physically interact, especially with proteins. The final challenge is to apply this knowledge in the diagnosis, monitoring, and cure of lipid-related diseases.

1. INTRODUCTION

The European Lipidomics Initiative was funded by the European Commission as a specific support action over the 2005 to 2006 period. Its task was to mobilize and network the European basic and medical researchers and industry by means of a series of dedicated workshops and a concluding conference on technology development, scientific challenges, and medical applications of lipidomics and the involvement of industry. One aim of the initiative was to create a European technology platform bringing together technological know-how and industry. The second aim was to raise the level of awareness of the potential of lipidomics and to shape the way in which metabolomics research and, in particular, lipidomics research is organized in Europe. Finally, the initiative aimed at defining the most suitable ways in which the new technologies can be applied in the clinical setting to assess, cure, and prevent membrane lipid disorders.

Towards these goals, four dedicated workshops and meetings have been organized: (1) ELIfe workshop on Lipid Mass Spectrometry, Dresden, Germany, May 20 to 21, 2005, (2) ELIfe/EMBO workshop "Dynamic Lipid Organization in Cells," Bilbao, Spain, June 3 to 7, 2006, (3) ELIfe/ICBL (International Conference on the Bioscience of Lipids) joint workshop "Lipidomics and Health," Pécs, Hungary, September 5 to 10, 2006, and (4) ELIfe/FEBS special meeting "New Concepts in Lipidology: From Lipidomics to Disease," Noordwijkerhout, the Netherlands, October 21 to 25, 2006. In addition, a number of papers were published to inventory the field and its challenges (Fahy et al., 2005; Griffiths, 2006; Helms, 2006; Spener et al., 2006a,b; van Meer, 2005, 2006; Varfolomeyev et al., 2005). Finally, to support the field, a lipidomics expertise platform was created as an interactive database on the web (www.lipidomics-expertise.de), with access to an interactive white paper based on the Wikipedia software on lipidomics.

Where lipidology is the general science of lipids, the term "lipidomics" has a different connotation. Lipidomics covers everything connected to the study of the lipidome, the total complement of lipids. As such, lipidomics covers not only the quantitative determination of all lipids in space and time, but also includes the study of the lipid-metabolizing enzymes and lipid transporters, which determine the local concentration, their genes, and regulation. Finally, lipidomics concerns lipid function, which cannot be understood without a thorough understanding of the physical basis of their behavior, especially of lipid–lipid and lipid–protein interactions. Because lipidomics is part of metabolomics, which aims at a complete description of the full metabolome, there is no need for a stringent definition of lipids (Fahy *et al.*, 2005). This description of lipidomics defines what types of methods are commonly applied to this field.

First of all, lipidomics requires methods that can quantitatively analyze the total complement of lipids. It is the precipitous development of the various mass spectrometric methods that is making this possible. These methods are highly sensitive and can be used in high-throughput mode. Data obtained this way must be correlated with related data obtained at the protein, mRNA, and gene level—the relevant enzymes of lipid synthesis, degradation, and transport have been identified using genetic, molecular biological, and data mining approaches. Their expression pattern and regulation by protein–protein interactions and posttranslational modifications are measured in time using genomics and proteomics approaches. Novel sophisticated bioinformatics approaches must be developed in a global harmonizing effort in each area and to cross-reference the different types of data. Standard operation protocols must be developed for applying these techniques in the clinic.

2. Methods of Lipidomics: Quantitative Analysis in Time

2.1. Ongoing technology development

The introduction of soft ionization techniques like electrospray ionization (ESI) revolutionized the field of lipid analysis by making polar lipids accessible to mass spectrometric analysis. Now, numerous methodologies have been developed for both the structural analysis of complex lipids as well as quantitative profiling of complex lipid mixtures (for reviews in phospholipids, see Han and Gross, 2005; Pulfer and Murphy, 2003; in sphingolipids, see Merrill *et al.*, 2005; in glycosphingolipids, see Levery, 2005; in prostaglandins, see Murphy *et al.*, 2005). In the late 1990s, it becomes clear that nano-ESI tandem mass spectrometry (ESI-MS/MS) permits a direct quantification of the major phospholipid classes from a minute amount of material without separation of crude lipid extracts (Brügger *et al.*, 1997). Fragments generated from the phospholipid headgroups during collision-induced dissociation (CID) were used to assign a species to a phospholipid class. However, such headgroup scanning only permits the detection of the sum of both fatty acid residues for glycerophospholipids. For example, a phosphatidylcholine 38:5 containing 38 carbons and 5 double bonds in the fatty acid moiety may correspond to various fatty acid combinations, such as 16:0/22:5, 18:0/20:5, 18:1/20:4, etc. The distinct fatty acids can be analyzed in negative ion mode using acyl anions (fatty acid scan). A combination of headgroup and fatty acid scanning facilitates the analysis of single glycerophospholipid species (Ekroos *et al.*, 2002). Although, in principle, triple quadrupole instruments would be suitable for such an analysis, numerous different scans would be necessary. An ideal platform for such an approach is a hybrid quadrupole time-of-flight mass spectrometer, which allows the parallel recording of all fragment ions (Ekroos *et al.*, 2002).

Mass spectrometric analysis could be performed with a high sample throughput either by direct flow injection analysis (Liebisch *et al.*, 2002, 2004, 2006) or using an automated nanospray chip ion source (Ejsing *et al.*, 2006; Schwudke *et al.*, 2006; Zamfir *et al.*, 2004). For performing high throughput quantification of lipid species the automated analysis of data is a prerequisite: First, it is necessary to correct the raw data for an overlap of isotopes resulting from adjacent species (e.g., for phosphocholine containing phosphatidylcholine and sphingomyelin in a precursor ion scan of m/z 184) (Liebisch *et al.*, 2004). Second, species assignment needs software assistance especially to handle the fragment information received from quadrupole time-of-flight mass spectrometry (Ekroos *et al.*, 2002). Third, to generate quantitative data, matrix suppression effects as well as the individual species response have to be addressed. The latter is determined by the lipid class and may depend on lipid concentration, chain length, and double bond number of the species (Brügger *et al.*, 1997; Koivusalo *et al.*, 2001).

Analysis of crude lipid extracts by mass spectrometry may be hampered by signal suppression effects due to other lipid components as well as lack of specificity in the presence of isobaric molecules or uncharacteristic molecule fragmentation. A useful approach to differentiate quasi-isobaric molecules, such as alkyl and acyl containing glycerophospholipids, in unseparated lipid extracts is high-resolution mass spectrometry including time-of-flight or Orbitrap analyzers (Ejsing *et al.*, 2006; Schwudke *et al.*, 2006). For lower resolution mass spectrometers like triple quadrupole machines, sample separation may be necessary prior to analysis, either off-line or by direct mass spectrometer coupling. Besides classical approaches like coupling liquid and gas chromatography to mass spectrometry, capillary electrophoresis is increasingly applied especially in the field of glycosphingolipid analysis (Levery, 2005). Another emerging method is the direct analysis (e.g., of glycosphingolipids, directly from thin-layer chromatography plates by matrix-assisted laser desorption/ionization mass spectrometry) (Dreisewerd et al., 2005; Ivleva et al., 2004).

Another valuable technology for lipidomic research is nuclear magnetic resonance (NMR). Applications have been described for the investigation of biomembranes as well as protein–lipid interactions (Gawrisch et al., 2002). Moreover, the analysis of lipoprotein subclasses using NMR allows the analysis of particle size, which may have higher disease relevance than cholesterol quantification in distinct lipoprotein classes (Otvos, 2002).

An approach to detect lipids involved in autoimmunity is based on lipid microarrays (Kanter et al., 2006). Arrays were produced using an automatic TLC sampler (Camag, Berlin, Germany), which sprays the lipids in spatially addressable locations on polyvinylidene fluoride (PVDF) membranes affixed to microscope slides. These arrays are incubated with diluted serum or cerebrospinal fluid samples, and chemiluminescence is used to detect autoantibody binding to specific lipids, mostly glycolipids, on the arrays. Lipid microarrays were 5 to 25 times more sensitive than conventional enzyme-linked immunosorbent assay (ELISA) for detecting lipid-specific antibodies, and up to 200 individual lipid species can be applied on a single slide.

A promising technique to investigate the role of gene regulation related to lipids is based on the recently presented pre-spotted TaqMan real-time PCR low-density arrays (TLDA). This method needs only minute amounts of mRNA concomitant with a high assay precision. An array for all known human ABC transporters was already validated in human primary monocytes, in vitro–differentiated macrophages, and cells stimulated with the LXR/RXR agonists, mimicking sterol loading (Langmann et al., 2006). Currently, TLDA assays for various lipid pathways such as fatty acid metabolism are being tested, which will allow a standardized comparison of whole pathways in different patient groups or in vitro experiments.

2.2. Need for further technology development

Despite the substantial progress in lipid analytical techniques, further technology development is necessary to advance the field of lipidomic research. A basic requirement is the purity of the analysis material. Analysis of tissue or blood cells may need further purification of cell types present. The procedures should be fast and non-activating for the cell type purified. A promising approach may be the use of magnet bead selection via specific antibodies either by positive or negative selection. Commercial platforms for such a cellomics approach are available (e.g., from Miltenyi Biotec GmbH, Auburn, CA) and are currently being evaluated to develop protocols for blood cell separation. To study cellular lipid function, the preparation of pure cellular subfractions may be crucial, since cellular compartments differ significantly in their lipid composition (van Meer, 2005). Furthermore, sample preparation methods used for lipid analysis should fix lipid compositions immediately, since lipid conversion may occur very fast. Related to this, lipid autoxidation may be a serious problem (Lütjohann, 2004).

3. IMAGING LIPIDS

In the upcoming era of clinical cellomics, biomarker analysis for diagnosis, patient stratification, and therapy monitoring will combine well-established analysis tools with techniques that transfer research methods into the clinical laboratory. In this regard, flow cytometry is already established as a valuable method to investigate cell surface as well as intracellular expression, activation, and colocalization of proteins, also in regard of signaling events (phosphorylation, calcium homeostasis). With the application of fluorescent lipids and the increasing availability of antibodies against the various lipids, it is likely to play a major role in the characterization of lipid parameters as well. This will become more and more important with the increase in the number of lipid biomarkers.

The physiological status of a cell is influenced by and reflected in the set of proteins and metabolites that are present at a given time point. However, simply knowing how much of a protein is expressed or how much of a metabolite like a lipid is present is not sufficient to understand their contribution to cell function. It is particularly important to know their subcellular location because changes in the subcellular distribution of components can have major effects on the cell behavior. Perhaps the most thoroughly studied example of this phenomenon is the change in the location of proteins and lipids associated with apoptosis. Accordingly, there is a need for the high-resolution, comprehensive analysis of the subcellular location of proteins and lipids. Therefore, an important goal in lipidomics is to define the localization of lipid species with high resolution in biological material. In addition, topological information must be obtained on the enzymes and transporters that define the lipid distribution. Next to histochemical, cytochemical, and physical imaging techniques, cell fractionation is commonly applied to study the subcellular localization of molecules.

The ultrastructural localization of lipids by electron microscopy is feasible but much more difficult than that of proteins (which is outside the scope of this paper), and the microscopy of lipids requires a whole different set of preparatory methods (Hoetzl *et al.*, 2007). Complementary to the localization of a lipid in a certain membrane, a variety of dedicated chemical and physical techniques is applied to study the sidedness of lipids in membranes (Sillence *et al.*, 2000). Another way to study how cells handle lipids is the live cell imaging of fluorescent lipid analogs. However, one problem is the interpretation of the data, as commonly used fluorescent probes may not accurately mimic the biophysical behavior of their natural counterparts. Thus, the specificity of lipid probes or the application of fluorescent lipids has to be critically evaluated, and new lipid probes have to be designed (Kuerschner et al., 2005; van Meer and Liskamp, 2005). Recently, first applications of mass spectrometry have been presented for lipid imaging. Thus, direct tissue analysis at light microscopic resolution was performed using time-of-flight secondary ion mass spectrometry (TOF-SIMS) (Sjövall et al., 2004), as well as desorption electrospray ionization (DESI) (Cooks et al., 2006).

With regard to the application of imaging in the clinic, high-throughput methods for protein (and lipid) imaging and automated methods for image analysis are mandatory. The feasibility of an automated classification system for subcellular distribution patterns has been demonstrated. The pattern analysis has been refined to the point that all major subcellular patterns can be recognized in two- and three-dimensional images of single cultured cells with high accuracy. An important conclusion from this work is that automated classification systems may perform better than visual examination in regard of standardised criteria. Although automated, these classification approaches still have the same limitation as the visual approach since they can recognize the major patterns with which they have been instructed. An important alternative, therefore, is to use unsupervised machine learning approaches (cluster analysis) to group molecules by their high-resolution patterns. Taken together, the combination of large-scale protein and lipid tagging, high-resolution imaging, and clustering by subcellular pattern is one way to enhance the capabilities of cellomics.

In addition to standardized, automated cellular imaging analysis, it will be necessary in the future to screen several different tissue specimens from healthy controls and patients for abnormalities in protein and lipid occurrence and distribution. In this regard, tissue microarrays were recently proven to be a valuable tool. Tissue microarrays are composed of hundreds of sections to allow for the parallel investigation of various histological entities or subtypes. Each section is linked to a well-characterized patient chart that enables the direct correlation of distribution patterns or expression signals to disease state or tumor type. This approach has been found to be very useful in the analysis of protein expression patterns (e.g., in breast carcinoma). Cellular imaging in addition to flow cytometry will also help to identify and visualize distinct plasma membrane compartments, so-called raft microdomains (Hoetzl et al., 2007).

4. METHODS TO STUDY THE PHYSICAL PROPERTIES OF LIPIDS

The physical properties of lipids are unique in that lipids are not water-soluble. In turn, the large hydrophobic moieties of the lipid molecules give rise to lipid–lipid and lipid–protein supramolecular structure through hydrophobic interactions. Even pure lipids in water can adopt a variety of aggregated structures ("phases") depending on concentration, temperature, and other factors. This important property of lipids is referred to as mesomorphism.

4.1. Lipids as amphiphiles

Many lipid molecules contain clearly separated hydrophobic and hydrophilic moieties. These lipids are said to be amphiphilic. Amphiphilic molecules are often surface-active (i.e., they decrease the water surface tension when spread at an air–water interface by orienting themselves with the hydrophobic moiety towards the aqueous solvent and the hydrophobic moiety towards the air). Lipid surface activity is best measured with the Langmuir balance (Marsh, 1996). Some lipid amphiphiles (e.g., lysophospholipids) can give rise to optically clear micellar solutions. These are called "soluble amphiphiles," and they often have detergent properties (Goñi and Alonso, 2000; Helenius and Simons, 1975).

4.2. Lipid–lipid interactions and mesomorphism

The different phases that lipids adopt in aqueous media (lamellar, hexagonal, cubic, etc.) are best defined by low-angle X-ray scattering, using the technology first established by Luzzati et al. (1968; Caffrey and Wang, 1995). Complementary techniques in the detection of lipid mesomorphism are ^{31}P-NMR (Cullis and De Kruijff, 1978; Sot et al., 2005) and cryotransmission electron microscopy (cryo-TEM) (Basáñez et al., 1997; Siegel et al., 1989).

Differential scanning calorimetry (DSC) provides information on thermotropic (i.e., heat-driven) phase transitions in lipid-water systems, including some cell membranes. In particular, lamellar-gel to lamellar-fluid and lamellar to hexagonal transitions are readily measured. These transitions are highly sensitive to the system composition (presence of other lipids, proteins, drugs, etc.); thus DSC is particularly useful in the study of lipid–lipid and lipid–protein interactions. DSC is nondestructive, requires micrograms of lipids, and can be automated for handling many samples (Goñi and Alonso, 2006).

4.3. Lipid structural functions

Lipids constitute the basic scaffold of important biological structures, chiefly the fat globules, the cell membranes, and the skin stratum corneum. Such lipid structures can be examined, to a different extent, using the techniques mentioned above (X-ray, ^{31}P-NMR, cryo-TEM). In addition, specific structural aspects of lipids may be studied with particular techniques.

4.3.1. Lipid bilayers and biomembranes

The basic structure of cell membranes is the lipid bilayer. Lipid bilayers can also be artificially formed by dispersing certain amphipathic lipids in water. Natural and synthetic lipid bilayers can be examined by several electron microscopy techniques, particularly cryo-TEM (Siegel *et al.*, 1989) and freeze-fracture electron microscopy (Sen *et al.*, 1981). The latter technique provides a direct view of the lipid matrix, since the two lipid leaflets making up the membrane are separated during sample preparation. Intrinsic membrane proteins are easily distinguished against the lipid background (Luna and McConnell, 1977). The combination of confocal fluorescence microscopy and giant unilamellar vesicles (GUV) formed in electric fields (Angelova *et al.*, 1992) provides new and often spectacular views of bilayer shape and deformations caused by a variety of agents (Sot *et al.*, 2006). When multiphoton fluorescence microscopy is used, certain fluorescent probes (e.g., Laurdan) can provide additional information on the physical state of the lipid (Bagatolli, 2006). This information has been successfully applied to cellular systems (Gaus *et al.*, 2006).

4.3.2. Membrane fluidity: lateral and transversal diffusion

Fluidity is an ill-defined parameter encompassing the overall molecular motions in the membrane. It can be more accurately represented by the lateral diffusion coefficients of lipids and proteins in the plane of the bilayer. Diffusion coefficients can be measured by fluorescence recovery after photobleaching (FRAP), a technique in which the fluorescent molecules in a microscope field are bleached by a laser flash, and then fluorescence recovery (due to the migration of unbleached fluorescent molecules towards the microscopic field) gives a measure of lateral diffusion (Axelrod *et al.*, 1976; Meyvis *et al.*, 1999). A number of NMR techniques (e.g., high-pressure proton NMR and excitation transfer ^{31}P-NMR) have been used to measure lateral diffusion coefficients. A more recent, useful method is the use of pulsed-field gradients for diffusion measurements by NMR, which is applied on macroscopically oriented bilayers (Oradd and Lindblom, 2004). Electron spin resonance (ESR) techniques, particularly field gradient ESR, have also been applied to measure molecular diffusion in membranes (Freed, 1994). Transbilayer ("flip-flop") diffusion is an important physiological process in cell membranes under steady-state conditions. It is mostly protein-mediated and specific. However, there are circumstances, such as membrane fusion, membrane protein insertion, the activation of a "scramblase," or the asymmetric generation of certain lipids (e.g., ceramide), when transbilayer lipid motion is observed. Methods for the observation of this kind of diffusion are available (Contreras *et al.*, 2005; López-Montero *et al.*, 2005; Sillence *et al.*, 2000).

4.3.3. Membrane domains: rafts

The idea that cell membranes are not homogeneous in the 100 to 1000 nm scale has been present in the field for a long time; only recently has this idea been tested with biophysical techniques. Nowadays, the existence of domains, or membrane regions with compositional and functional peculiarities, is widely accepted. Domains are known to exist in both model and cell membranes. Rafts are a particular kind of transient microdomains, that is, enriched in sphingolipids and cholesterol and is involved, among others, in intracellular membrane traffic (Simons and Ikonen, 1997; Simons and van Meer, 1988). Domains can be visualized by electron microscopy using immunogold or other staining techniques (Hoetzl et al., 2007; Strzelecka-Kiliszek et al., 2004) or by fluorescence microscopy, again using fluorescent probes that are specific for a given protein in which a domain is enriched or selective for a particular property (e.g., high or low fluidity) (Veatch and Keller, 2005). The terms "domains" and "rafts" are sometimes used synonymously, leading to some confusion. True rafts, because of their transient nature, are difficult to observe. Pulse saturation recovery ESR has been applied to measure lipid exchange between raft and non-raft domains in membranes (Kawasaki et al., 2001). Submicron-sized raft-like domains in model membranes have been detected by small-angle neutron scattering (Pencer et al., 2005) and fluorescence correlation spectroscopy (Lenne et al., 2006). Lagerholm et al. (2005) have reviewed the methods for microdomain detection in cell membranes.

A special form of lamellar phase, the liquid-ordered phase, was described by Ipsen et al. (1987) as allowing high lateral mobility of the lipid molecules (i.e., diffusion in the plane of the membrane) together with high order of the lipid acyl chains (i.e., low proportion of *gauche* conformers). Liquid-ordered phases are believed to exist in nature in the membrane raft microdomains (London and Brown, 2000). The observation of liquid-ordered phases requires the independent assessment of fluidity and order. Fluidity (lateral diffusion) can be measured as outlined in the "Membrane fluidity" section (4.3.2). Lipid chain order can be measured through the "order parameters" derived from ^{2}H-NMR (Hsueh et al., 2002) or ESR (Collado et al., 2005) spectroscopies. Infrared spectroscopy provides a direct procedure for the estimation of the proportion of *gauche* and *anti* conformers in membranes (Arrondo and Goñi, 1998). Certain properties of the fluorescent probe Laurdan also allow the detection of liquid-ordered phases by two-photon fluorescence microscopy (Bagatolli, 2006; Gaus et al., 2006).

4.3.4. Membrane fusion

Membrane fusion is defined as the merging of two membranes (each of them limiting a vesicle) with the formation of an intervesicular pore. The mechanism involves the transient interruption of the two membrane bilayers and the formation of a non-lamellar intermediate. Whether the fusion pore in cells is mainly lipidic or rather proteo-lipidic in nature is a topic of debate. From the point of view of lipids, proper description of membrane fusion requires the independent assessment of vesicle aggregation, intervesicular lipid mixing, intervesicular mixing of aqueous contents, and vesicle leakage. In principle, fusion should be defined by the concurring detection of vesicle aggregation, intervesicular lipid mixing, and intervesicular mixing of aqueous contents in the absence of vesicle efflux. Techniques for the assessment of these phenomena have been described (Goñi et al., 2003). A technique for the detection of inner monolayer lipid mixing, a useful complement of contents mixing evaluation, has been developed (Villar et al., 2000). More recently, an electron microscopy technique (thin sectioning and conical electron tomography) has allowed the observation of fusion pores and other fusion intermediates (hemifusion) in synaptic vesicles docked to the presynaptic membrane (Zampighi et al., 2006).

4.4. Lipid signaling functions

In addition to their structural role, lipids are important metabolic signals. Again, their mostly hydrophobic, or amphiphilic, nature explains their peculiar properties as signal molecules. Lipid signaling occurs mainly in the hydrophobic environment of cell membranes, and the lipids themselves have to be transported in a protein-bound form along the cytosol, or in the blood plasma. Some lipid signals (e.g., ceramides) are believed to act in both modulating the membrane physical properties and binding to specific protein receptors (Kolesnick et al., 2000; van Blitterswijk et al., 2003).

At the membrane level, signaling lipids interact with the other lipids and the proteins just as any other amphiphilic molecule; thus, the techniques mentioned in the "Lipid–lipid interactions and mesomorphism" section are as appropriate for the signaling lipids as they are for the structural ones. Signaling lipids travel in the blood plasma bound to the main plasma lipoproteins. Vast amounts of literature on the methodology of plasma lipoproteins exists, some of it published in this series (e.g., Volumes 128, 129, and 263). Within the cytosol, signaling lipids are transported by specific lipid transfer proteins. Methods for the study of these proteins are also found in Volume 98 of this series.

Finally, some lipids exert their signaling function by covalently binding certain proteins. They are most often saturated fatty acids (e.g., myristic or palmitic), bound through amide or ester bonds to amino acid side chains (Mikic et al., 2006) or isoprenoid chains (Berzat et al., 2005). Lipid binding often targets the protein to the cell membranes. A particular kind of lipid anchor is glycosylphosphatidylinositol (GPI), which is attached to the protein through a carbohydrate chain. Reviews on GPI-anchored proteins with methodological details can be found in Bütikofer et al. (2001) and Maeda et al. (2006). Villar et al. (1999) described the asymmetric incorporation of GPI into liposomal membranes.

5. Data Handling and Standardization

5.1. Software requirements

The novel lipidomic techniques, especially mass spectrometry–based methods, generate a vast amount of data. As mentioned above, a high sample throughput requires software assistance for data analysis as well as species assignment. Commercial software for lipidomic analysis is thus far not available. However, tools developed in academic environments, such as LipidInspector or LIMSA (Haimi et al., 2006), may be very useful. Such tools are required even more when assessing the metabolism of single lipid species (e.g., when several stable isotope labels are used in parallel, like for monitoring glycerophospholipid metabolism) (Binder et al., 2006). The final challenge on this side is to develop improved information technology solutions for platform integration in the various areas of genomics, proteomics, and metabolomics that will allow cross-referencing for the various types of data.

To forward the field of lipidomics, it is necessary to make data exchangeable and accessible worldwide. This accessibility requires the generation of standards:

1. The nomenclature of lipids has to be unified to a system that can be easily incorporated into databases. Three organizations involved in lipidomics—ELIfe, LIPID MAPS, and Lipid Bank—already agreed on a comprehensive classification system for lipids (Fahy et al., 2005).
2. Although the classification scheme allows a specification of single lipid species, methods like mass spectrometry need an additional, analysis-specific nomenclature. As already pointed out, MS/MS using a headgroup-specific scan cannot differentiate the fatty acids attached to the glycerol backbone. Moreover, using fatty acid–specific scans, the position and geometry of double bonds cannot be assigned. Therefore, a standardized data format should be developed that fits to instrument software as well as lipid profile databases.
3. To compare different analytical setups, commercially available reference material is necessary (i.e., lipid species standards [calibrators] and biological samples [controls] with known quantity). Moreover, for quantitative mass spectrometry, internal standards are beneficial either as stable isotope-labeled or non–naturally occurring species.

5.2. Existing lipid-related databases

Another important issue is lipid-related databases. Already, several useful websites exist that provide information about lipid molecules, lipid analysis, and lipid metabolism (the following list is only a small selection):

- Cyberlipid (http://www.cyberlipid.org/): contains information on methods for lipid analysis, a link collection as well as lipid structures linked to bibliography
- Kyoto Encyclopedia of Genes and Genomes (KEGG) (http://www.genome.jp/kegg/): collection of databases for pathways, genes, ligands, and functional hierarchies of biological systems; contains numerous pathways related to lipid metabolism, including enzymes and ligands
- Lipid Bank (http://lipidbank.jp/): contains more than 6000 entries of lipid species with their chemical and physical properties, information on biological activities, metabolism, and genetic information, as well as graphic data including structural formula, chromatographic data, ultraviolet (UV) data, infrared (IR) data, NMR data, and mass spectra; the lipids are grouped into lipid classes and advanced search tools are available
- Lipid Data Bank (LDB) (http://www.caffreylabs.ul.ie): includes the following databases: LIPIDAT (thermodynamic and associated information on lipids with about 20,000 records), LIPIDAG (lipid miscibility and associated information with almost 1600 phase diagrams), and LMSD (includes almost 13,000 molecular structures)
- LIPID MAPS (http://www.lipidmaps.org/): consortium supported by a National Institutes of Health (NIH) "Glue" grant; focuses on the lipid section of the metabolome by developing an integrated metabolomic system capable of characterizing the global changes in lipid metabolites ("lipidomics"); recently (with others) published its lipid classification system based on a 12-digit code reflecting the lipid class and subclasses (Fahy et al., 2005); includes following features:
  - LIPID MAPS Structure Database (LMSD), with currently more than 10,000 lipid species, including chemical structure, as well as links to LIPID ID, LipidBase ID and to the NCBI PubChem Substance ID (SID)
  - Proteome Database (LMPD), with lipid-associated protein sequences with annotations from UniProt, EntrezGene, ENZYME, Gene Ontology GO, and KEGG (Cotter et al., 2006)
  - Various searching capabilities; tools for drawing lipid structures; mass spectrometry tools (search algorithms based on mass spectrometric information, calculation of isotope pattern)
- Lipid Library (http://www.lipidlibrary.co.uk/): created by W.W. Christie from the Scottish Crop Research Institute; great source of information on lipid analysis (especially fatty acid analysis), including up-to-date bibliography, hundreds of mass spectra, ^{13}C-NMR chemical shifts of fatty acid derivatives, and practical advice on methods; includes biochemistry and chemistry of lipids and a valuable collection of links related to lipids

- "Seed oil fatty acids" (SOFA) Database (http://www.bagkf.de/sofa/): created by Federal Research Centre for Nutrition and Food, Germany; collection of data concerning plant oils and their lipid composition

5.3. Lipidomics expertise platform

In order to advance the field of lipidomics, a close dialog has to be established between application scientists and scientists in biology, medicine, and industry. Therefore, we launched the European Lipidomics Expertise Platform (http://www.lipidomics-expertise.de), which will serve as a forum to collect and provide information on labs involved in lipidomics. As a first step to fully mobilize the field, an awareness of the strength and expertise is created via a survey, including the areas of interest, the lipid classes, the applied technologies, and the organisms. These electronic surveys will collect contact information and give an overview of the expertise present in the various labs. Also, existing networks and plans for new networks will be mapped. This information is the source for establishing task forces and for initiating national and international networks and grants. Moreover, databases for lipid standards and for methods are included, which should improve methodology and standard material exchange. Currently, the Wikipedia software is integrated into the Lipidomics Expertise Wiki Portal (LEP-Wiki), which will provide information on lipids from basic information to advanced aspects, like the biology and pathophysiology on which lipids are involved. Editing LEP-Wiki is open to all members of the LEP, who include anybody who is interested in lipid science and registered for the site. Moreover, the website will provide channels for an optimal contact within the field and allow an efficient discussion on selected topics.

6. PERSPECTIVES

A major goal for the future will be to move basic to applied lipidomics; basic lipidomics stands for the technical requirements to analyze the lipidome (i.e., sample preparation, lipid synthesis, analytical techniques, and bioinformatics methods). In a next step, these technologies are applied in a functional context. Theoretical fundaments are lipid species maps as well as the genome, proteome, metabolome, and signalome related to lipids. A key task for rapid development of lipidomics will be to build a common, open-access repository of these theoretical fundaments. The databases already existing at LIPID MAPS may comprise a good foundation. Such a central database would be beneficial for all different disciplines of applied lipidomics, which include functional, evolutionary, clinical, pharmaceutical, and toxico- and nutri-lipidomics. Conversely, these approaches contribute data and knowledge to this central repository, increasing the basic knowledge of lipids. In addition, an integrative database is a prerequisite for other promising strategies to gain further insight into the regulation of lipid metabolism, such as the correlation of lipid metabolic flux rates and mRNA and/or protein expression levels based on pathway models. Recently, a theoretical model of the sphingolipid metabolism was validated by lipid analysis in yeast (Alvarez-Vasquez *et al.*, 2005). Moreover, the application of a combination of lipid analysis and gene expression studies to investigate the function of sphingolipids in the yeast heat-stress response has been discussed (Cowart and Hannun, 2005).

Another important aim of lipidomics is to map and quantify lipid species involved in signaling, a field of lipidomics termed mediator-lipidomics (Serhan, 2005). Both precursors, which also may act as structural components, as well as the signaling molecules are of relevance. For instance, the analysis of glycerophospholipids as the precursor pool has to be integrated into a complex network with the quantification of prostaglandins, leukotrienes, and lipoxins.

In summary, to increase our knowledge of lipids, especially of distinct lipid species, it is necessary to introduce immediately nomenclature and data standards in the emerging field of lipidomics. A major issue would be an integrative, open-access database for lipidomics that collects basic knowledge and provides the basis for bioinformatic tools that allow us to integrate all chemical, physical, biological, and medical information that is relevant for a specific lipid-related issue. This in turn would help us to discover novel lipid biomarkers or patho-mechanisms and utilize this knowledge for the diagnosis, monitoring, and cure of lipidome-related diseases (Helms and van Meer, 2006).

ACKNOWLEDGMENTS

This work was supported by specific support action LSSG-CT-2004-013032 of the European Commission.

REFERENCES

Alvarez-Vasquez, F., Sims, K. J., Cowart, L. A., Okamoto, Y., Voit, E. O., and Hannun, Y. A. (2005). Simulation and validation of modelled sphingolipid metabolism in *Saccharomyces cerevisiae*. *Nature* **433**, 425–430.

Angelova, M. I., Soleau, S., Meleard, P., Faucon, J. F., and Bothorel, P. (1992). Preparation of giant vesicles by external AC fields. Kinetics and applications. *Progr. Colloid Polym. Sci.* **89**, 127–131.

Arrondo, J. L., and Goñi, F. M. (1998). Infrared studies of protein-induced perturbation of lipids in lipoproteins and membranes. *Chem. Phys. Lipids* **96**, 53–68.

Axelrod, D., Koppel, D. E., Schlessinger, J., Elson, E., and Webb, W. W. (1976). Mobility measurement by analysis of fluorescence photobleaching recovery kinetics. *Biophys. J.* **16,** 1055–1069.
Bagatolli, L. A. (2006). To see or not to see: Lateral organization of biological membranes and fluorescence microscopy. *Biochim. Biophys. Acta* **1758,** 1541–1556.
Basáñez, G., Ruiz-Argüello, M. B., Alonso, A., Goñi, F. M., Karlsson, G., and Edwards, K. (1997). Morphological changes induced by phospholipase C and by sphingomyelinase on large unilamellar vesicles: A cryo-transmission electron microscopy study of liposome fusion. *Biophys. J.* **72,** 2630–2637.
Berzat, A. C., Brady, D. C., Fiordalisi, J. J., and Cox, A. D. (2005). Using inhibitors of prenylation to block localization and transforming activity. *Methods Enzymol.* **407,** 575–597.
Binder, M., Liebisch, G., Langmann, T., and Schmitz, G. (2006). Metabolic profiling of glycerophospholipid synthesis in fibroblasts loaded with free cholesterol and modified low density lipoproteins. *J. Biol. Chem.* **281,** 21869–21877.
Brügger, B., Erben, G., Sandhoff, R., Wieland, F. T., and Lehmann, W. D. (1997). Quantitative analysis of biological membrane lipids at the low picomole level by nano-electrospray ionization tandem mass spectrometry. *Proc. Natl. Acad. Sci. USA* **94,** 2339–2344.
Bütikofer, P., Malherbe, T., Boschung, M., and Roditi, I. (2001). GPI-anchored proteins: Now you see 'em, now you don't. *FASEB J.* **15,** 545–548.
Caffrey, M., and Wang, J. (1995). Membrane-structure studies using X-ray standing waves. *Annu. Rev. Biophys. Biomol. Struct.* **24,** 351–377.
Collado, M. I., Goñi, F. M., Alonso, A., and Marsh, D. (2005). Domain formation in sphingomyelin/cholesterol mixed membranes studied by spin-label electron spin resonance spectroscopy. *Biochemistry* **44,** 4911–4918.
Contreras, F. X., Basáñez, G., Alonso, A., Herrmann, A., and Goñi, F. M. (2005). Asymmetric addition of ceramides but not dihydroceramides promotes transbilayer (flip-flop) lipid motion in membranes. *Biophys. J.* **88,** 348–359.
Cooks, R. G., Ouyang, Z., Takats, Z., and Wiseman, J. M. (2006). Detection technologies. Ambient mass spectrometry. *Science* **311,** 1566–1570.
Cotter, D., Maer, A., Guda, C., Saunders, B., and Subramaniam, S. (2006). LMPD: LIPID MAPS proteome database. *Nucleic Acids Res.* **34,** D507–D510.
Cowart, L. A., and Hannun, Y. A. (2005). Using genomic and lipidomic strategies to investigate sphingolipid function in the yeast heat-stress response. *Biochem. Soc. Trans.* **33,** 1166–1169.
Cullis, P. R., and De Kruijff, B. (1978). Polymorphic phase behaviour of lipid mixtures as detected by ^{31}P NMR. Evidence that cholesterol may destabilize bilayer structure in membrane systems containing phosphatidylethanolamine. *Biochim. Biophys. Acta* **507,** 207–218.
Dreisewerd, K., Muthing, J., Rohlfing, A., Meisen, I., Vukelic, Z., Peter-Katalinic, J., Hillenkamp, F., and Berkenkamp, S. (2005). Analysis of gangliosides directly from thin-layer chromatography plates by infrared matrix-assisted laser desorption/ionization orthogonal time-of-flight mass spectrometry with a glycerol matrix. *Anal. Chem.* **77,** 4098–4107.
Ejsing, C. S., Moehring, T., Bahr, U., Duchoslav, E., Karas, M., Simons, K., and Shevchenko, A. (2006). Collision-induced dissociation pathways of yeast sphingolipids and their molecular profiling in total lipid extracts: A study by quadrupole TOF and linear ion trap-orbitrap mass spectrometry. *J. Mass Spectrom.* **41,** 372–389.
Ekroos, K., Chernushevich, I. V., Simons, K., and Shevchenko, A. (2002). Quantitative profiling of phospholipids by multiple precursor ion scanning on a hybrid quadrupole time-of-flight mass spectrometer. *Anal. Chem.* **74,** 941–949.

Fahy, E., Subramaniam, S., Brown, H. A., Glass, C. K., Merrill, A. H., Jr., Murphy, R. C., Raetz, C. R., Russell, D. W., Seyama, Y., Shaw, W., Shimizu, T., Spener, F., et al. (2005). A comprehensive classification system for lipids. *J. Lipid Res.* **46,** 839–861.
Freed, J. H. (1994). Field gradient ESR and molecular diffusion in model membranes. *Annu. Rev. Biophys. Biomol. Struct.* **23,** 1–25.
Gaus, K., Zech, T., and Harder, T. (2006). Visualizing membrane microdomains by Laurdan 2-photon microscopy. *Mol. Membr. Biol.* **23,** 41–48.
Gawrisch, K., Eldho, N. V., and Polozov, I. V. (2002). Novel NMR tools to study structure and dynamics of biomembranes. *Chem. Phys. Lipids* **116,** 135–151.
Goñi, F. M., and Alonso, A. (2000). Spectroscopic techniques in the study of membrane solubilization, reconstitution and permeabilization by detergents. *Biochim. Biophys. Acta* **1508,** 51–68.
Goñi, F. M., and Alonso, A. (2006). Differential scanning calorimetry in the study of lipid structures. *In* "Chemical Biology" (B. Larijani, C. A. Rosser, and R. Woscholski, eds.), pp. 47–66. Wiley, Chichester, England.
Goñi, F. M., Villar, A. V., Nieva, J. L., and Alonso, A. (2003). Interaction of phospholipases C and sphingomyelinase with liposomes. *Methods Enzymol.* **372,** 3–19.
Griffiths, W. (2006). Why steroidomics in brain? *Eur. J. Lipid Sci. Technol.* **108,** 707–708.
Haimi, P., Uphoff, A., Hermansson, M., and Somerharju, P. (2006). Software tools for analysis of mass spectrometric lipidome data. *Anal. Chem.* **78,** 8324–8331.
Han, X., and Gross, R. W. (2005). Shotgun lipidomics: Electrospray ionization mass spectrometric analysis and quantitation of cellular lipidomes directly from crude extracts of biological samples. *Mass Spectrom. Rev.* **24,** 367–412.
Helenius, A., and Simons, K. (1975). Solubilization of membranes by detergents. *Biochim. Biophys. Acta* **415,** 29–79.
Helms, B. (2006). Host-pathogen interactions: Lipids grease the way. *Eur. J. Lipid Sci. Technol.* **108,** 895–897.
Helms, B., and van Meer, G., eds. (2006). Lipidome and disease. *FEBS Lett.* **580**(Special issue), 5429–5610.
Hoetzl, S., Sprong, H., and van Meer, G. (2007). The way we view (glyco)sphingolipids. *J. Neurochem.* In press.
Hsueh, Y. W., Giles, R., Kitson, N., and Thewalt, J. (2002). The effect of ceramide on phosphatidylcholine membranes: A deuterium NMR study. *Biophys. J.* **82,** 3089–3095.
Ipsen, J. H., Karlstrom, G., Mouritsen, O. G., Wennerstrom, H., and Zuckermann, M. J. (1987). Phase equilibria in the phosphatidylcholine-cholesterol system. *Biochim. Biophys. Acta* **905,** 162–172.
Ivleva, V. B., Elkin, Y. N., Budnik, B. A., Moyer, S. C., O'Connor, P. B., and Costello, C. E. (2004). Coupling thin-layer chromatography with vibrational cooling matrix-assisted laser desorption/ionization Fourier transform mass spectrometry for the analysis of ganglioside mixtures. *Anal. Chem.* **76,** 6484–6491.
Kanter, J. L., Narayana, S., Ho, P. P., Catz, I., Warren, K. G., Sobel, R. A., Steinman, L., and Robinson, W. H. (2006). Lipid microarrays identify key mediators of autoimmune brain inflammation. *Nat. Med.* **12,** 138–143.
Kawasaki, K., Yin, J. J., Subczynski, W. K., Hyde, J. S., and Kusumi, A. (2001). Pulse EPR detection of lipid exchange between protein-rich raft and bulk domains in the membrane: Methodology development and its application to studies of influenza viral membrane. *Biophys. J.* **80,** 738–748.
Koivusalo, M., Haimi, P., Heikinheimo, L., Kostiainen, R., and Somerharju, P. (2001). Quantitative determination of phospholipid compositions by ESI-MS: Effects of acyl chain length, unsaturation, and lipid concentration on instrument response. *J. Lipid Res.* **42,** 663–672.

Kolesnick, R. N., Goñi, F. M., and Alonso, A. (2000). Compartmentalization of ceramide signaling: Physical foundations and biological effects. *J. Cell. Physiol.* **184,** 285–300.
Kuerschner, L., Ejsing, C. S., Ekroos, K., Shevchenko, A. J., Anderson, K. I., and Thiele, C. (2005). Polyene-lipids: A new tool to image lipids. *Nat. Meth.* **2,** 39–45.
Lagerholm, B. C., Weinreb, G. E., Jacobson, K., and Thompson, N. L. (2005). Detecting microdomains in intact cell membranes. *Annu. Rev. Phys. Chem.* **56,** 309–336.
Langmann, T., Mauerer, R., and Schmitz, G. (2006). Human ATP-binding cassette transporter TaqMan low-density array: Analysis of macrophage differentiation and foam cell formation. *Clin. Chem.* **52,** 310–313.
Lenne, P. F., Wawrezinieck, L., Conchonaud, F., Wurtz, O., Boned, A., Guo, X. J., Rigneault, H., He, H. T., and Marguet, D. (2006). Dynamic molecular confinement in the plasma membrane by microdomains and the cytoskeleton meshwork. *EMBO J.* **25,** 3245–3256.
Levery, S. B. (2005). Glycosphingolipid structural analysis and glycosphingolipidomics. *Methods Enzymol.* **405,** 300–369.
Liebisch, G., Drobnik, W., Lieser, B., and Schmitz, G. (2002). High-throughput quantification of lysophosphatidylcholine by electrospray ionization tandem mass spectrometry. *Clin. Chem.* **48,** 2217–2224.
Liebisch, G., Lieser, B., Rathenberg, J., Drobnik, W., and Schmitz, G. (2004). High-throughput quantification of phosphatidylcholine and sphingomyelin by electrospray ionization tandem mass spectrometry coupled with isotope correction algorithm. *Biochim. Biophys. Acta* **1686,** 108–117.
Liebisch, G., Binder, M., Schifferer, R., Langmann, T., Schulz, B., and Schmitz, G. (2006). High throughput quantification of cholesterol and cholesteryl ester by electrospray ionization tandem mass spectrometry (ESI-MS/MS). *Biochim. Biophys. Acta* **1761,** 121–128.
London, E., and Brown, D. A. (2000). Insolubility of lipids in Triton X-100: Physical origin and relationship to sphingolipid/cholesterol membrane domains (rafts). *Biochim. Biophys. Acta* **1508,** 182–195.
López-Montero, I., Rodriguez, N., Cribier, S., Pohl, A., Velez, M., and Devaux, P. F. (2005). Rapid transbilayer movement of ceramides in phospholipid vesicles and in human erythrocytes. *J. Biol. Chem.* **280,** 25811–25819.
Luna, E. J., and McConnell, H. M. (1977). The intermediate monoclinic phase of phosphatidylcholines. *Biochim. Biophys. Acta* **466,** 381–392.
Lütjohann, D. (2004). Sterol autoxidation: From phytosterols to oxyphytosterols. *Br. J. Nutr.* **91,** 3–4.
Luzzati, V., Gulik-Krzywicki, T., and Tardieu, A. (1968). Polymorphism of lecithins. *Nature* **218,** 1031–1034.
Maeda, Y., Ashida, H., and Kinoshita, T. (2006). CHO glycosylation mutants: GPI anchor. *Methods Enzymol.* **416,** 182–205.
Marsh, D. (1996). Lateral pressure in membranes. *Biochim. Biophys. Acta* **1286,** 183–223.
Merrill, A. H., Jr., Sullards, M. C., Allegood, J. C., Kelly, S., and Wang, E. (2005). Sphingolipidomics: High-throughput, structure-specific, and quantitative analysis of sphingolipids by liquid chromatography tandem mass spectrometry. *Methods* **36,** 207–224.
Meyvis, T. K., De Smedt, S. C., Van Oostveldt, P., and Demeester, J. (1999). Fluorescence recovery after photobleaching: A versatile tool for mobility and interaction measurements in pharmaceutical research. *Pharm. Res.* **16,** 1153–1162.
Mikic, I., Planey, S., Zhang, J., Ceballos, C., Seron, T., von Massenbach, B., Watson, R., Callaway, S., McDonough, P. M., Price, J. H., Hunter, E., and Zacharias, D. (2006). A live cell, image-based approach to understanding the enzymology and pharmacology of 2-bromopalmitate and palmitoylation. *Methods Enzymol.* **414,** 150–187.

Murphy, R. C., Barkley, R. M., Zemski Berry, K., Hankin, J., Harrison, K., Johnson, C., Krank, J., McAnoy, A., Uhlson, C., and Zarini, S. (2005). Electrospray ionization and tandem mass spectrometry of eicosanoids. *Anal. Biochem.* **346,** 1–42.

Oradd, G., and Lindblom, G. (2004). Lateral diffusion studied by pulsed field gradient NMR on oriented lipid membranes. *Magn. Reson. Chem.* **42,** 123–131.

Otvos, J. D. (2002). Measurement of lipoprotein subclass profiles by nuclear magnetic resonance spectroscopy. *Clin. Lab.* **48,** 171–180.

Pencer, J., Mills, T., Anghel, V., Krueger, S., Epand, R. M., and Katsaras, J. (2005). Detection of submicron-sized raft-like domains in membranes by small-angle neutron scattering. *Eur. Phys. J. E Soft Matter* **18,** 447–458.

Pulfer, M., and Murphy, R. C. (2003). Electrospray mass spectrometry of phospholipids. *Mass Spectrom. Rev.* **22,** 332–364.

Schwudke, D., Oegema, J., Burton, L., Entchev, E., Hannich, J. T., Ejsing, C. S., Kurzchalia, T., and Shevchenko, A. (2006). Lipid profiling by multiple precursor and neutral loss scanning driven by the data-dependent acquisition. *Anal. Chem.* **78,** 585–595.

Sen, A., Williams, W. P., Brain, A. P., Dickens, M. J., and Quinn, P. J. (1981). Formation of inverted micelles in dispersions of mixed galactolipids. *Nature* **293,** 488–490.

Serhan, C. N. (2005). Mediator lipidomics. *Prostaglandins Other Lipid Mediat.* **77,** 4–14.

Siegel, D. P., Burns, J. L., Chestnut, M. H., and Talmon, Y. (1989). Intermediates in membrane fusion and bilayer/nonbilayer phase transitions imaged by time-resolved cryo-transmission electron microscopy. *Biophys. J.* **56,** 161–169.

Sillence, D. J., Raggers, R. J., and van Meer, G. (2000). Assays for transmembrane movement of sphingolipids. *Methods Enzymol.* **312,** 562–579.

Simons, K., and van Meer, G. (1988). Lipid sorting in epithelial cells. *Biochemistry* **27,** 6197–6202.

Simons, K., and Ikonen, E. (1997). Functional rafts in cell membranes. *Nature* **387,** 569–572.

Sjövall, P., Lausmaa, J., and Johansson, B. (2004). Mass spectrometric imaging of lipids in brain tissue. *Anal. Chem.* **76,** 4271–4278.

Sot, J., Aranda, F. J., Collado, M. I., Goñi, F. M., and Alonso, A. (2005). Different effects of long- and short-chain ceramides on the gel-fluid and lamellar-hexagonal transitions of phospholipids: A calorimetric, NMR, and X-ray diffraction study. *Biophys. J.* **88,** 3368–3380.

Sot, J., Bagatolli, L. A., Goñi, F. M., and Alonso, A. (2006). Detergent-resistant, ceramide-enriched domains in sphingomyelin/ceramide bilayers. *Biophys. J.* **90,** 903–914.

Spener, F., Kohlwein, S. D., and Schmitz, G. (2006a). Lipid droplets and lamellar bodies—from innocent bystanders to prime targets of lipid research for combating human diseases. *Eur. J. Lipid Sci. Technol.* **108,** 541–543.

Spener, F., Zechner, R., and Borlak, J. (2006b). Is lipotoxicity an oxymoron? *Eur. J. Lipid Sci. Technol.* **108,** 625–627.

Strzelecka-Kiliszek, A., Korzeniowski, M., Kwiatkowska, K., Mrozinska, K., and Sobota, A. (2004). Activated FcγRII and signaling molecules revealed in rafts by ultra-structural observations of plasma-membrane sheets. *Mol. Membr. Biol.* **21,** 101–108.

van Blitterswijk, W. J., van der Luit, A. H., Veldman, R. J., Verheij, M., and Borst, J. (2003). Ceramide: Second messenger or modulator of membrane structure and dynamics? *Biochem. J.* **369,** 199–211.

van Meer, G. (2005). Cellular lipidomics. *EMBO J.* **24,** 3159–3165.

van Meer, G. (2006). How do sphingolipids and lipid rafts relate to pathology? *Eur. J. Lipid Sci. Technol.* **108,** 799–801.

van Meer, G., and Liskamp, R. M. (2005). Brilliant lipids. *Nature Methods* **2,** 14–15.

Varfolomeyev, S., Efremenko, E., Beletskaya, I., Bertini, I., Blackburn, G. M., Bogdanov, A., Cunin, R., Eichler, J., Galaev, I., Gladyshev, V., O'Hagan, D.,

Haertle, T., et al. (2005). Postgenomic chemistry (IUPAC Technical Report). *Pure Appl. Chem.* **77**, 1641–1654.

Veatch, S. L., and Keller, S. L. (2005). Seeing spots: Complex phase behavior in simple membranes. *Biochim. Biophys. Acta* **1746**, 172–185.

Villar, A. V., Alonso, A., and Goñi, F. M. (2000). Leaky vesicle fusion induced by phosphatidylinositol-specific phospholipase C: Observation of mixing of vesicular inner monolayers. *Biochemistry* **39**, 14012–14018.

Villar, A. V., Alonso, A., Paneda, C., Varela-Nieto, I., Brodbeck, U., and Goñi, F. M. (1999). Towards the *in vitro* reconstitution of caveolae. Asymmetric incorporation of glycosylphosphatidylinositol (GPI) and gangliosides into liposomal membranes. *FEBS Lett.* **457**, 71–74.

Zamfir, A., Vukelic, Z., Bindila, L., Peter-Katalinic, J., Almeida, R., Sterling, A., and Allen, M. (2004). Fully-automated chip-based nanoelectrospray tandem mass spectrometry of gangliosides from human cerebellum. *J. Am. Soc. Mass Spectrom.* **15**, 1649–1657.

Zampighi, G. A., Zampighi, L. M., Fain, N., Lanzavecchia, S., Simon, S. A., and Wright, E. M. (2006). Conical electron tomography of a chemical synapse: Vesicles docked to the active zone are hemi-fused. *Biophys. J.* **91**, 2910–2918.

CHAPTER TEN

LIPIDOMIC ANALYSIS OF SIGNALING PATHWAYS

Michael J. O. Wakelam,[*,†] Trevor R. Pettitt,[†] and Anthony D. Postle[‡]

Contents

1. General Considerations	235
2. Lipid Extraction	235
2.1. General lipid extraction (Folch procedure)	235
2.2. Acidified phosphoinositide/lysolipid extraction	236
3. HPLC-MS Analysis	237
4. General Phospholipid HPLC	237
5. Ceramide, Diradylglycerol, and Monoradylglycerol Separation	239
6. Phosphoinositide Separation	240
7. Analysis of Phospholipid Synthesis by ESI-MS/MS Using Stable Isotopes	241
8. Labeling Protocols	242
8.1. In vivo	242
8.2. In cultured cells	242
9. Lipid Extraction for ESI-MS/MS	243
10. LC-MS Analysis (methyl-d_9)-Choline	243
11. MRM Analysis of (methyl-d_9)-Choline Enrichment in Cell and Tissue Phosphorylcholine	244
12. ESI-MS/MS Analysis of Native and Newly Synthesized Phospholipids	244
13. Data Analysis	245
References	246

Abstract

This chapter outlines methods that can be applied to determine the levels of lipids in cells and tissues. In particular, the methods focus upon the extraction and analysis of those lipids critical for monitoring signal transduction pathways.

[*] The Babraham Institute, Babraham Research Campus, Cambridge, United Kingdom
[†] Institute for Cancer Studies, Birmingham University, Birmingham, United Kingdom
[‡] School of Chemistry, University of Southampton, Southampton, United Kingdom

The methods address the analysis of the phosphoinositides, the lipid agonists lysophosphatidic acid and sphingosine 1-phosphate, and the neutral lipid messengers diacylglycerol and ceramide. Additionally, because of the increasing need to determine the dynamics of signaling, the analysis of phospholipids synthesis using stable isotope methods is described. The use of these methods as described or adaptation to permit both approaches should allow investigators to determine changes in signaling lipids and to better understand such processes in most cell types.

The increasing appreciation of the central roles played by lipid signaling pathways has dispelled the misconception that lipids are inert structural components that are involved solely in keeping a cell intact. Advances in our understanding of cell-signaling pathways have identified particular lipids that act to regulate the functions of a number of proteins either by controlling enzyme activity directly, or by localizing proteins to particular intracellular compartments where they perform a specialized role. These lipid-binding domains (e.g., PH, PX, FYVE) have been found in many proteins, and considerable detail is recorded of the structural basis of lipid protein interaction. Additional lipid-binding domains exist, which remain less well characterized (e.g., those that bind phosphatidic acid [PA] or ceramide); however, the important regulatory roles that these lipids play and the pathways involving these messengers are increasingly appreciated.

While the downstream targets are thus being defined, the actual changes in lipid concentration in a stimulated cell or membrane are less characterized. The primary reason for this lack has been a deficiency in methodology. Much of the reported studies of lipid messengers in stimulated cells have depended upon monitoring changes in radio-labeled cells. Many well-documented problems are associated with this type of methodology, including lack of isotopic equilibrium, distinct pools with different turnover rates, and inadequate separation of radio-labeled metabolites; however, much important information has been generated. The second approach has been to make use of the lipid-binding properties of the target protein domains and to generate a tagged fusion protein, generally GFP, which permits identification of a region rich in a signaling lipid (Guillou *et al.*, 2007). This has proved useful in monitoring PI-3-kinase activation in stimulated cells; however, considerable caveats must be raised, not least the problems associated with lipid specificity and the fact that many of these domains have associated protein-binding regions that can compromise the findings. A further problem associated with these two methodologies is that they tend to group lipids together and take no account of the multiple acyl chain structures that occur in all lipids.

These concerns point to the need to determine actual changes in lipid compositions. Until relatively recently, such an analysis was unachievable; however, advances in both chromatographic separation and mass spectrometry (MS) have permitted the development of lipidomic analysis. This chapter outlines a number of methods that allow determination of changes in signaling lipids. Adaptation of the methods here for the analysis of other molecules should be relatively straightforward in the future. Much of the lipidomic research in the United

Kingdom is focused upon signaling lipidomics, with particular foci upon phosphoinositide-related signaling in Birmingham and Cambridge (Wakelam) and London (Larijani), upon eicosanoids in Cardiff (O'Donnell), and steroids in London (Griffiths). Meanwhile, the use of stable isotopes has been particularly developed in Southampton (Postle).

1. General Considerations

In an ideal situation, it would be possible to extract all the lipids from a cell or tissue, inject them into a mass spectrometer, and generate a detailed analysis. Unfortunately, this is not achievable since different solvents are necessary to extract different lipids; indeed, acidification, which is necessary in some cases, destroys other lipids of interest. Thus, a general extraction method is outlined together with one for the phosphoinositides and the lysolipids, such as lysophosphatidic acid (LPA).

Classically, lipids are extracted into glass rather than plastic tubes; however, lipids can stick to glass, and it is essential that, if used, all glassware is silanized, especially when working with more polar lipids, such as the phosphoinositides. Simply infusing a mixture of lipids into a MS instrument will also generate incomplete data because of the different concentrations of lipid; their distinct ionizations will lead to ion suppression and, thus, a false analysis of the mixture. At high lipid concentrations, further suppression of the signal will occur. Thus, the use of high performance liquid chromatography (HPLC) with a silica column is adopted prior to online injection to the MS.

2. Lipid Extraction

Two methods are outlined. The first is for the more general lipids and is described for a sample of cells in suspension. The second is for the polar lipids and is described for a tissue sample and suspension cells. It is possible, of course, to adapt these methods for the appropriate sample being used.

2.1. General lipid extraction (Folch procedure)

1. Place 0.25 ml of the sample (assume 0.25 g of cellular material contains 0.25 ml water; add more water if necessary) in a glass screw-capped tube.
2. Add 2 ml of ice-cold methanol (this kills cells and denatures protein). Add internal standards if required (200 ng of 12:0/12:0-DAG, PA, PC, PE, PG, PS).
3. Vortex the mix and let it stand on ice for 10 min.
4. Add 4 ml of chloroform, vortex, and then sonicate in ultrasonic bath for 5 min (keep cool with ice). Let stand for 10 min.

5. Add 1.25 ml of 0.88% KCl to split the phases. Mix vigorously, then let stand on ice for 10 min.
6. Complete the phase split by centrifugation (5 min 200×g); alternatively, let the mix stand overnight in a cold room.
7. Carefully remove the upper aqueous phase, together with interfacial material, using a glass Pasteur pipette. Transfer the lower phase into a clean glass tube.
8. Dry under a stream of nitrogen or on a vacuum dryer.
9. Resuspend the phase in 50 μl of chloroform/methanol/water (90:9.5:0.5). Resuspend in chloroform if analyzing diradylglycerols (DRG). Transfer into an autosampler vial.
10. Cap tightly and store at −20° if necessary, although it is generally best to analyze as soon as possible to minimize any sample degradation.

2.2. Acidified phosphoinositide/lysolipid extraction

To minimize acid hydrolysis, extraction should be performed on ice without delays.

1. Freeze tissue (~50 mg) in liquid nitrogen. Pulverize in a liquid nitrogen–cooled pulverizer.
2. Transfer the frozen, pulverized sample into a 3-ml glass homogenizer. Add 1 ml of ice-cold methanolic HCl (0.1 M). Add internal standards (400 ng of 16:0/16:0-PI, PI5P, PI4P, PI45P$_2$, PI35P$_2$, PI345P$_3$, plus 500 ng of 18:1-LPA, c17-S1P, d31-16:0-LPC).
3. Homogenize tissue until fully disrupted. Transfer suspension into a 15-ml polypropylene centrifuge tube (e.g., from Sarstedt, Nümbrecht, Germany) using a silanized Pasteur pipette. Rinse the homogenizer with 2 ml of methanolic 0.1 M HCl/chloroform (1:1), and then with 3 ml of chloroform, and combine washes with the homogenized sample.
4. Vortex vigorously for 30 s, and then sonicate in an ice-cooled sonicating bath for 5 min.
5. If cells are being used for the analysis, pellet ~5 × 10^6 cells in a 15-ml polypropylene tube (e.g., from Sarstedt), add 2 ml of ice-cold methanolic 0.2 M HCl, and add internal standards as previously described. Vortex vigorously for 30 s and sonicate in an ice-cooled sonicating bath for 10 min.
6. For either sample type, add 1.5 ml of 0.1 M HCl, 2 mM EDTA, and 5 mM of tetrabutylammonium sulphate 0.9% NaCl to split phases. Vortex vigorously for 30 s, and then sonicate in an ice-cooled sonicating bath for 5 min.
7. Let stand on ice for 10 min.
8. Complete the phase split by centrifugation (250×g, 5 min). Transfer the lower phase into a clean tube using a silanized pipette. Re-extract the remaining upper phase with 4 ml of ice-cold, synthetic lower phase.

9. Carefully collect the lower phase with a silanized pipette. Neutralize the combined lower phases with 3 ml of 100 mM sodium EDTA pH 6.0/methanol (1:1). Vortex vigorously.
10. Complete the phase split by centrifugation (250×g, 5 min). Remove and discard the upper phase using a silanized pipette.
11. Dry the lower phase using a vortex evaporator.
12. Resuspend the phase in 50 μl of chloroform/methanol/water (49:49:2). Transfer into silanized auto-sampler vials. Analyze quickly to minimize any losses on standing.

3. HPLC-MS Analysis

HPLC separation of lipid classes permits MS detection of pseudomolecular ions by electrospray ionization (ESI) on a single quadrupole MS (Pettitt and Wakelam, 2006). However, this type of instrument cannot provide unambiguous identification of the exact fatty acid composition, only the total number of carbons and double bonds (e.g., C38:4 for a 18:0/20:4 or a 18:2/20:2 structure). Identification of exact fatty acid compositions requires fragmentation using a triple quadrupole or ion-trap MS (Taguchi *et al.*, 2005). If HPLC separation cannot resolve different lipid classes with identical molecular ions, and then triple quad or ion-trap MS can identify the class on the basis of characteristic fragmentation (e.g., m/z 184 for phosphocholine-containing lipids such as PC or SM). Equally, the resolution of structural isomers such as PI3P and PI4P requires the controlled fragmentation capabilities of an ion-trap (Pettitt *et al.*, 2006). Attomole LC-MS detection limits (signal-to-noise [S/N] > 5) can be achieved for some lipid structures on certain instruments; however, femtomole limits are more usual. The major signaling lipid ions are [M–H]$^-$-PA, PEtOH, PE (and [M+H]$^+$), PG, PI, PS, LPA, S-1-P, FFA, PC, LysoPC; SM are detected as [M+H]$^+$, [M+Na]$^+$, [M–CH$_3$]$^-$, DRG, monoradylglycerols (MRG), and ceramide as [M+Na]$^+$, [M–OH]$^+$; phosphoinositides are [M–H]$^-$, [M–H]$^{2-}$, [M–2H+Na]$^-$.

4. General Phospholipid HPLC

Unmodified silica remains the column material of choice for the separation of most lipids. High-grade, uniformly spherical silicas with low pore size, high surface area, and very low metal content (e.g., Luna silica from Phenomenex, Torrance, CA) generally give good separation with minimal peak tailing. For general phospholipids separations, use 100% chloroform/dichloromethane/methanol/water (45:45:9.5:0.5) containing 7.5-mM ethylamine as solvent A, changing to 20% acetonitrile/chloroform/methanol/water (30:30:32:8) containing 10 mm of ethylamine (solvent B) over 1 min, 35%

solvent B over 9.5 min, 70% solvent B over 9.5 min, and then recycle back to 100% solvent A over 1 min and hold for a further 14 min to re-equilibrate (total program time, 35 min). The flow rate is 100 µl/min for 20 min, and then set to 130 µl/min over 5 min, held for 9 min, and then back to 100 µl/min over 1 min. The sample injection volume was 0.1 to 1 µl (Fig. 10.1).

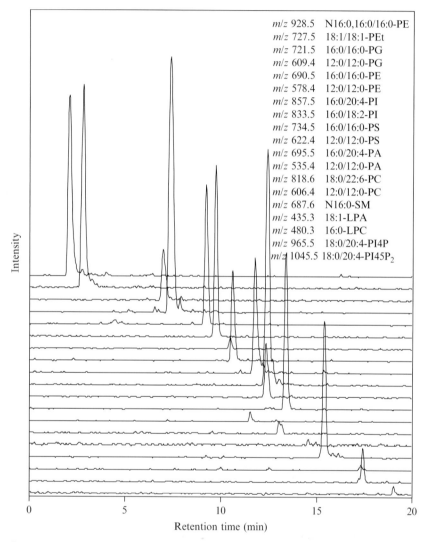

Figure 10.1 Separation of lipid standards. The traces are in the order listed in the figure. Detection is as [M−H]$^-$ ions, except PC, LPC, and SM (very small double peak at 14.5 to 15.0 min), which are detected as [M−CH$_3$]$^-$ (best detected as [M+H]$^+$). Extracted ion chromatograms correspond to the listed lipid structures. Although not shown, free fatty acids elute at 4 min, dimethylPE at 6 min, monomethyl PE at 9 min, S1P at 17 min, and sphingosinephosphoryl choline at 20 min.

Signaling Lipidomics

Table 10.1 Diagnostic ions for tandem mass spectrometric (MS/MS) fragmentation

Lipid class	Precursor ion	MS/MS	Fragment
Glycerophospholipid	[M-H]⁻	153.0	Glycerophosphate backbone
Sphingolipids	[M+H]⁺	264.4	Sphingosine backbone
PC, LysoPC, SM	[M+H]⁺	184.1	Phosphocholine
PE	[M-H]⁻	196.0	Glycerophosphoethanolamine
PI	[M-H]⁻	241.1	Cyclic inositol phosphate
PS	[M-H]⁻	Neutral loss of 87.0	Serine
PIP	[M-H]⁻	321.1	Phosphoinositol phosphate
PIP$_2$	[M-H]⁻	401.1	Diphosphoinositol phosphate
PIP$_3$	[M-H]⁻	481.1	Triphosphoinositol phosphate
All	[M-H]⁻	227.2	14:0 FA
All	[M-H]⁻	255.2	16:0 FA
All	[M-H]⁻	279.2	18:2 FA
All	[M-H]⁻	281.2	18:1 FA
All	[M-H]⁻	283.2	18:0 FA
All	[M-H]⁻	303.3	20:4 FA
All	[M-H]⁻	305.3	20:3 FA
All	[M-H]⁻	307.3	20:2 FA
All	[M-H]⁻	309.3	20:1 FA
All	[M-H]⁻	311.3	20:0 FA
All	[M-H]⁻	327.3	22:6 FA
All	[M-H]⁻	329.3	22:5 FA
All	[M-H]⁻	331.3	22:4 FA

MS detection was by ESI as negative (PA, PE, PG, PI, PS, LPA, S-1-P) or positive ions (PC, SM, LPC). The optimum desolvation temperature was 300°; probe voltage, ±3.5 kV; and nebulizing N$_2$, 3.5 l/min (values used with a Shimadzu QP8000α MS). Diagnostic ions for MS/MS fragmentation are presented in Table 10.1.

5. Ceramide, Diradylglycerol, and Monoradylglycerol Separation

As the gradient, use 100% chloroform/hexane/propan-2-ol (80:20:1) as solvent A, hold for 2 min, change to 100% chloroform/methanol/water (49:49:2) as solvent B for over 17 min, hold for 2 min, and then recycle over

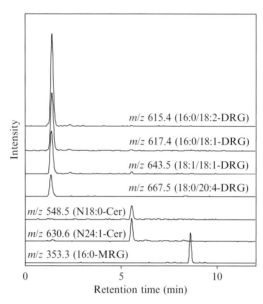

Figure 10.2 Neutral lipid separation. Diradylglycerols and monoradylglycerols (MRG) are detected as $[M+Na]^+$. Ceramide detected is as $[M-OH]^+$, although it also gives a strong $[M+Na]^+$ ion. Very low polarity lipids, such as triacylglycerol and sterol esters, elute in the solvent front. Phospholipids elute after 10 min, but with poor resolution—this part of the gradient is to prevent lipid carryover into subsequent runs.

1 min and hold for a further 19 min to re-equilibrate (total program time of 40 min). Run the HPLC with a 100 µl/min flow rate.

Add 20 mM ammonium formate in methanol at 50 µl/min as a post-column addition in order to enable MS detection primarily as $[M+Na]^+$. Optimum responses are obtained with a low probe voltage (0.75 kV with a Shimadzu QP8000α MS), as seen in Fig. 10.2.

6. Phosphoinositide Separation

As the gradient, use 100% chloroform/methanol/water (90:9.5:0.5) containing 7.5-mM ethylamine as solvent A, hold for 2 min, change to 45% acetonitrile/chloroform/methanol/water (30:30:32:8) containing 15 mm of ethylamine as solvent B over a 1-min period, change to 50% solvent B over 17 min, and recycle back to solvent A over 1 min and hold for a further 14 min to re-equilibrate. PIP, PIP$_2$, and PIP$_3$ are detected as $[M–H]^-$ and $[M–2H]^{2-}$; the stronger $[M–2H]^{2-}$ signals are seen at lower temperatures, with 240° being optimum (Fig. 10.3).

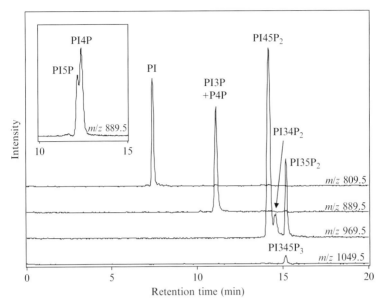

Figure 10.3 Separation of 16:0/16:0 phosphoinositide standards. Detection is as [M−H]⁻ ions. Identical molecular species of PI3P and PI4P co-elute, but PI5P will partially resolve (inset).

7. ANALYSIS OF PHOSPHOLIPID SYNTHESIS BY ESI-MS/MS USING STABLE ISOTOPES

A variety of tandem MS (MS/MS) are employed for the analysis of phospholipid dynamics. Synthesis analysis by ESI-MS relies on quantification of the relative ion intensities of a subset of intact lipid species that have incorporated substrate labeled with the relevant isotope. This is dependent on the formation by MS/MS of diagnostic fragment ions containing the stable isotope. The overall sensitivity of the approach is then a function of the number of isotope atoms within this diagnostic fragment. In practice, three or more isotope atoms are required to overcome the effect of the M+1, M+2, M+3 isotope due to the approximately 1% natural abundance of [^{13}C].

The methodology has been best characterized for phosphatidylcholine (PC) synthesis; incorporation of (*methyl-d₉*)-choline by the CDP:choline Kennedy pathway generates PC species 9 mass units higher than the corresponding native species (Hunt *et al.*, 2002). MS/MS fragmentation of this ion generates a choline phosphate product ion of m/z 184⁺ for native and m/z 193⁺ for newly synthesized PC. Precursor scans of m/z 184⁺ and 193⁺ then provide direct comparison of the patterns of native and newly synthesized PC species. Quantification of the ion intensities of these two precursor scans provides an estimate of the fractional enrichment of each PC species with the (*methyl-d₉*)-choline substrate. Comparison of the modification of the pattern of incorporation of *(methyl-d₉)*-choline with time in comparison with the molecular species composition of the endogenous native PC gives an indication of the acyl remodeling.

Phosphatidylcholine can also be synthesized by sequential N-methylation of PE using S-adenosylmethionine (Adomet) as the methyl donor. The molecular specificity of PE N-methylation in liver can be determined from the incorporation of *(methyl-d₉)*-choline. A portion of these methyl groups are then recycled back into choline by PE N-methylation, generating PC with one or two *methyl-d₃* moieties in the phosphorylcholine headgroup of PC, monitored by precursor scans of m/z 187$^+$ and m/z 190$^+$, respectively.

8. Labeling Protocols

8.1. In vivo

1. Inject a mouse (20–25 g) i.p. with 1 mg *(methyl-d₉)*-choline chloride and/or 1 mg of d_6-*myo*-inositol in 200 μl of 0.9% saline. Record body weight and sacrifice mice at intervals from 1.5 to 24 h.
2. Take approximately 250 μl of blood by heart puncture into 1-ml EDTA blood tube, mix, and stand on ice. Rapidly remove organs as required and freeze in liquid nitrogen.
3. Centrifuge blood samples at 400×g for 10 min at 4°; aspirate the plasma.
4. Store all tissue and plasma samples at −80°.

8.2. In cultured cells

1. Add stable isotopes dissolved in filter-sterilized medium to cells; use 100-μl stock solutions of 10 mg/ml of *(methyl-d₉)*-choline, d_6-*myo*-inositol, and d_3-serine to achieve enrichments, respectively, of 96, 74, and 76%.
2. End incubations at intervals from 10 min to 6 h by centrifuging suspension cultures at 400×g for 10 min at 4° followed by the addition of 1 ml of ice-cold methanol or aspirating medium, adding 1 ml of ice-cold methanol and scraping cells with a cell scraper. Acyl remodeling mechanisms in cultured cells are generally complete by 3 h.
3. Transfer cell suspensions to glass screw-cap extraction tubes and combine with a second wash of tube or flask with 1 ml methanol; add 800 μl phosphate-buffered saline (PBS) containing 20 μg/ml butylated hydroxytoluene (BHT).

9. Lipid Extraction for ESI-MS/MS

1. Pulverize frozen tissue samples in a liquid nitrogen–cooled pulverizer.
2. Homogenize powdered tissue samples (up to 100 mg) using an Ultraturrex in 5 ml of buffered saline containing 20 μg/ml of BHT.
3. Remove 800 μl of tissue homogenate to a disposable glass screw-cap extraction tube.
4. For plasma extraction, add 50 μl of to 750 μl PBS containing 20 μg/ml of BHT in a glass screw-cap tube.
5. Add 2 ml of methanol to the 800 μl of saline homogenates of tissue or plasma. At this stage, tissue, plasma, and cell samples are then processed identically. Add internal standards to the methanol:saline extracts followed by 1 ml chloroform. For 10^7 cells, use 15 nmol of PC14:0/14:0 or PC20:0/20:0, 4 nmol PE14:0/14:0, 2 nmol PS14:0/14:0, 2 nmol PI16:0/16:0, and 0.5 nmol PA14:0/14:0 (check that there are no significant amounts of these in a typical sample spectrum).
6. Add 1 ml of chloroform, mix vigorously for 1 min to form a single phase, add 1 ml of chloroform and 1 ml of water, mix vigorously for 1 min to form a biphasic system, and centrifuge at $400 \times g$ for 10 min to separate the phases.
7. Remove the chloroform-rich lower phase with a Pasteur pipette to a glass tube and dry at 40° under nitrogen. Dissolve the lower phase in 1 ml of chloroform:methanol (9:1,v:v), transfer to a 2-ml HPLC sample vial, and dry under nitrogen.
8. Immediately before analysis, dissolve in methanol:chloroform:water: ammonia (70:20:10:2, v:v). Use sufficient solvent, 0.3 to 2 ml, for a final concentration of 1 μM for the most abundant components.

10. LC-MS Analysis *(methyl-D$_9$)*-Choline

1. Add 2 nmol of *(N,N,N-trimethyl-d$_9$;1,1,2,2-d$_4$)*-choline (200 μl of 10 μM in water) as an internal recovery standard to 200-μl plasma aliquots. Extract with chloroform and methanol, as previously shown.
2. Remove upper aqueous layer, dilute to 4 ml with acetonitrile, and transfer 1 ml to a 2-ml HPLC sample vial. Inject sample aliquots (50 μl) on a Chromasil® C4–3μ column (30 × 2 mm) (Phenomenex, Maccelsfield, UK) at 300 μl/min using a linear gradient from solvent A (95% acetonitrile:0.2% formic acid) to solvent B (5% acetonitrile: 0.2% formic acid) over 6 min.
3. Quantify choline, d_9-choline, and d_{13}-choline in the eluate using, for example, a Micromass Quattro Micro triple quadrupole MS (Waters, Elstree, UK), operated in positive ionization in the MRM mode with

transitions of m/z 104.1→59.9 u, m/z 113.1→68.9 u, and m/z 117.1→68.9 u for choline, d_9-choline, and d_{13}-choline, respectively.
4. Calculate concentrations of choline and *(methyl-d_9)*-choline by reference to external standard curves, correcting back for recovery of the d_{13}-choline internal standard. The linear range is from 2.6 to 4000 nM for all components.

11. MRM Analysis of *(Methyl-D_9)*-Choline Enrichment in Cell and Tissue Phosphorylcholine

1. Freeze-dry the aqueous upper phase from the Bligh/Dyer lipid extracts of tissues and cells. Dissolve in 1-ml 0.05% ammonia:methanol (1:9), sonicate in a water bath for 2 min, and centrifuge (13,000×g, 30 sec) to remove particulates.
2. Apply the supernatant to an aminopropyl BondElut sample cartridge previously conditioned with 1-ml of the same solvent, followed by a further 1-ml wash.
3. Elute phosphorylcholine with 1.5 ml of 0.05% ammonia: 6% (w/v) NH$_4$HCO$_3$:methanol (1:1:4). Dry the eluate under nitrogen gas, resuspend in 100 μl of methanol, and inject directly into the LC-MS system as previously shown with methanol at 5 μl/min.
4. The Quattro Micro mass spectrometer is operated in the MRM mode with transitions of 183→82 u and 193→95 u for native and deuterated phosphoroylcholine, respectively.

12. ESI-MS/MS Analysis of Native and Newly Synthesized Phospholipids

1. Introduce samples dissolved in methanol:chloroform:water:ammonia (70:20:10:2) into the Z-spray ESI interface of a Micromass Quatro Ultima triple quadrupole mass spectrometer (Micromass, Wythenshaw, UK) by direct injection using a Hamilton syringe at a flow rate of 5 μl/min.
2. Acquire the appropriate diagnostic scans in the MCA mode on a recycling sequential basis for each diagnostic scan (precursor or neutral loss). Obtain 10 scans each of 12 to 36 s, with longer acquisition times for the less intense signal. Mass spectrometry conditions for each phospholipid class are optimized for the relevant dimyristoyl species. Collision-induced fragmentation using argon gas is then optimized to give a maximal response of the relevant fragment ion.
3. The relevant diagnostic scans are detailed in Table 10.2.

Table 10.2 Diagnostic electrospray ionization tandem mass spectrometric scans for the major phospholipid classes

Lipid class	Diagnostic scan (native)	Synthesis substrate	Diagnostic scan (synthetic)[a]
Phosphatidylcholine (PC)	P184$^+$	(methyl-d_9) choline Cl	P193$^+$ (CDP:choline pathway) P187$^+$ (PE-N-methylation)
Phosphatidylethanolamine (PE)	NL141$^+$	d_4-ethanolamine	NL145$^+$
Phosphatidylserine (PS)	NL87$^-$	d_3-serine	NL90$^-$
Phosphatidylinositol	P241$^-$	d_6-myo-inositol	P247$^-$
Phosphatidic acid (PA) and phosphatidylglycerol (PG)	P153$^-$	d_5-glycerol	P158$^-$
Phosphatidylinositol-4,5-bisphosphate (PIP$_2$)	P401$^-$	d_6-myo-inositol	P407$^-$

[a] Precursor (P) and neutral loss (NL) scans are indicated for both native phospholipid species and those with the incorporated appropriate deuterium-labeled substrate.

13. Data Analysis

Calculate the fractional and absolute synthetic rates of phospholipid synthesis, convert mass spectra to centroid format, and export to an Excel spreadsheet. Correct for formation of adduct ions where appropriate. Correct for the isotope distribution due to the approximately 1% natural abundance of ^{13}C in biological compounds. This correction eliminates the contribution of the M+2 ion peak due to the presence of two ^{13}C atoms from the species 2 mass units higher. Correct for the progressive decrease of the collision-induced ionization response with increasing m/z. The formulae are generated empirically under analytical conditions using a range of saturated and unsaturated synthetic phospholipid standards of known concentration. Calculate fractional synthetic rates for each individual molecular species from the ratio of corrected ion intensities from the diagnostic scans for the native and newly synthesized species. Calculate the absolute rates of synthesis from the recovery of the internal standard.

REFERENCES

Guillou, H., Lecureuil, C., Anderson, K. E., Suire, S., Ferguson, G. J., Ellson, C. D., Gray, A., Divecha, N., Hawkins, P. T., and Stephens, L. R. (2007). Use of the GRP1 PH domain as a tool to measure the relative levels of PtdIns(3,4,5)P3 through a protein-lipid overlay approach. *J. Lipid Res.* **48,** 726–732.

Hunt, A. N., Clark, G. T., Neale, J. R., and Postle, A. D. (2002). A comparison of the molecular specificities of whole cell and endonuclear phosphatidylcholine synthesis. *FEBS Lett.* **530,** 89–93.

Pettitt, T. R., and Wakelam, M. J. O. (2006). Identification and quantification of lipids using mass spectrometry. *In* "Chemical Biology" (B. Larijani, R. Woscholski, and C. A. Rosser, eds.), pp. 85–93. John Wiley and Sons, New York.

Pettitt, T. R., Dove, S. K., Lubben, A., Calaminus, S. D. J., and Wakelam, M. J. O. (2006). The analysis of intact phosphoinositides in biological samples. *J. Lipid Res.* **47,** 1588–1596.

Taguchi, R., Houjou, T., Nakanishi, H., Yamazaki, T., Ishida, M., Imagawa, M., and Shimizu, T. (2005). Focused lipidomics by tandem mass spectrometry. *J. Chromatog. B* **823,** 26–36.

CHAPTER ELEVEN

BIOINFORMATICS FOR LIPIDOMICS

Eoin Fahy,* Dawn Cotter,* Robert Byrnes,* Manish Sud,* Andrea Maer,* Joshua Li,* David Nadeau,* Yihua Zhau,* *and* Shankar Subramaniam*,[†,‡]

Contents

1. Introduction	248
2. Lipid Structure Databases	249
2.1. Ontology, classification, and nomenclature	249
2.2. Lipid structure databases	251
2.3. Lipid structure representation	253
3. Lipid-Associated Protein/Gene Databases	254
3.1. LIPID MAPS genome/proteome database	254
3.2. Extracting lipid genes and proteins from databases and legacy knowledge	255
4. Tools for Lipidomics	256
4.1. LIPID MAPS website and user interface	256
4.2. Tools for automated drawing and naming of lipids	260
4.3. Tools for prediction of lipid mass spectra	262
4.4. Tools for managing lipidomic metadata	263
4.5. Tools for collection, display, and analysis of experimental lipidomic data	264
4.6. Tools for lipid profiling	267
5. Lipid Pathways	268
5.1. Bioinformatics resources for lipid-related pathways	268
5.2. Tools for drawing pathways	269
6. Challenges for Future Lipid Informatics	270
Acknowledgments	272
References	272

* San Diego Supercomputer Center, University of California, San Diego, La Jolla, California
[†] Departments of Bioengineering and Chemistry and Biochemistry, University of California, San Diego, La Jolla, California
[‡] Graduate Program in Bioinformatics and Systems Biology, University of California, San Diego, La Jolla, California

Abstract

Lipids are recognized as key participants in the regulation and control of cellular function, having important roles in signal transduction processes. The diversity in lipid chemical structure presents a challenge for establishing practical methods to generate and manage high volumes of complex data that translate into a snapshot of cellular lipid changes. The need for high-quality bioinformatics to manage and integrate experimental data becomes imperative at several levels: (1) definition of lipid classification and ontologies, (2) relational database design, (3) capture and automated pipelining of experimental data, (4) efficient management of metadata, (5) development of lipid-centric search tools, (6) analysis and visual display of results, and (7) integration of the lipid knowledge base into biochemical pathways and interactive maps. This chapter describes the recent contributions of the bioinformatics core of the LIPID MAPS consortium toward achieving these objectives.

1. Introduction

The LIPID MAPS consortium is an exemplary systems biology project that quantitatively measures cell-wide lipid changes upon stimulus and attempts to reconstruct biochemical pathways associated with lipid processing and signaling. The cell-wide measurements of components of these pathways include mass spectrometric measurements of lipid changes in response to stimulus in mammalian cells, changes in transcription profiles in response to stimulus, and, in select cases, proteomic changes in response to stimulus. Reconstruction efforts will rely on organization, analysis, and integration of these data, and this warrants a strong bioinformatics effort. The UCSD Bioinformatics Laboratory is involved in the creation of the infrastructure of LIPID MAPS and in the methodology needed to carry out systematic reconstruction leading to new biological hypotheses. This infrastructure involved the development of the following:

- A completely systematic and universal classification and nomenclature system
- A LIPID MAPS website that provides a comprehensive introduction, description of the project, and reports of developments in the project
- A series of databases pertinent to lipids
- Web interfaces that provide easy navigation and query capabilities for LIPID MAPS
- Analysis tools for lipid-related molecules
- A laboratory information management system
- A pathway representation and visualization system

The following sections provide a comprehensive, albeit synoptic, description of these facets of the LIPID MAPS bioinformatics infrastructure for lipidomics.

2. Lipid Structure Databases

2.1. Ontology, classification, and nomenclature

The first step toward classification of lipids is the establishment of an ontology that is extensible, flexible, and scalable. When efforts to sequence genomes and to measure proteomes were initiated, significant investments were made in bioinformatics infrastructures that could accommodate the huge amounts of data. Structured vocabularies were developed to classify these macromolecules, database schemas were developed to house the data in relational tables, and numerous algorithms were developed to carry out extensive gene and protein sequence analysis (Jorde et al., 2005). Similarly, in the case of lipids, one must first be able to classify, name, and represent these molecules in a logical manner that is amenable to databasing and computational manipulation. A number of online resources (Table 11.1) outline comprehensive classification schemes for lipids. In view of the fact that lipids comprise an extremely heterogeneous collection of molecules from a structural and functional standpoint, it is not surprising that significant differences exist with regard to the scope and organization of current classification schemes. The Lipid Library (http://www.lipidlibrary.co.uk) site defines lipids as "fatty acids and their derivatives, and substances related biosynthetically or functionally to these compounds." The Cyberlipids (http://www.cyberlipid.org) website has a broader approach and includes isoprenoid-derived molecules such as steroids and terpenes. Both of these resources classify lipids into "simple" and "complex" groups, with simple lipids being those yielding at most two types of distinct entities upon hydrolysis (e.g., acylglycerols: fatty acids and glycerol) and complex lipids (e.g., glycerophospholipids: fatty acids, glycerol, and headgroup) yielding three or more products upon hydrolysis. The LipidBank (http://lipidbank.jp) database in Japan defines 17 top-level categories in their classification scheme, covering a wide variety of animal and plant sources. The LIPID MAPS group (http://www.lipidmaps.org) has taken a more chemistry-based approach and defines lipids as hydrophobic or amphipathic small molecules that may originate entirely or in part by carbanion-based condensations of thioesters (fatty acids, polyketides, etc.) and/or by carbocation-based condensations of isoprene units (prenols, sterols, etc.). Lipids are divided into eight categories (fatty acyls, glycerolipids, glycerophospholipids, sphingolipids, sterol lipids, prenol lipids, saccharolipids,

Table 11.1 Online resources for lipid classification and databasing

Resource	URL	Country	Comments
Lipid Library	http://www.lipidlibrary.co.uk	U.K.	Multiple reference topics and examples, mass spectrometry (MS) and nuclear magnetic resonance (NMR) libraries, literature service
Cyberlipid Center	http://www.cyberlipid.org	France	Descriptions of lipid classes with examples and literature references
LipidBank	http://lipidbank.jp	Japan	Lipid database with a wide variety of categories, including plant lipids
LIPIDAT	http://www.lipidat.chemistry.ohio-state.edu	U.S.	Database composed mostly of phospholipids and associated thermodynamic data
LIPID MAPS	http://www.lipidmaps.org	U.S.	Classification scheme, lipid and protein databases, MS libraries, search tools

and polyketides) containing distinct classes and subclasses of molecules. A 12-digit unique identifier is associated with each distinct lipid molecule.

For each lipid, the unique 12-character identifier based on this classification scheme (Fahy et al., 2005) is an important database field in the Lipids Database. The format of the LIPID ID provides a systematic means of assigning a unique identification to each lipid molecule and allows for the addition of large numbers of new categories, classes, and subclasses in the future. The last four characters of the ID comprise a unique identifier within a particular subclass and are randomly assigned. By initially using numeric characters, this allows 9999 unique IDs per subclass; however, with the additional use of 26 uppercase alphabetic characters, a total of 1.68 million possible combinations can be generated, providing ample scalability within each subclass. In cases where lipid structures were obtained from other sources, such as LipidBank or LIPIDAT, the corresponding IDs for those databases are included to enable cross-referencing. The LIPID MAPS classification scheme has been adopted by KEGG (Kyoto Encyclopedia of Genes and Genomes), where functional hierarchies involving lipids, reactions, and pathways have been constructed (http://www.genome.ad.jp/brite/). The classification system may conveniently be browsed online from the LIPID MAPS website (http://www.lipidmaps.org), in which the various classes and subclasses are linked to the LIPID MAPS structure database (LMSD).

Nomenclature of lipids falls into two main categories: systematic names and common, or trivial, names. The latter includes abbreviations, which are a convenient way to define acyl/alkyl chains in acylglycerols, sphingolipids, and glycerophospholipids. The generally accepted guidelines for systematic names have been defined by the International Union of Pure and Applied Chemists and the International Union of Biochemistry and Molecular Biology (IUPAC-IUBMB) Commission on Biochemical Nomenclature (http://www.chem.qmul.ac.uk/iupac/). The use of core structures such as prostanoic acid, cholestane, or phosphocholine is strongly recommended as a way of simplifying systematic nomenclature; commercially available software packages that perform structure-to-name conversions generally create overly complicated names for several categories of lipids.

Many lipids, in particular the glycerolipids, glycerophospholipids, and sphingolipids, may be conveniently described in terms of a shorthand name, where abbreviations are used to define backbones, headgroups, and sugar units, and the radyl substituents are defined by a descriptor indicating carbon chain length and number of double bonds. These shorthand names lend themselves to fast, efficient, text-based searches and are used widely in lipid research as compact alternatives to systematic names. The use of a shorthand notation for selected lipid categories that incorporates a condensed text nomenclature for glycan substituents has also been deployed by LIPID MAPS. The abbreviations for the sugar units follow the current IUPAC-IUBMB recommendations (Chester, 1998).

2.2. Lipid structure databases

Modern bioinformatics has become increasingly sophisticated, permitting complex database schemas and approaches that enable efficient data accrual, storage, and dissemination. A large number of repositories, such as GenBank, SwissProt, and ENSEMBL (http://www.ensembl.org), support nucleic acid and protein databases; however, only a few specialized databases (e.g., LIPIDAT [Caffrey and Hogan, 1992] and LipidBank [Watanabe *et al.*, 2000]) are dedicated to cataloging lipids. The LIPIDAT database, developed by Martin Caffrey's group at Ohio State University, focuses on the biophysical properties of glycerolipids and sphingolipids and contains over 12,000 unique molecular structures. The LipidBank online database in Japan contains over 6000 structures across a broad range of lipid classes and is a rich resource for associated spectral data, biological properties, and literature references.

Given the importance of these molecules in cellular function and pathology, it is essential to have a well-organized database of lipids with a defined ontology that is extensible, flexible, and scalable. The ontology of lipids must incorporate classification, nomenclature, structure representations, definitions, related biological/biophysical properties, cross-references, and structural features (formula, molecular weight, number of carbon atoms, number of various functional groups, etc.) of all objects stored in the database. This ontology is then transformed into a well-defined schema that forms the foundation for a relational database of lipids. An object-relational database of lipids, based on the previously mentioned classification scheme and containing structural, biophysical, and biochemical characteristics, is available on the LIPID MAPS website with browsing and searching capabilities. The database (Sud et al., 2007) currently contains over 10,000 structures, which are obtained from these four sources: the LIPID MAPS consortium's core laboratories and partners; lipids identified by LIPID MAPS experiments; computationally generated structures for appropriate lipid classes; and biologically relevant lipids manually curated from LipidBank, LIPIDAT, and other public sources. All structures have been classified and redrawn according to LIPID MAPS guidelines. After lipids have been selected for inclusion into LMSD, they are classified following the LIPID MAPS classification scheme as explained earlier under the "Ontology, Classification, and Nomenclature" section. Structures of the lipids are drawn either manually or generated automatically by computational structure drawing tools developed by the LIPID MAPS consortium; the structure representation is consistent and adheres to the rules proposed by the LIPID MAPS consortium. LMSD implements storage of lipid structure representations using the following three formats: Binary Large Object (BLOB), ChemDraw Exchange (CDX) format (www.cambridgesoft.com/services/documentation/sdk/chemdraw/cdx/), and Graphics Interchange Format (GIF). LMSD uses BLOB format to store MDL MOL file structural representation via the Accord Chemical Cartridge. CDX format, a richer and more flexible format with support for not only structure data but also for visual characteristics and annotations, is stored in LMSD as Character Large Object (CLOB) data; CDX format objects are used to support viewing of structures via ChemDraw viewer. GIF images representing structures are stored in the database table as BLOB objects. The structures are either drawn manually using ChemDraw or generated automatically by structure drawing tools developed by the LIPID MAPS consortium. Based on its classification, each lipid structure in LMSD is assigned a unique 12-character LIPID MAPS identifier (LM ID). The format of the LM ID not only maintains uniqueness of ID, but also provides the capability to add new categories, classes, and subclasses as the need arises. In addition to manual curation of biologically relevant lipids from LIPIDAT and LipidBank according to LIPID MAPS classification and structure representation schemes, LMSD also maintains their original IDs to enable cross-referencing.

2.3. Lipid structure representation

In addition to having rules for lipid classification and nomenclature, it is important to establish clear guidelines for drawing lipid structures. These guidelines are a prerequisite for any useful lipid molecular database in terms of consistent molecular structure presentation. Large and complex lipids are difficult to draw, which leads to the use of shorthand and unique formats that often generate more confusion than clarity among lipid researchers. The LIPID MAPS consortium has chosen a consistent format for representing lipid structures where, in the simplest case of the fatty acid derivatives, the acid group (or equivalent) is drawn on the right and the hydrophobic hydrocarbon chain is on the left (Fig. 11.1). For fatty acids and derivatives, the acid group or its equivalent is drawn on the right side and the

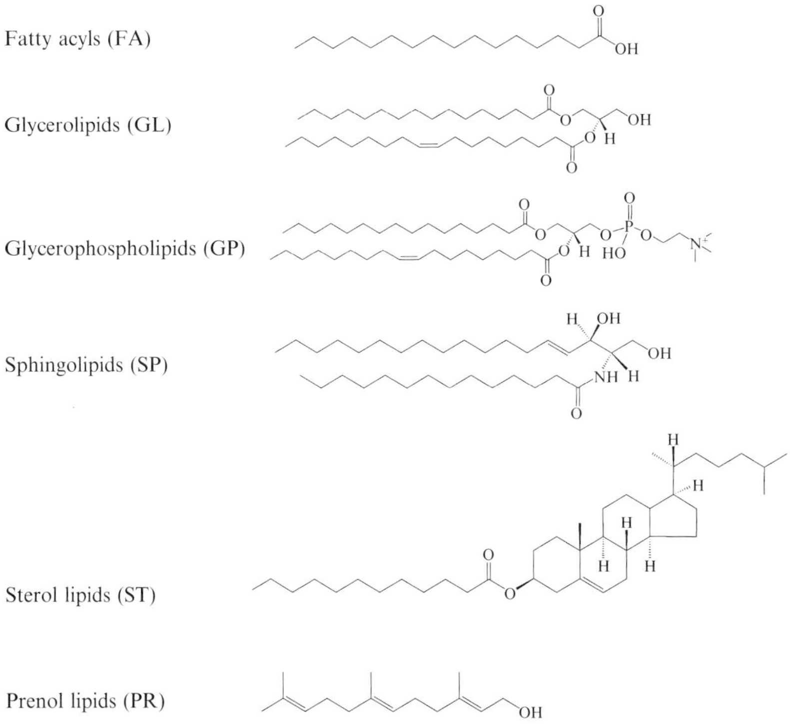

Figure 11.1 Consistent format for representing lipid structures.

hydrophobic chain on the left, except for the eicosanoid class, in which the hydrocarbon chain wraps around in a counter clockwise fashion to produce a more compact structure. For glycerolipids and glycerophospholipids, the radyl hydrocarbon chains are drawn to the left; the glycerol group is drawn horizontally with stereochemistry defined at *sn* carbons; the headgoups for glycerophospholipids are depicted on the right. For sphingolipids, the C1 hydroxyl group of the long-chain base is placed on the right and the alkyl portion on the left; the headgroup of sphingolipids ends up on the right. The linear prenols and isoprenoids are drawn like fatty acids, with the terminal functional groups on the right. Several structurally complex lipids—acylaminosugar glycans, polycyclic isoprenoids, and polyketides—cannot be drawn using these simple rules. Rather, these structures are drawn using commonly accepted representations. This approach, adopted by the LIPID MAPS consortium, has the potential to standardize drawings of lipids and eliminate ambiguity in structural representations to a reasonable extent.

3. Lipid-Associated Protein/Gene Databases

3.1. LIPID MAPS genome/proteome database

Although the public domain EntrezGene and UniProt databases contain sequences and annotations for most of the known lipid-associated genes and proteins, no unique gene and protein database of lipid-associated proteins contains comprehensive summary- and context-dependent annotation of these molecules. In order to fill this void, the LIPID MAPS Proteome Database (LMPD) was developed to provide a catalog of genes and proteins involved in lipid metabolism and signaling. The initial release of LMPD establishes a framework for creating a lipid-associated protein list, collecting relevant annotations, databasing this information, and providing a user interface. The LMPD is an object-relational database of lipid-associated protein sequences and annotations (Cotter *et al.*, 2006). The current version contains approximately 3000 records, representing human and mouse proteins involved in lipid metabolism and signaling. A new release, which is currently undergoing extensive curation, is planned for spring of 2007 with substantially more records. Users may search LMPD by database ID or keyword, and filter by species and/or lipid class associations. From the search results, one can then access a compilation of data relevant to each protein of interest, cross-linked to external databases. LMPD is publicly available from the LIPID MAPS consortium website (http://www.lipidmaps.org/data/proteome/index.cgi).

3.2. Extracting lipid genes and proteins from databases and legacy knowledge

In order to populate the LMPD, a list of lipid-related GO (Gene Ontology, http://www.geneontology.org) (Harris et al., 2004) terms and KEGG (Kanehisa and Goto, 2000) pathway data was compiled, using lipid-specific keywords, such as trivial names of classes, subclasses, and individual lipid compounds (Fig. 11.2). In the new release, the list of lipid-specific keywords was expanded, and, in addition to GO and KEGG terms, the description field of UniProt records and the EntrezGene names were scanned (for human and mouse species). The UniProt (Apweiler et al., 2004) proteins annotated with those terms were then collected, and these proteins were classified (based on their substrates/products or interactions) according to the lipid classification scheme previously described. Annotations are organized by category: record overview, Gene/GO/KEGG Information, UniProt annotations, related proteins, and experimental data. The record overview contains LMPD_ID, species, description, gene symbols, lipid categories, EC number, molecular weight, sequence length, and protein sequence. Gene information includes EntrezGene ID, chromosome, map location, primary name, primary symbol, and alternate names and symbols; GO IDs and descriptions; and KEGG pathway IDs and descriptions. UniProt annotations include primary accession number, entry name, and comments, such as catalytic activity, enzyme

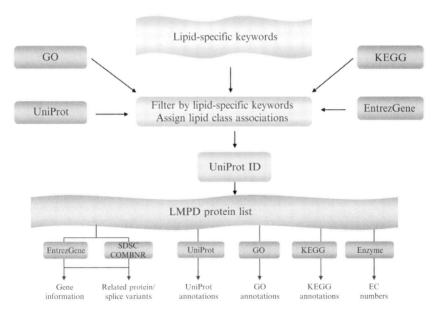

Figure 11.2 Overview of the LIPID MAPS Proteome Database (LMPD). (See color insert.)

regulation, function, and similarity. Efforts are underway to curate the current database of lipid-associated genes/proteins by (1) selecting entries with annotations from multiple sources (GO, UniProt, KEGG, EntrezGene), (2) ruling out entries whose annotations consist only of similarity-based terms ("like," "similar to," "hypothetical," etc.), (3) conducting PubMed and other literature searches, and (4) consulting with experts in the lipidomics field. The resulting "high-confidence" set of lipid-associated genes/proteins will be useful as sources of information for interfacing with studies on lipid-related pathways and networks.

4. TOOLS FOR LIPIDOMICS

4.1. LIPID MAPS website and user interface

In an effort to increase public awareness of lipidomics research and resources, the LIPID MAPS consortium has developed a website (http://www.lipidmaps.org) that provides a comprehensive introduction and description of the project, a set of databases and online query interfaces pertinent to lipids, access to a wide range of experimental data relating to lipids, and a collection of analysis tools for lipid-related molecules. In addition to providing background information about the goals of the LIPID MAPS project, participants, and literature references, the website (Fig. 11.3) is divided into three main sections:

1. **Data**: Interfaces to lipid and gene/protein databases, lipid classification system, lipid standards, experimental lipid time-course data, and gene-array data.
2. **Tools**: Interfaces to lipidomics tools, such as drawing programs and mass spectrometry (MS) prediction tools
3. **Protocols**: Information on procedures and protocols relating to lipidomic research

A good example of the type of online browsing and querying capability of the website is provided by the interface to the LMSD. The LMSD browsing page provides multiple search formats, including the capability to retrieve lipids based on the LIPID MAPS classification scheme. After the user selects one of the main categories of lipids, a listing of all lipids present in the selected category, along with a link to the set of lipids in each main class and subclass, is provided. The user may then select all lipids that belong to either a main class or a subclass and display the results as a result summary page. In the case of lipids containing multiple functional groups, assignment of a structure to a particular subclass may be somewhat subjective. For example, a fatty acid containing both epoxy and hydroxy groups could be assigned to either epoxy or hydroxy

LIPID MAPS Bioinformatics of Lipids

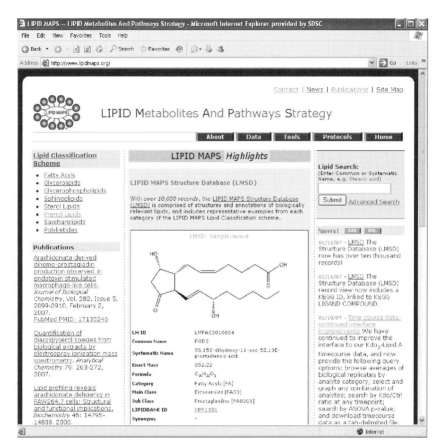

Figure 11.3 LIPID MAPS public website (http://www.lipidmaps.org). (See color insert.)

fatty acids subclass. To address this situation, an ontology-based search is also provided. The user may choose to search for lipids containing similar functionality, and all the lipids with the specified functionality, irrespective of their subclass designation, will be retrieved.

The text-based query page (Fig. 11.4) allows the user to search LMSD by any combination of the following data fields: LM ID, common or systematic name, mass along with a tolerance value, formula, category, main class, and subclass. Selecting a category from the category drop-down menu causes the corresponding set of main classes to appear in the main class drop-down menu. Selecting a main class then shows the corresponding set of subclasses in the subclass menu.

The structure-based query page provides the capability to search LMSD by performing a substructure or exact match using the structure drawn by the user. Supported structure drawing tools are Chemdraw, MarvinSketch

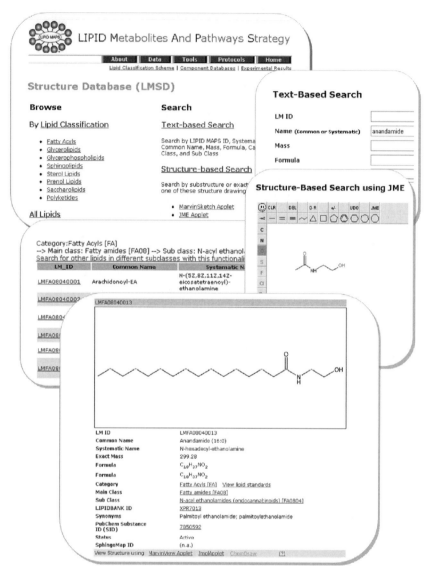

Figure 11.4 Online interfaces for searching the LIPID MAPS Structure Database (LMSD). (See color insert.)

(www.chemaxon.com/marvin/), and JME (www.molinspiration.com/jme/index.html). The latter two structure drawing tools are Java applets and require only applet support in the browser. In addition to structure, the user can also specify an LM ID or a common or systematic name for the search.

The results summary page displays a table containing all the lipids matching specified search criteria. The results table contains the following columns for text-based and structure-based queries: LM ID, common name, systematic name, main class, subclass, and exact mass. The record details page, in addition to displaying the structure for the selected lipid, also contains the following data fields: LM ID, common name, systematic name, formula, category, main class, subclass, synonyms, and status. Appropriate links to other external databases and collections, such as LipidBank, KEGG, LIPIDAT, PubChem, and SphingoMap, are also provided, where applicable.

The default page uses a GIF image for representing the structure of the lipid. The following additional structure viewing tools are also supported: MarvinView, (www.chemaxon.com/products.html), JMol (http://jmol.sourceforge.net/), and ChemDraw.ActiveX/Plugin Viewer (www.cambridgesoft.com/software/ChemDraw/). LIPID MAPS lipid structures are now available on NCBI's PubChem website, where they have been assigned PubChem Substance ID's. LMSD lipid structures are deposited into the PubChem database (http://pubchem.ncbi.nlm.nih.gov/) periodically, and a link to PubChem Substance ID (SID) is also maintained within LMSD. All the deposited LIPID MAPS lipids on the PubChem website have hyperlinks back to the LIPID MAPS site. The entire LMSD dataset in now available for download in SD file format from the LIPID MAPS website at http://www.lipidmaps.org/data/structure/index.html.

The online interface (Fig. 11.5) to LMPD also has multiple browsing/query formats. The default query form allows the user to browse the protein list with the ability to browse by associated lipid category. The "Advanced Search" query form provides choices for conducting a more focused search, including options to search by database ID or keyword and to filter by species and/or lipid class association. Database ID fields searched include UniProt accession, UniProt entry name, gene symbols, GenBank GI, EC number, GO ID, and KEGG pathway ID. Keyword search fields include UniProt description and Swiss-Prot comments. The results summary page presents a sortable list of proteins matching the query criteria, along with selected summary information, including LMPD_ID, accession, protein name, protein symbol, and associated lipid categories. From the summary page, the user may display complete LMPD annotations for each protein. For each record selected from the results summary, all LMPD data relevant to that protein are displayed, with external database IDs linked to their respective resources. Annotations are organized by category: record overview, Gene/GO/KEGG information, UniProt annotations, and related proteins. The record overview contains LMPD_ID, species, description, gene symbols, lipid categories, EC number, molecular weight, sequence length, and protein sequence. Gene information includes EntrezGene ID, chromosome, map location, primary name, primary symbol, and alternate names and symbols; GO IDs and descriptions; and KEGG pathway IDs and descriptions. UniProt annotations include primary

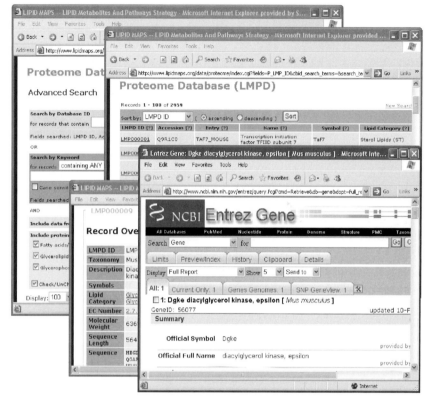

Figure 11.5 Online interface to LIPID MAPS Proteome Database (LMPD) at the LIPID MAPS website. (See color insert.)

accession number, entry name, and comments, such as catalytic activity, enzyme regulation, function, and similarity.

4.2. Tools for automated drawing and naming of lipids

In addition to having rules for lipid classification and nomenclature, it is important to establish clear guidelines for drawing lipid structures. Large and complex lipids are difficult to draw, which leads to the use of many unique formats that often generate more confusion than clarity among the lipid research community. For example, the use of SMILES (Simplified Molecular Input Line Entry System) strings to represent lipid structures, while being very compact and accurate in terms of bond connectivity, valence, and chirality, causes problems when the structure is rendered. This is due to the fact that the SMILES format does not include two-dimensional (2D) coordinates. Therefore, the orientation of the structure is arbitrary, making visual recognition and comparison of related structures much more difficult. Additionally, the structure drawing step is typically the most time-consuming process in creating molecular databases of lipids; however, many classes of lipids lend themselves well as targets for automated structure drawing due to their consistent 2D layout. A suite of structure-drawing tools has been developed and deployed that dramatically increases the efficiency of data entry into lipid-structure databases and permits "on-demand" structure generation in conjunction with a variety of MS–based informatics tools. Online versions of the structure drawing tools for fatty acyls, glycerolipids, glycerophospholipids, sphingolipids, and sterols are available in the "Tools" section of the LIPID MAPS website. Current versions of this software are now capable of drawing chiral centers and ring structures. In addition, a set of standalone drawing tools, written in Perl, are available for download from the LIPID MAPS website (Fig. 11.6) (http://www.lipidmaps.org/tools/).

Concurrently, a generalized lipid abbreviation format has been developed that enables structures, systematic names, and ontologies to be

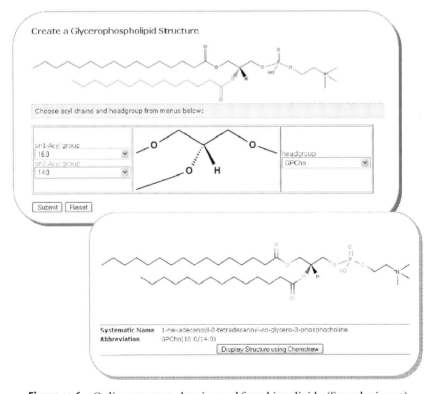

Figure 11.6 Online structure-drawing tool for sphingolipids. (See color insert.)

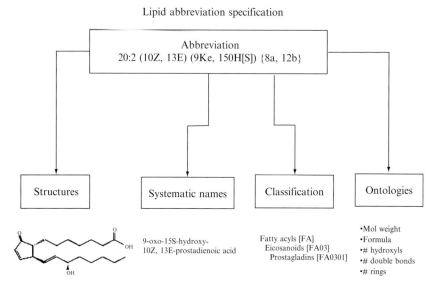

Figure 11.7 Flowchart showing structure/name/ontology generation from abbreviation.

generated automatically from a single source format (Fig. 11.7). Lipid structure, nomenclature, classification, and ontology creation may be performed computationally, using well-defined, unique abbreviations as a starting point. Figure 11.7 demonstrates the conversion of a text abbreviation for a bile acid into a dataset containing structure (Molfile), systematic name, classification, and various molecular attributes, such as formula, molecular weight, number of functional groups, double bonds, and rings.

4.3. Tools for prediction of lipid mass spectra

Certain classes of lipids, such as acylglycerols and phospholipids, composed of an invariant core (glycerol and headgroups) and one or more acyl/alkyl subsituents are good candidates for MS computational analysis since they fragment more predictably in electrospray ionization mass spectrometry (ESI-MS), leading to loss of acyl side-chains, neutral loss of fatty acids, and loss of water and other diagnostic ions depending on the phospholipid headgroup (Murphy et al., 2001). It is possible to create a virtual database of permutations of the more common side chains for glycerolipids and glycerophospholipids and to calculate "high-probability" product ion candidates in order to compare experimental data with predicted spectra. The LIPID MAPS group has developed a suite of search tools allowing the user to enter a mass-to-charge (m/z) value of interest and view a list of matching structure candidates, along with a list of calculated neutral-loss ions and "high-probability" product ions. These search interfaces have been integrated with structure-drawing and isotopic-distribution tools (Fig. 11.8).

Currently, MS prediction tools (http://www.lipidmaps.org/tools/) for three different categories of lipids (glycerolipids, cardiolipins, and glycerophospholipids) can be found online. In each case, all possible structures corresponding to a list of likely headgroups and acyl, alkyl-ether, and vinyl-ether chains were expanded and enumerated by computational methods to generate a table containing the nominal and exact mass for each discrete structure as well as additional ontological information, such as formula, abbreviation, and numbers of chain carbons and double bonds. This tabular data was then uploaded into category-specific database tables, making it amenable for online tools.

4.4. Tools for managing lipidomic metadata

To accomplish transfer, centralized storage, sharing, and integration of "metadata" (information on experimental procedures, protocols, reagents, and resources) among LIPID MAPS member laboratories, a Laboratory Information Management System (LIMS) customized for lipidomics research has been developed. The LIPID MAPS LIMS is written in the Java programming language and is implemented as a Java Web Start program that runs on a user's desktop computer. The modules allow users to enter information and browse the LIMS data, which is connected to a central Oracle database. The LIMS also allows tracking of laboratory materials and protocols via printed labels that may be scanned into modules using barcode readers.

The LIPID MAPS LIMS is a two-tier client-server window form-based application that, at its core, enables entry and tracking of information pertinent to lipidomics experiments. It is composed of 13 modules, selectable from a main form. The LIMS is primarily devoted to cell treatments and MS experiments, and the essential process of a simplified pattern of usage is shown in Fig. 11.9.

The LIPID MAPS LIMS is a specialized, yet powerful data tracking system for lipidomics studies. It provides functions that encompass many of the goals of systems biology, where each LIMS module represents an idealized abstraction of an experimental step performed in a laboratory and records information pertinent for that step. It provides a means of accurately associating experimental conditions and protocols (metadata) with the corresponding experimental results (data) in a multi-user, multi-center research environment.

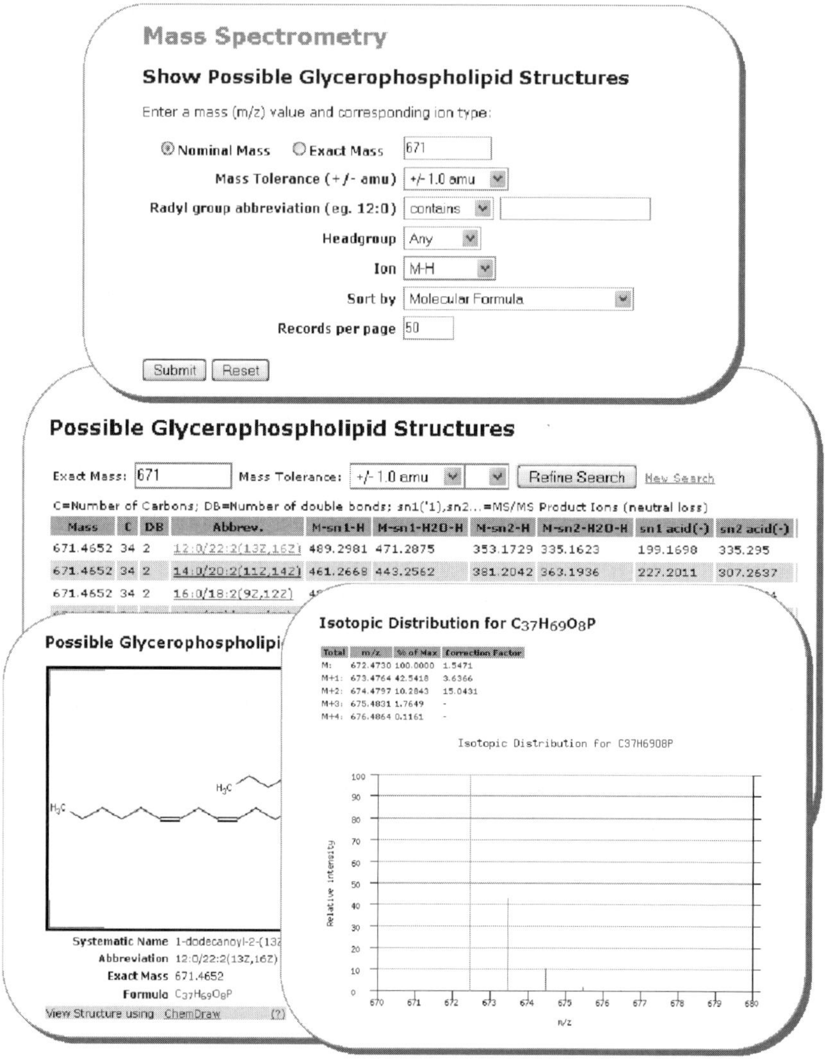

Figure 11.8 Online utility to predict possible lipid structures from mass spectrometry (MS) data. (See color insert.)

4.5. Tools for collection, display, and analysis of experimental lipidomic data

Data analysis of lipidomics experiments represents significant challenges both in volume and complexity. Typical experiments yield component lists of lipids with quantitative content data and a catalog of interactions and networks involving lipids. A truly comprehensive "lipidomics" approach must

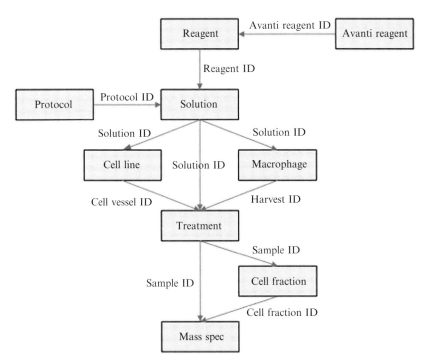

Figure 11.9 Simplified sequential flow of Laboratory Information Management System (LIMS) usage. Rectangles represent modules of functionality as described in the text. Arrow labels indicate sets of one or more identifiers that are generated by the module at the base of the arrow and entered into the immediately following target module. (See color insert.)

incorporate multiple separation and identification techniques, maintaining sufficient sensitivity to distinguish closely related metabolites while remaining robust enough to cope with the wide variety of heterogeneous classes of cellular lipid species. ESI-MS, with its extraordinary sensitivity and its capacity for high throughput, has become the technique of choice for the analysis of complex mixtures of lipids from biological samples. From a bioinformatics standpoint, one must record and store all the experimental conditions and protocols pertaining to a particular set of experiments (metadata) via the aforementioned LIMS, as well as the actual experimental measurements (data).

With the advent of sensitive analytical instrumentation, such as MS, it is now possible to obtain quantitative data on large numbers of lipid species under a variety of experimental conditions, allowing us to investigate time-dependent changes in lipid-associated processes in response to a variety of stimuli. For example, the LIPID MAPS consortium has embarked on a time-dependent study of a wide range of lipid classes in mouse macrophage cells in response to stimulation of the Toll-like receptor 4 (TLR4) with

Kdo$_2$-Lipid A. This required the establishment of an informatics pipeline to convert MS experimental data into a format suitable for viewing by end-users. In this case, MS quantitative time-course measurements in heterogeneous formats on a variety of lipids over a 24-hour time-course were first converted into a common data format prior to processing and import into an Oracle database. A middleware layer composed of Apache webserver and PHP scripting was used to create a web-based user interface with the MS database data. All online data displays were integrated with the LIMS system (via sample barcodes) and the LIPID MAPS structure database, allowing seamless navigation across both data and metadata. A software drawing component was used to generate "on-demand" online graphs in response to user input. In addition, a set of online query and display tools was developed to allow the end-user to view MS time-course data in a number of different formats (Fig. 11.10). These include tabular and graphical displays of data as averages of technical and biological replicates, as well as "drill-down" links to the corresponding LIMS metadata (cell samples) and structure/classification information (analytes).

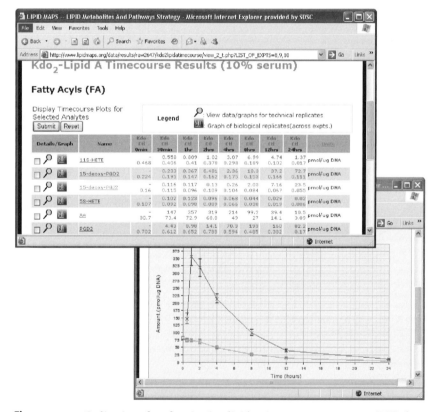

Figure 11.10 Online interface for viewing lipidomics mass spectrometry (MS) data. (See color insert.)

Query tools include a feature that allows the end-user to select any combination of analytes for graphing with a view to comparing time-course profiles of lipids that are part of the same biochemical or signaling pathways. Quantitative data from these experiments are being used to validate existing lipid networks and to elucidate novel interactions.

The structural heterogeneity of many classes of lipids precludes accurate and reliable prediction of MS fragmentation in tandem spectra (MS/MS) in contrast to the situation for MS interpretation of proteins, where facile cleavage of the peptide bonds has led to the development of search algorithms, which compare MS/MS fragmentation data to a list of "virtual spectra" computed from a protein database. In this situation, comparison of spectral data with libraries of lipid standards can be very informative. An online library of lipid standards, including MS/MS data generated by the LIPID MAPS core facilities, has recently been made available to the public. This database currently consists of approximately 200 analytes spanning seven lipid categories. An online search interface (Fig. 11.11) allowing end-users to search by precursor or product ion mass has also been developed.

4.6. Tools for lipid profiling

With the advent of high-sensitivity liquid chromatography MS (LC-MS) systems, one increasingly utilized type of LC-MS application is differential profiling in which the extraction, LC methods, and MS instrument setup are set to provide a broad coverage of compounds in order to enable relative quantitative comparisons for individual compounds across multiple samples. The applications of such an approach can be found in domains of systems biology, functional genomics, and biomarker discovery. Within the last couple of years, a number of software packages have been developed by various companies and research groups to analyze data generated by MS profiling of metabolites, including lipids. The data processing for differential profiling usually proceeds through several stages, including input file manipulation, spectral filtering, peak detection, chromatographic alignment, normalization, visualization, and data export. Examples of metabolic profiling software are the MarkerView package (ABI) and the freely-available Java-based Mzmine (Katajamaa and Oresic, 2005) application (Turku Centre for Biotechnology, Finland). The LIPID MAPS bioinformatics group has set up the Mzmine software to run in parallel-processing, creating a high-throughput computational cluster devoted to the analysis of lipid profiling experiments of the LIPID MAPS cores.

Within the LIPID MAPS consortium, in-house software has been developed for more specific profiling tasks, such as processing and comparing control and deuterium-supplemented MS data sets from LC-MS experimental runs. This involves a stepwise process of (1) extracting data from raw MS data files, (2) integrating peaks in each scan over a user-defined mass

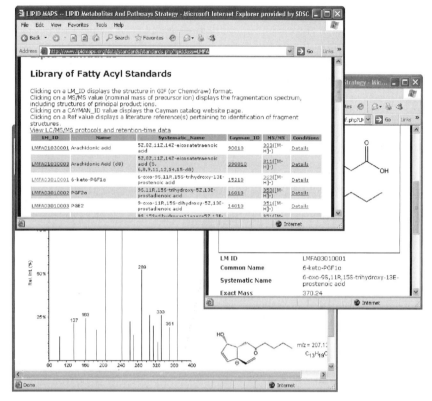

Figure 11.11 Online interface to Lipid Standards Library and associated data. (See color insert.)

window to generate a formatted array of quantitated data, (3) applying an algorithm to identify deuterated metabolites, (4) performing data normalization and subtractive analysis, and (5) displaying and listing peaks of interest according to user-defined criteria. This software has been utilized to investigate the biosynthesis of eicosanoids derived from arachidonic acid in RAW cells (Harkewicz et al., 2007).

5. Lipid Pathways

5.1. Bioinformatics resources for lipid-related pathways

Lipids play central roles in energy storage, cell membrane structure, cellular communication, and regulation of biological processes, such as inflammatory response, neuronal signal transmission, and carbohydrate metabolism.

Organizing these processes into useful, interactive pathways and networks represents a great bioinformatics challenge. The KEGG consortium maintains a collection of manually drawn pathway maps representing current knowledge on the molecular interaction and reaction networks, several of which pertain to lipids (http://www.genome.jp/kegg/pathway.html), including fatty acid biosynthesis and degradation, sterol metabolism, and phospholipids pathways. Additionally, the KEGG Brite (http://www.genome.jp/kegg/brite.html) collection of hierarchical classifications includes a section devoted to lipids where the user can select a lipid of interest and view reactions and pathways involving that molecule. A number of category-specific lipid pathways have been constructed, notably the SphinGOMAP (http://www.sphingomap.org), a pathway map of approximately 400 different sphingolipid and glycosphingolipid species (Sullards et al., 2003), by a member of the LIPID MAPS consortium.

5.2. Tools for drawing pathways

LIPID MAPS has developed a comprehensive graphical pathway display, editing, and analysis program, LIPID MAPS Biopathways Workbench (LMBP), with the capability to display a variety of lipid-related entities and states, such as lipids, enzymes, genes, pathways, activation states, and experimental results. This graphical tool is tightly integrated with the underlying LIPID MAPS relational databases and will play an important role in the reconstruction of lipid networks using emerging data from LIPID MAPS core facilities and other public sources.

The central theme for LMBP data is a reaction network graph, or pathway, with chemical compounds represented as nodes, and chemical interactions represented as links, or edges, between the nodes. Numeric, text, and structured data attributes may be bound to nodes and edges, including molecule names, synonyms, locations, concentrations, charges, interaction types, molecule roles in those interactions, and so forth. For example, experimental data such as time-course measurements for a series of lipids may be displayed in graphical form in a pathway window (Fig. 11.12).

The LMBP software tools may be downloaded from the http://www.biopathwaysworkbench.orgwebsite. The current alpha/beta release is comprised of two components: (1) The Biopathways Workbench (BPW; a Java programming toolkit for managing pathway data) and (2) the Pathway Editor (a Java application built using the BPW). It presents a pathway drawing area, menu bar, toolbar, and operations to open, edit, and save pathways and search databases of pathways, compounds, and time-course data.

The current version contains an extensive list of metabolic pathways that have been imported from KEGG as well as tools allowing the user to create pathway diagrams from scratch.

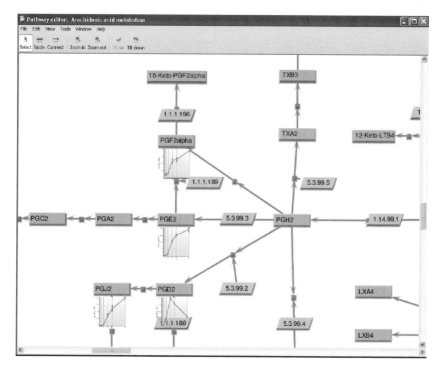

Figure 11.12 Screenshot of the Biopathways editor showing a portion of the arachidonate pathway with overlaid timecourse data. (See color insert.)

The freely available VANTED application (http://vanted.ipk-gatersleben.de/) has also been used by the bioinformatics core to import existing KEGG lipid-specific pathways, to modify and customize existing KEGG pathways, and to draw new pathways and networks. The application enables the user to import and graph experimental data from Excel templates, to store and export data-containing pathways, and to perform many types of statistical analyses on the networks.

Additionally, custom VANTED templates (Fig. 11.13) have been created in which the nodes corresponding to lipids and enzymes have been hyperlinked to the LIPID MAPS structure and protein databases, respectively, enabling a high level of integration between key bioinformatics resources.

6. Challenges for Future Lipid Informatics

Despite the significant role of lipids in normal and disease physiology, lipidomics has not reached the same stage of advancement and knowledge as genomics and proteomics. With the advent of precise MS methods, we are

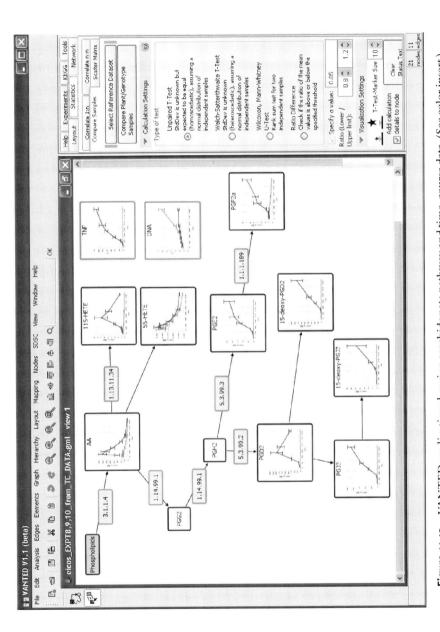

Figure 11.13 VANTED application showing arachidonate pathway and time-course data. (See color insert.)

now in a position to measure dynamical changes in lipids in cells subject to stimulus. However, no tools are currently available that relate these changes to phenotypic changes in cellular states or physiology. The context-specific identification of lipid components, their integration with the proteome and transcriptome, and the derivation of lipid networks in normal and pathological states remain as major challenges for lipidomics. Initiatives such as LIPID MAPS are early steps in this direction.

ACKNOWLEDGMENTS

The authors wish to express their sincere gratitude to LIPID MAPS core directors H. Alex Brown (Vanderbilt University), Christopher K. Glass (University of California, San Diego), Alfred H. Merrill Jr. (Georgia Institute of Technology), Robert C. Murphy (University of Colorado Health Sciences Center), Christian R. H. Raetz (Duke University Medical Center), David W. Russell (University of Texas Southwestern Medical Center), and LIPID MAPS principal investigator Edward A. Dennis (University of California, San Diego). This work was supported by the National Institutes of Health (NIH) National Institute of General Medical Sciences (NIGMS) Glue Grant NIH/NIGMS Grant 1 U54 GM69558.

REFERENCES

Apweiler, R., Bairoch, A., Wu, C. H., Barker, W. C., Boeckmann, B., Ferro, S., Gasteiger, E., Huang, H., Lopez, R., Magrane, M., Martin, M. J., Natale, D. A., et al. (2004). UniProt: The Universal Protein knowledgebase. *Nucleic Acids Res.* **32,** D115–D119.

Caffrey, M., and Hogan, J. (1992). LIPIDAT: A database of lipid phase transition temperatures and enthalpy changes. DMPC data subset analysis. *Chem. Phys. Lipids* **61,** 1–109.

Chester, M. A. (1998). IUPAC-IUB Joint Commission on Biochemical Nomenclature (JCBN). Nomenclature of glycolipids LIPIDAT: A database of lipid phase transition temperatures and enthalpy changes. *Eur. J. Biochem* **257,** 293–298. Accessed at http://www.chem.qmul.ac.uk/iupac/misc/glylp.html.

Cotter, D., Maer, A., Guda, C., Saunders, B., and Subramaniam, S. (2006). LMPD: LIPID MAPS proteome database. *Nucleic Acids Res.* **34,** D507–D510.

Fahy, E., Subramaniam, S., Brown, H. A., Glass, C. K., Merrill, A. H., Jr., Murphy, R. C., Raetz, C. R. H., Russell, D. W., Seyama, Y., Shaw, W., Shimizu, T., Spener, F., et al. (2005). A comprehensive classification system for lipids. *J. Lipid Res.* **46,** 839–862.

Harkewicz, R., Fahy, E., Andreyev, A., and Dennis, E. A. (2007). Arachidonate-derived dihomoprostaglandin production observed in endotoxin-stimulated macrophage-like cells. *J. Biol. Chem.* **282,** 2899–2910.

Harris, M. A., Clark, J., Ireland, A., Lomax, J., Ashburner, M., Foulger, R., Eilbeck, K., Lewis, S., Marshall, B., Mungall, C., et al. (2004). The Gene Ontology (GO) database and informatics resource. *Nucleic Acids Res.* **32,** D258–D261.

Jorde, L. B., Little, P. F. R., Dunn, M. J., and Subramaniam, S. (2005). "Encyclopedia of Genetics, Genomics, Proteomics and Bioinformatics." Wiley Press, Hoboken, NJ.

Kanehisa, M., and Goto, S. (2000). KEGG: Kyoto encyclopedia of genes and genomes. *Nucleic Acids Res.* **28,** 27–30.

Katajamaa, M., and Oresic, M. (2005). Processing methods for differential analysis of LC/MS profile data. *BMC Bioinformatics* **6**, 179.

Murphy, R. C., Fiedler, J., and Hevko, J. (2001). Analysis of nonvolatile lipids by mass spectrometry. *Chem. Rev.* **101**, 479–526.

Sud, M., Fahy, E., Cotter, D., Brown, A., Dennis, E. A., Glass, C. K., Merrill, A. H., Jr., Murphy, R. C., Raetz, C. R., Russell, D. W., and Subramaniam, S. (2007). LMSD: LIPID MAPS structure database. *Nucleic Acids Res.* **35**, D527–D532.

Sullards, M. C., Wang, E., Peng, Q., and Merrill, A. H., Jr. (2003). *Cell Mol. Biol.* **49**, 789–797.

Watanabe, K., Yasugi, E., and Ohshima, M. (2000). How to search the glycolipid data in "LipidBank for web," the newly-developed lipid database in Japan. *Trends Glycosci. Glycotechnol.* **12**, 175–184.

CHAPTER TWELVE

MEDIATOR LIPIDOMICS: SEARCH ALGORITHMS FOR EICOSANOIDS, RESOLVINS, AND PROTECTINS

Charles N. Serhan, Yan Lu, Song Hong, *and* Rong Yang

Contents

1. Introduction: Metabolomics and Mediator Lipidomics–Informatics	276
2. Engineered Animals and Human Tissues	278
3. New Chemical Mediator Pathways in Resolution of Inflammation	279
4. Logic Diagram to Identify PUFA-Derived Lipid Mediators: Eicosanoids, Resolvins, and Protectins	284
4.1. Databases	286
5. Cognoscitive-Contrast-Angle Algorithm	286
5.1. Identification of mass spectral ions	286
5.2. Modification of MS/MS ion intensities according to identities	289
5.3. Cognoscitive-contrast-angle algorithm and databases contrast angle	290
5.4. Integrating MS/MS spectra with UV spectra and chromatograms in COCAD to identify LMs	291
6. Theoretical Database for Novel LM and Search Algorithm	292
6.1. RvD1, 0,0,0,0-d_4-RvD1, and 21,21,22,22,22-d_5-RvD1	301
6.2. PD1, 0,0,0-d_3-PD1, 21,21,22,22,22-d_5-PD1, and 0,0,0, 21,21,22,22,22-d_8-PD1	306
7. Discussion	308
7.1. Impact of a functional group and enes on MS/MS fragmentation of various lipid mediators	308
8. Conclusions and Next Steps	311
8.1. Materials and sample preparation	312
8.2. Instrumentation	313
Acknowledgments	313
References	313

Center for Experimental Therapeutics and Reperfusion Injury, Department of Anesthesiology, Perioperative, and Pain Medicine, Brigham and Women's Hospital and Harvard Medical School, Boston, Massachusetts

Abstract

Within the domain of lipidomics, lipid mediator lipidomics and informatics is an exciting area because of the important roles of lipid-derived mediators in health and disease. It is well-appreciated that arachidonic acid is a precursor to potent bioactive mediators, such as prostaglandins, leukotrienes, and lipoxins. Recent experiments employing a mediator-lipidomic approach have uncovered that other major essential polyunsaturated fatty acids, such as docosahexaenoic acid (DHA) and eicosapentaenoic acid (EPA), serve as precursors for the production of potent bioactive mediator families, coined "resolvins" and "protectins." These omega-3 polyunsaturated fatty acids have long been noted to have beneficial effects in human systems; however, molecular evidence for their beneficial actions has been lacking and/or subject to debate. In this chapter, we review the databases and search algorithms used to identify these novel lipid mediators using liquid chromatography-ultraviolet-tandem mass spectrometry (LC-UV-MS/MS). Cognoscitive-contrast-angle algorithms and databases (COCAD) and systematic naming and empirical fragmentation rules useful in MS/MS ion identification were developed. Examples of identifying eicosanoids, resolvins, and protectins are given in human and murine tissues. The findings reviewed in this chapter demonstrate the advantages of COCAD in profiling and identification of bioactive lipid mediators and documenting activation of their biosynthetic pathways as well as in establishing relationships between lipid mediator pathways with agonists and antagonists present in the biological system.

1. Introduction: Metabolomics and Mediator Lipidomics–Informatics

Powerful uses of mediator lipidomics and informatics are illustrated from our studies with transgenic animals, specifically, transgenic rabbits (Serhan et al., 2003), and mice (Devchand et al., 2003), where the relative balance of specific class of eicosanoid lipid mediators (LMs) has a dramatic outcome on the phenotype of the organism. In general, unsaturated double bonds present in polyunsaturated fatty acids, such as arachidonic acid, EPA, or DHA are nonconjugated, making the fatty acids essentially devoid of a characteristic UV spectrum (Fig. 12.1). During the release of arachidonic acid and its transformation to bioactive mediators, specific stereoselective hydrogen abstraction leads to formation of conjugated diene-, triene-, or tetraene-containing chromophores, particularly in linear eicosanoid groups of mediators, such as the lipoxygenase pathway products leukotrienes (LT) and lipoxins (LX). The presence of specific UV chromophore and characteristic MS/MS spectra, the fragmentation patterns of these compounds and retention time, and the potency of their bioactions are the basis for their

Lipid Mediator Informatics and Profiling

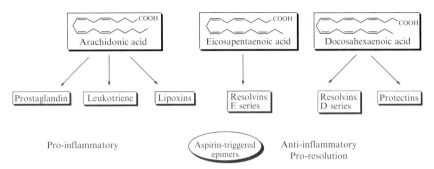

Figure 12.1 Lipid-derived mediators in programmed resolution of acute inflammation. *Precursors of Lipid Mediators*: Arachidonic acid is the precursor to eicosanoids that have distinct roles as proinflammatory mediators. The prostaglandins and leukotrienes each play specific actions pivotal to the progression of inflammation. Arachidonic acid–derived epoxyeicosatetraenoic acids (EETs) produced via P450 (Funk, 2001; Samuelsson et al., 1987) and ω-3 PUFA P450 epoxides may also play roles (Arita et al., 2005a; Capdevila et al., 1990). Cell–cell interactions, exemplified by platelets–leukocytes within blood vessels and/or PMN–mucosal interactions, enhance generation of lipoxins that serve as endogenous anti-inflammatory mediators self-limiting the course of inflammation (Serhan and Savill, 2005). The essential omega-3 fatty acids eicosapentaenoic acid and docosahexaenoic acid (C20:5 and C22:6) are converted to two novel families of lipid mediators, resolvins and protectins, that play pivotal roles in promoting resolution. Resolvins of the E series are generated from eicosapentaenoic acid (e.g., RvE1), and resolvins of the D series (e.g., resolvin D1) are generated from DHA as well as the protectins, such as neuroprotectin D1. *Aspirin-triggering epimers of lipid mediators*: Aspirin impacts the formation of lipoxins and resolvins by acetylating COX-2 (e.g., in human vascular endothelial cells that stereoselectively can generate, in the case of RvE1 biosynthesis, 18R-HPEPE, which is picked up via transcellular cell–cell interactions by leukocytes and converted in a lipoxygenase-like mechanism to RvE1). The complete stereochemistry of RvE1 and at least one of its receptors were established (Arita et al., 2005c; see Fig. 12.2). The biosynthesis of RvE1 can also be initiated by P450-like enzymes in microbes (Serhan et al., 2000). Aspirin also influences the biosynthesis of D-series resolvins. Aspirin catalytically switches COX-2 to a 17R-lipoxygenase-like mechanism that generates 17R-containing series of resolvin D and protectins (e.g., neuroprotectin D1/protectin D1; see text). (See color insert.)

identification in biologic material. Eicosanoids (derived from the Greek word *eicosa*, meaning "twenty") display profound stereoselective bioactions, which are usually active in the nano- to picomolar range. In general, eicosanoids are rapidly formed within seconds to minutes, act on cells locally, and then rapidly become inactive. Hence, eicosanoids usually act as extracellular mediators (paracrine or autocrine) within their local milieu and, therefore, are grouped according to these physiologic and pathophysiologic role(s), in some cases with the broader group of autocoids or chemical mediators, such as serotonin and histamine.

The specific physical and biological properties of each of these related structures permit identification and profiling from complex cellular milieu.

Unlike phospholipids or other structural lipids that keep a barrier function, those derived from arachidonic acid, such as prostaglandins, LT, and LX, have potent stereoselective bioactions and serve as local mediators. This characteristic makes it very important for profiling efforts to clearly chromatographically separate these related structures for their identification because, in some cases, closely related structures can be devoid of bioactions or can display opposing bioactions. Thus, an accurate profiling of the relationships between individual lipid mediators within a snapshot of a biological process or disease state can give valuable information on the events *in vivo* and their potential outcome (Levy *et al.*, 2001; Freire-de-Lima *et al.*, 2006). Moreover, when specific drugs are taken, such as aspirin, the relationship between individual pathway(s) and lipid mediators can change (Fig. 12.1), and their relationship may be directly linked to the drug's action *in vivo* (Serhan *et al.*, 2000, 2002). In view of these very different scenarios, mediator-lipidomics and informatics can give valuable means to assess the phenotype in many prevalent diseases, particularly those in which inflammation has an important pathologic basis.

2. Engineered Animals and Human Tissues

The now widely used approach of transgenics, namely deletion and/or overexpression of a gene product, provides exciting opportunities for lipidomics and, specifically, LM informatics, which can give valuable information and an assessment of pathways and their interactions *in vivo* (Hudert *et al.*, 2006; Xia *et al.*, 2006). For example, we used this LM informatics-lipidomics approach earlier to evaluate transgenic rabbits overexpressing the human 15-lipoxygenase (LOX) type 1 in their leukocytes (Serhan *et al.*, 2003) and in mice overexpressing human GPCR (Chiang *et al.*, 1999; Devchand *et al.*, 2003). When these profiles are compared directly to nontransgenic rabbits where the key enzyme is not overproduced but rather in its normal (wild-type) state, we can take a snapshot upon cell activation and examine the difference(s) between the nontransgenic versus overexpression of key enzyme(s) or receptors in a given biosynthetic pathway or signaling route. In this example, human 15-LOX overexpression gave enhanced lipoxin A_4 and 5S,15S-dihydroxyeicosatetraenoic acid (5S,15S-diHETE) formation with an apparent reduction in leukotriene B_4 formation (Fig. 12.1). Since leukotriene B_4 is a potent chemoattractant and lipoxin A_4 is a counter-regulatory antiinflammatory from within the eicosanoid family, their interrelationship(s) and/or overproduction of lipoxin A_4 are key indices in appreciating the *in vivo* role of the 15-LOX type 1 in inflammation. Overexpression of 15-LOX type 1 *in vivo* up-regulates lipoxin A_4, a potent bioactive product of this mediator pathway. Of interest, the transgenic rabbits display overall reduced inflammation and protection from tissue damage (Jain et al., 2003; Serhan et al., 2003; Shen et al., 1996). Also, overexpression of the LXA$_4$ receptor ALX reduced the magnitude of the acute inflammatory response and shortened the time required to resolve (Devchand et al., 2003).

3. New Chemical Mediator Pathways in Resolution of Inflammation

Inflammation plays a key role in many prevalent diseases. In addition to the chronic inflammatory diseases such as arthritis, psoriasis, and periodontitis (Van Dyke and Serhan, 2006), it is now increasingly apparent that diseases such as asthma, Alzheimer's disease, and even cancer have an inflammatory component associated with the disease process (Majno and Joris, 2004). Therefore, it is important for us to gain more detailed information on the molecules and mechanisms controlling local tissue inflammation (Calder, 2001) and its resolution. In order to accomplish this, we recently identified new families of lipid mediators (Fig. 12.1) that are generated from fatty acids during resolution of inflammation, termed resolvins and protectins (Serhan, 2005, 2004).

Using a systematic approach and analysis of resolving inflammatory exudates, we harvested exudates during resolution phase, namely, as leukocytic infiltrates were declining in cell number, to determine whether new chemical mediators biosynthesized during resolution (Serhan et al., 2000, 2002). We used a functional mediator-lipidomics approach employing LC-UV-MS/MS analysis to evaluate and profile temporal profiles of compounds at defined time intervals during experimental inflammation and its resolution. The focus is on bioactions of potential novel mediators; to achieve this, libraries of physical properties and MS/MS fragmentation for known mediators (i.e., prostaglandins, epoxyeicosatetraenoic acids (EETs), leukotrienes, and lipoxins) were constructed (as later shown), as well as theoretical compounds and their potential diagnostic tandem mass fragments that were used as signatures for specific enzymatic pathways. When we encountered novel compounds pinpointed within chromatographic profiles, complete structural elucidation and retrograde chemical analyses were carried out, which involved both biogenic and total organic synthesis. This permitted us to scale up compounds of interest in order to evaluate them both *in vitro* and *in vivo*. The *in vivo* models included a murine air pouch model of inflammation and peritonitis; the *in vitro* cell systems employed human isolated cells and focused on regulation of cytokines and leukocyte migration across transepithelial and/or transendothelial monolayers (Serhan et al., 2000, 2002) (Figs. 12.2 and 12.3).

Mediator-lipidomic databases containing MS/MS along with UV spectra and chromatographic profiles with appropriate search algorithms are

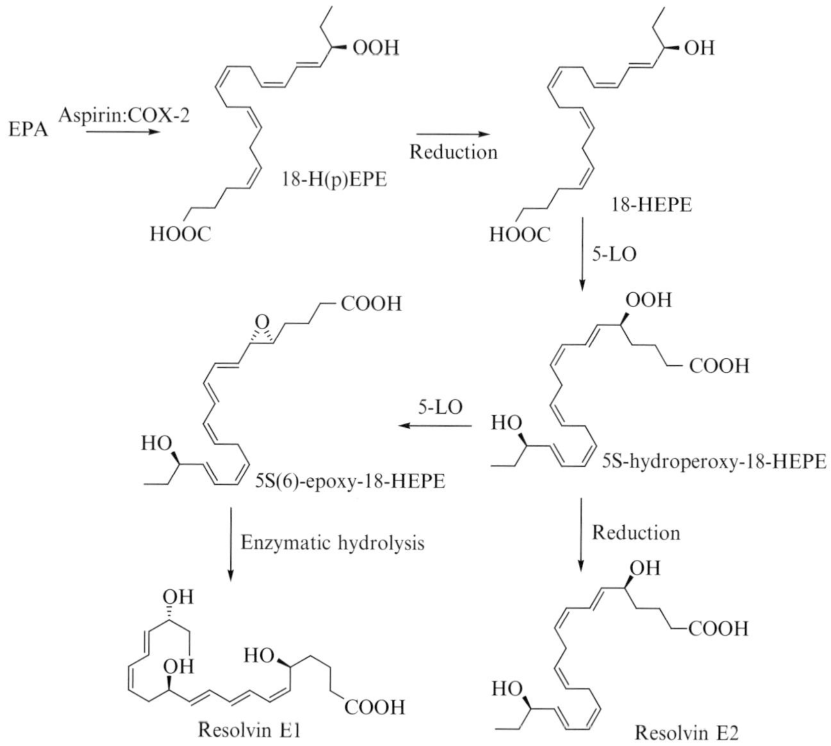

Figure 12.2 Biosynthesis of E series resolvins. RvE1 was identified in Serhan et al. (2000) and the biosynthesis of RvE1 with human cells was established with isolated cells in vitro. For example, hypoxic human endothelial cells expressing COX-2 treated with aspirin transform eicosapentaenoic acid (EPA). The mechanism involves abstracting hydrogen at carbon 16 in eicosapentaenoic acid to give R insertion of molecular oxygen, yielding 18R-hydroperoxy-EPE that is reduced to 18R-HEPE. They are further converted via sequential actions of human leukocyte 5-LO, leading to formation of the trihydroxy bioactive product resolvin E1 (Serhan et al., 2000, 2002). Using matching criteria, both physical and bioactions, the complete stereochemistry of RvE1 was assigned as 5S,12R,18R-trihydroxy-6Z,8E,10E,14Z,16E-eicosapentaenoic acid and one of its receptors identified as a GPCR (Arita et al., 2005c).

critical for accurate and timely analysis of the families of bioactive lipid mediators (LMs) biosynthesized from polyunsaturated fatty acids (PUFA) in mammalian systems (Lu et al., 2005). Although gas chromatography–mass spectrometry (GC-MS) is useful in many cases for identification of specific mediator structures, the required high-column temperatures limit its use because many of the PUFA-derived LMs (i.e., eicosanoids) are thermolabile (Murphy et al., 2001). In this regard, the search algorithms used for GC-MS electron-impact-ionization spectra are well studied (Ausloos et al., 1999; Stein, 1995; Stein and Scott, 1994), as are those for both GC-MS chromatograms and mass spectra (Mallard and Reed, 1997).

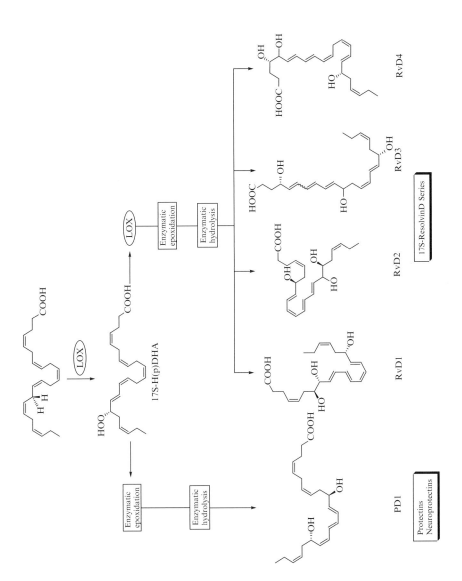

The contrast-angle (or dot-product, which equals cosine of contrast-angle) algorithm is widely used for these studies. Here, the contrast-angle is the angle between two spectra represented as vectors composed of ordered peak intensities (Stein and Scott, 1994). Because databases (i.e., MS-MS fragments) and search algorithms for LMs (such as the eicosanoids and the new resolvins and protectins) employing LC-UV-MS/MS were not available at the time of our studies, we used MassFrontier™ (ThermoFinnigan, Waltham, MA), GC-MS mass spectral commercial software (Lu et al., 2005), to assemble and construct an LM database and the necessary search algorithm.

The MassFrontier search algorithm, which was developed by Stein and Scott (1994; Stein, 1995), is a dot-product. The dot-product algorithm in MassFrontier uses the intensities of all peaks in the MS/MS as a vector for computation and does not establish the identities of ions. It does not distinguish whether the ions are molecular ions, ions derived from molecular ions, or, for example, ions from interfering substances in the samples (Ausloos et al., 1999; Mallard and Reed, 1997; Stein, 1995; Stein and Scott, 1994). Using the ions generated from interfering background can decrease the percent correct for the best match in a given search. Even the ions generated from unknown LMs can contribute differently to the identification. For LMs, ions generated by cleavage of the carbon–carbon bond inside the carbon chain have greater diagnostic yield for determining the structures than the ions formed by loss of water or CO_2, which are common features for most of the PUFA-derived LMs (Chiang et al., 1998; Griffiths et al., 1996; Hong et al., 2003; Murphy et al., 2001; Serhan et al., 2002).

The relationship(s) between ESI-MS/MS spectra and LM structures within the eicosanoid family are extensively studied (Griffiths et al., 1996; Hong et al., 2003; Murphy et al., 2001; Serhan et al., 2000, 2002; Wenk

Figure 12.3 Biosynthesis of protectins and D series resolvins. Docosahexaenoic acid (DHA) is precursor to the lipoxygenase product 17S-H(p)DHA that is converted to a 16 (17)-epoxide, converted to the 10,17-dihydroxy bioactive product (Hong et al., 2003; Serhan et al., 2002), denoted earlier as 10, 17S-docosatriene (DT) (Marcheselli et al., 2003), and more recently coined neuroprotectin D1/protectin D1 because of its potent actions in neural systems in vivo (Marcheselli et al., 2003; Mukherjee et al., 2004). The complete stereochemistry of NPD1/PD1 (Serhan et al., 2006) and RvD1 (Sun et al., 2007) was recently established. The complete stereochemical assignments for RvD3 and RvD4 are tentative and based on biogenic considerations (Serhan et al., 2002). *Aspirin-triggered 17R epimers:* The 17R series resolvins are produced from DHA in the presence of aspirin. Human endothelial cells expressing COX-2 treated with aspirin transform DHA to 17R-HpDHA. Also, recombinant COX-2 treated with aspirin converts DHA to 17R-HpDHA. Human PMN convert 17R-HDHA to two compounds via 5-lipoxygenation; each is rapidly transformed into two epoxide intermediates. One of these is a 7(8)-epoxide and the other a 4(5)-epoxide (Serhan et al., 2002). These two novel epoxide intermediates can be enzymatically opened to bioactive products denoted 17R series ATRvD1 through ATRvD6 (Serhan et al., 2002). Also, the confirmation of the assigned structure and synthesis of RvD2 was reported by others (Rodriguez and Spur, 2004).

et al., 2003). The elucidation of unknown structures of novel LMs employing LC-UV-MS/MS is based on the relationships between structural features (such as functional groups and double bonds), characteristics of the spectra, and chromatographic behavior (i.e., MS/MS ions, UV spectra, and LC retention times in relation to other known structures). With C18 reversed-phase LC, chromatographic retention times (CRTs) generally increase in the following order: tri-hydroxy–containing, di-hydroxy–containing, and mono-hydroxy–containing LMs, followed by the native or nonoxygenated PUFA precursors (such as arachidonic acid, EPA, or DHA), which are eluted depending on the mobile phase employed. For positional isomers of hydroxy-containing LMs, the CRTs decrease as the hydroxy groups locate closer to the methyl group at the omega-end of the long-chain PUFA. For instance, the CRT of LXB$_4$ (5S,14R,15S-trihydroxy-6E,8Z,10E, 12E-eicosatetraenoic acid) is usually shorter than that of LXA$_4$ (5S,6R, 15S-trihydroxy-7E,9E,11Z,13E-*trans*-11-*cis*-eicosatetraenoic acid). The CRT of native LXA$_4$ (containing a 15S-hydroxy at carbon 15) is shorter than that of the aspirin-triggered 15-epi-LXA$_4$, but greater than that of 11-*trans*-LXA$_4$, which is the natural, all-*trans*-containing isomer of LXA$_4$ (Serhan, 1989).

The UV spectrum, namely the band multiplicity and λ_{max}, is an additional signature for identifying specific LMs. For example, the presence of an asymmetric singlet band with λ_{max} ~235 nm is diagnostic for the conjugated diene present within the compounds associated with lipoxygenase pathways. These UV chromatophores include mono-HETEs (mono-hydroxy-eicosatetraenoic acids—conjugated dienes); a triplet with λ_{max} ~270 nm for the leukotrienes (e.g., the conjugated triene within LTB$_4$ [5S,12R-dihydroxy-6E,8Z,10Z,14E-eicosatetraenoic acid]); a triplet with λ_{max} ~300 nm for conjugated tetraene, as present within LXA$_4$ and LXB$_4$; and an asymmetric singlet with λ_{max} ~242 nm from the two conjugated dienes interrupted by a methylene group as present in the double lipoxygenation product 5S,15S-diHETE (Serhan, 1989). Some compounds of biologic significance and interest do not possess specific chromophores, such as those with λ_{max} in vacuum UV range (as in the 1,4-cis-pentadiene–containing PUFA [AA, EPA, DHA]), as well as some of the prostaglandins, such as PGE$_2$ (11α,15S-dihydroxy-9-oxo-prosta-5Z,13E-dien-1-oic acid) and PGF$_{2\alpha}$ (9α,11α,15S-trihydroxy-prosta-5Z,13E-dien-1-oic acid). Both PGE$_2$ and PGF$_{2\alpha}$ are potent prostanoids (Bergström, 1982; Samuelsson, 1982); neither possesses conjugated double-bond systems (Kiss *et al.*, 1998). When the conjugated tetraene of LXA$_4$ isomers is in the all-*trans*-geometry, the λ_{max} is shifted 1 or 2 nm upward to 302 nm rather than 300 nm (Serhan, 1989). The relationships between MS/MS spectra and the novel LM families resolvins and protectins are presented (*vide infra*).

Using current analytical approaches, most LMs are identified by comparing spectra and chromatographic behaviors acquired from tissue samples with those of authentic standards for the known LMs. When authentic standards are not available, as in the case of previously unknown LMs and their further metabolites, basic chemical structures can be obtained on the basis of the relationship between structures and features of their spectra and chromatographic behaviors compared to those of synthetic and biogenic products prepared to assist in evaluation and assignment. Routinely, this laboratory identifies LMs by matching spectra of "unknowns" (MS/MS, GC-MS, and UV spectra) and CRTs to those of authentic and synthetic standards, if available, and, importantly, bioactions, or with a theoretical database that consists of virtual UV, MS/MS spectra, and CRTs for identifying potentially novel chemical mediators (Hong *et al.*, 2003; Lu *et al.*, 2005; Serhan *et al.*, 2000, 2002).

We initially developed a theoretical database and algorithm according to the relationships between LM structures and their spectral and chromatographic characteristics (Lu *et al.*, 2005). The proposed structures of novel potential LMs in the theoretical databases were based on putative PUFA precursors and results of biosynthetic studies. To this end, we constructed mediator-lipidomic databases and search algorithms to assist in the identification of LM structures employing LC-UV ion trap MS/MS with the following objectives: (1) assemble a database using available mass-spectral software; (2) construct a cognoscitive-contrast-angle algorithm and databases to improve identification of LMs using MS/MS ions (currently not performed with available software); (3) develop a theoretical database and algorithm for assessing potential new and/or previously unknown structures derived from PUFA precursors; and (4) identify further metabolites of bioactive LM in complex tissue matrices. It proved useful to develop these mediator-lipidomic databases and algorithms using ion trap mass spectrometers, since they are relatively cheaper and widely used. Moreover, the fragmentation rules and patterns for collision-induced dissociation (CID) spectra from triple-quadrupole mass spectrometers—another popular MS instrumentation—are similar to those encountered using ion trap MS (Aliberti *et al.*, 2002; Murphy *et al.*, 2001; Serhan *et al.*, 2003; Takano *et al.*, 1998).

4. Logic Diagram to Identify PUFA-Derived Lipid Mediators: Eicosanoids, Resolvins, and Protectins

The routes for LM identification and structure elucidation of potentially novel LMs followed in mediator-lipidomic databases, and search algorithms are illustrated in Fig. 12.4. Two types of lipidomic databases for LMs were used; one contains LC-UV-MS/MS spectra and chromatograms acquired on LM standards and the other is based on theoretically

Lipid Mediator Informatics and Profiling

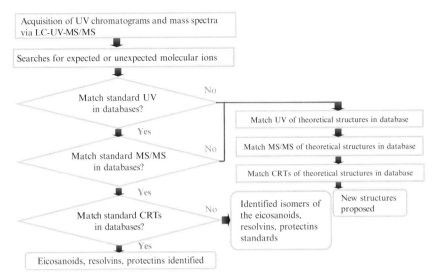

Figure 12.4 Logic diagram for developing LC-UV-MS/MS-based mediator lipidomic databases and search algorithms for polyunsaturated fatty acids (PUFA)-derived lipid mediators (LMs): eicosanoids, resolvins, and protectins (Lu et al., 2005). CRT is the chromatographic retention time. (See color insert.)

generated LC-UV-MS/MS spectra and chromatograms (Lu et al., 2005). Searches were conducted stepwise against either standards or the theoretical databases to increase the search speed. The search of MS/MS spectra was carried out only against the MS/MS subdatabase with the molecular ion of interest (i.e., M-1) and matched UV spectra (e.g., conjugated diene, triene, or tetraene chromophores). Subsequently, the matching of CRTs was performed. The standard error for CRTs for the chromatographic conditions used in the present set of experiments was ~0.3 min. Thus, only the unknown CRTs within plus or minus 2 × 0.3 min of the CRT (for 95% confidence intervals) in the databases were taken as a correct match. If the UV spectral pattern was unclear, the MS/MS and CRTs were still searched to avoid potential errors in assignment. A standard LM or theoretical fragmentation/ion fragmentation pattern that fulfilled the above match criteria was then assigned to the unknown set. If the match was a "hit" only with UV and MS/MS spectra but not with its CRT, then the LM in the sample was likely to be a geometric isomer (*vide infra*).

The UV absorbance patterns were presented in the database as maximum absorbance wavelength λ_{max} and divided into six clusters: ~301, 278, 270, 242, 235 nm, and vacuum UV. In contrast, each MS/MS spectrum has dozens of ions and therefore is a much larger data set than a UV λ_{max}. If the UV λ_{max} is matched first, then only the MS/MS subdatabase having LM standards with this UV λ_{max}, which is a fraction of the whole MS/MS database, needs to be searched. This type of narrowed search is faster than searching through the entire MS/MS database. This will become more significant when more and more standard MS/MS data are entered into the database. We tested the route with MS/MS spectra before UV; the search results are the same as having MS/MS after UV data. If the UV pattern is unclear, we can search the MS/MS and CRT to avoid assignment errors.

4.1. Databases

We assembled a lipid mediator informatics database composed of LC-UV-MS/MS spectra and chromatograms acquired from authentic LMs using the available GC-MS spectral software, MassFrontier. The UV λ_{max} of authentic LMs were written into the subdatabase names, and the CRTs were written into the LM names so that MassFrontier could handle the acquired UV spectral results and CRTs used in identification of known/unknown LMs (see the logic diagram outlined in Fig. 12.4). The demonstration of the efficacy of this database was assessed with the identification of eicosanoids biosynthesized by rabbit leukocytes using LC-UV-MS/MS (Lu et al., 2005). For example, to identify PGE_2, a subdatabase "mTOz 351Vacuum UV" was selected with the molecular ion of interest and matched UV λ_{max}. The search showed that the MS/MS spectrum matched best (highest matching score, 907) with PGE_2. Also, the CRT of peak I and that of synthetic PGE_2 matched. Therefore, the material was assigned as PGE_2, consistent with the manual identification and assessment of MS and LC chromatographic features.

5. Cognoscitive-Contrast-Angle Algorithm

5.1. Identification of mass spectral ions

The system with cognoscitive-contrast-angle algorithm and databases (COCAD) that we used (Hong et al., 2003; Levy et al., 2001; Lu et al., 2005) helped to elucidate the fragmentation of LMs in MS and to match unknown MS/MS spectra to those of synthetic and/or authentic standards. In this process of matching, the intensity of each peak is treated differently based on the ion identity. MS/MS ions are clustered into three types: "peripheral-cut" ions, formed by neutral loss of water, CO_2, amino acid, or amines derived from functional groups linking to LM carbon chain as hydroxy, hydroperoxy, carbonyl, epoxy, carboxy, amino acid group, or amino group, etc.; "chain-cut" ions, formed by cleavage of a carbon–carbon bond along the LM carbon chain; and "chain- plus peripheral-cut" ions, formed by combination of chain-cut and peripheral-cut ions. Molecular ions formed during ESI can easily be converted to peripheral-cut ions in the

MS/MS process. Similarly, chain-cut ions can also be readily converted to chain- plus peripheral-cut ions (Fig. 12.5).

Typical chain-cut ions for LMs in MS/MS are formed by α-cleavage of the carbon–carbon bonds connecting to the carbon with a functional group directly attached (Griffiths *et al.*, 1996; Hong *et al.*, 2003; Murphy *et al.*,

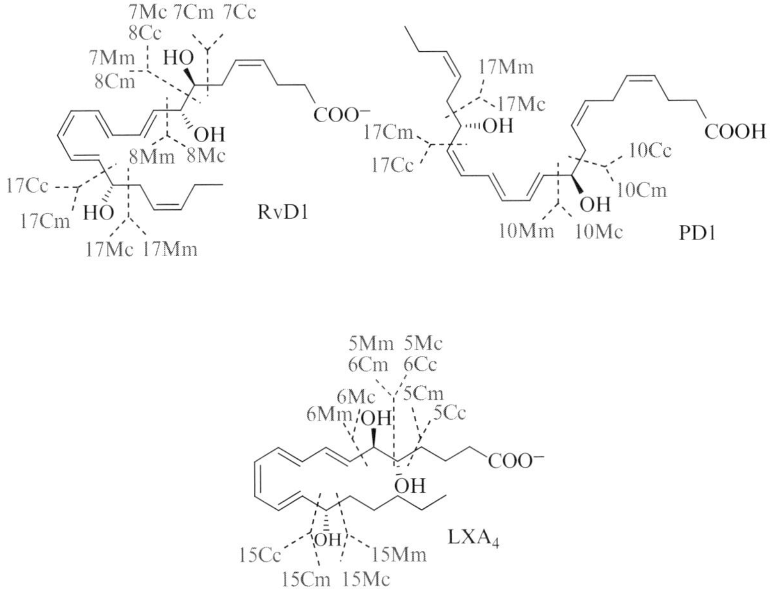

$UV\lambda_{max}$

Retention time

LM name | Molecularion | Peripheral-cut segments | Chain-cut segments

LXA_4 RT: 5.2min 300nm	[M-H=351.3]	5Cc=86	5Cm=265	5Mc=116	5Mm=235	6Cc=116	6Cm=235	6Mc=146	6Mm=205	15Cc=250	15Cm=101	15Mc=280	15Mm=71
$-H_2O$	333		247	98	217	98	217	128	187	232	83	262	
$-2H_2O$	315		229		199		199	110		214		244	
$-3H_2O$	297		211									226	
$-CO_2$	307	42		72		72		102		206		236	
$-H_2O-CO_2$	289			54		54		84		188		218	
$-2H_2O-CO_2$	271							66		170		200	
$-3H_2O-CO_2$	253											182	

Neutral loss derived from LM functional groups

Chain-plus-peripheral-cut segments

Figure 12.5 Liquid chromatographic ultraviolet tandem mass spectrometry (LC-UV-MS/MS) database layout: naming lipid-mediated segments for PD1, RvD1, and LXA_4. These examples depict RvD1, LXA_4, and PD1, formed via chain-cut, peripheral-cut, and chain- plus peripheral-cut for interpretation of MS/MS fragmentation (see text for further details). (See color insert.)

2001; Serhan et al., 2000, 2002). LMs readily undergo α-cleavage (Murphy et al., 2001). We proposed the nomenclatures illustrated with LXA$_4$, PD1, and RvD1 structures (Fig. 12.5) to systematically name segments formed via chain-cut and chain- plus peripheral-cut without concern for hydrogen shift occurring during MS analysis of PUFA-derived products. Each of these LMs derived from PUFA (Fig. 12.1) has a carboxyl terminus and a methyl terminus. An Arabic number was used to designate the position of the functional group on the LM carbon chain where the cleavage occurs. The uppercase letter right after the number indicates the side of the functional group on which the cleavage occurs: "C" is for cleavage on the carboxyl side of the functional group, and "M" is for cleavage on the methyl side of the functional group. Each cleavage can directly generate two segments. The lowercase letter that follows indicates the side of the cleavage on which the segment forms: "c" is for a segment formed on the carboxyl side of the cleavage, and "m" is for a segment formed on the methyl side of the cleavage. Segments formed through β-cleavage or γ-cleavage toward the functional group are named by adding β or γ between the uppercase and lowercase letters. Because three hydroxy groups of LXA$_4$ are located at C$_5$, C$_6$, and C$_{15}$ of the 20-carbon linear chain, α-cleavages can occur on the bonds at C$_4$-C$_5$, C$_5$-C$_6$, C$_6$-C$_7$, C$_{14}$-C$_{15}$, and C$_{15}$-C$_{16}$ to generate 10 chain-cut segments (Fig. 12.5). Each of the possible chain-cut, peripheral-cut, and chain- plus peripheral-cut segments for LXA$_4$, PD1, and RvD1 is shown in Fig. 12.5.

An MS/MS ion detected from LM samples in negative-ion mode generally is formed from a specific segment with the addition or subtraction of hydrogen(s) caused by hydrogen shift during the cleavage. The charge (z) of the LM negative ion is usually equal to one; therefore, the mass-to-charge ratio (m/z) of a LM ion is usually equal to its mass (m). Previous reports (Chiang et al., 1998; Griffiths et al., 1996; Murphy et al., 2001) and our results (Gronert et al., 2004; Hong et al., 2003; Levy et al., 2001; Serhan et al., 2000, 2002, 2003) demonstrated that the MS/MS fragmentation observes the empirical rules on the addition or subtraction of hydrogen(s) for the chain-cut segments to form the chain-cut MS/MS ions.

For the segment Cc, the detected MS/MS ions are Cc and Cc + H; for the segment Cm, the detected MS/MS ions are Cm, Cm ± H, and Cm ± 2H; for the segment Mc, the detected MS/MS ions are Mc and Mc − H; and for the segment Mm, the detected MS/MS ions are Mm, Mm ± H and Mm ± 2H.

These ions were used to identify the segments for instrument-detected MS/MS ions. The interpreted ions, such as M − H − H$_2$O, Cc, Cc + H, or Mc − H (identified as the MS/MS ions detected in the mass spectrometer) are called virtual ions. One detected MS/MS ion can be interpreted as one or several virtual ions. Through loss of H$_2$O, CO$_2$, NH$_3$, and/or amino acids, the chain-cut ions can form chain- plus peripheral-cut ions. For the chain-cut and chain- plus peripheral-cut ions in the present report, we focused on those formed by α-cleavages. Detected MS/MS ions that are uninterpretable via the general rules noted above and neutral loss are taken as unidentified ions.

5.2. Modification of MS/MS ion intensities according to identities

In nature, the isotope ^{13}C is 1.1% of elemental carbon. Therefore, if the intensity of an ion with mass M (ion charge z is 1) containing only carbons, hydrogens, and oxygen(s) is 100, the contribution of ^{13}C to the intensity of the ion with mass (M + 1) (the ion with mass 1 unit higher) is statistically equal to $100 \times 1.1\% \times$ number of carbons in the ion M (McLafferty and Turecek, 1993). The contribution of ^{13}C to the ion (M + 2) is much less than $100 \times 1.1\% \times$ carbon-number. Thus, the ^{13}C contribution of ion M to the intensities of ions (M + 1) and (M + 2) was subtracted in the computation of the matching score for COCAD and theoretical algorithms. If ion M could not be identified, the carbon number in the ion (M + 1) or (M + 2) is used, because M, (M + 1), and (M + 2) generally have the same carbon-number.

Chain-cut ions are most informative and could be diagnostic for determining specific LM structures, such as the position of functional groups and double bonds. Generally, more chain- plus peripheral-cut ions are found than chain-cut ions because the former are derived from the latter by one or several neutral losses. It is likely that some chain- plus peripheral-cut ions have the same m/z even though they originate by different mechanisms and represent different structural features. Therefore, chain-cut ions are more specific than chain- plus peripheral-cut ions for defining the LM structure. Peripheral-cut ions in MS/MS spectra are similar among LM isomers and, therefore, were not specific enough for differentiation of individual LM isomers.

According to the general fragmentation rule noted above, the nth MS/MS peak can be identified as one or several chain-cut (C) ions, peripheral-cut (P) ions, and/or chain- plus peripheral-cut (CP) ions. The weighted intensity $^{y}I_n$ of each identified ion is as follows:

$$^{y}I_n = I'_n \div \left(^{C}_{n}M + ^{CP}_{n}M + ^{P}_{n}M \times \rho \right) \times ^{y}W \qquad (1)$$

where y is the MS/MS ion type identified as C, P, or CP; I'_n is the relative intensity of the nth peak in the MS/MS spectrum; $^{C}_{n}M$ is the number of chain-cut ions identified for the nth MS/MS peak; $^{CP}_{n}M$ is the number of chain- plus peripheral-cut ions identified for the nth MS/MS peak; $^{P}_{n}M$ is the number of peripheral-cut ions identified for the nth MS/MS peak; and ^{y}W is the weight measuring the importance of the identified ion to determine the LM structure. It is 10 as ^{C}W and 1 as ^{CP}W or ^{P}W (for peripheral-cut ions). The fingerprint features of chain-cut ions are used to define LM structure by multiplying their intensities by 10, which was determined to be the best among values 2, 10, 20, and 100 tested. Weighted MS/MS ion intensities are used for COCAD and the theoretical system described later in this manuscript. ρ represents the contribution of peripheral-cut ions to I'_n ($\rho = 3$ for peripheral-cut ions formed via loss of one CO_2 from each molecular ion, $\rho = 10$ for peripheral-cut ions formed via loss of one H_2O from each molecular ion, and $\rho = 1$ for other peripheral-cut ions formed via multiple loss of CO_2 and/or H_2O from each molecular ion). The assignment of ρ values is arbitrary and based on the observation of relative intensities of peripheral-cut ions in MS/MS spectra of LMs.

5.3. Cognoscitive-contrast-angle algorithm and databases contrast angle

Cognoscitive-contrast-angle algorithm and databases used a contrast-angle algorithm to match an MS/MS spectrum between sample and standards. For this approach, the contrast angle is calculated as follows:

$$C_v = \sum_{n=1}^{N} \left({}_n^{C}B_v \times {}^{C}I_n\right) \qquad (2)$$

$$CP_v = \sum_{n=1}^{N} \left({}_n^{CP}B_v \times {}^{CP}I_n\right) \qquad (3)$$

$$P_v = \sum_{n=1}^{N} \left({}_n^{P}B_v \times {}^{P}I_n\right) \qquad (4)$$

$$D_C(D_{CP}, \text{or } D_P) = \frac{\sum_{v=1}^{V} U_v S_v}{\sqrt{\sum_{v=1}^{V} U_v^2 \sum_{v=1}^{V} S_v^2}} \qquad (5)$$

U_v is equal to C_v, CP_v, or P_v for unknown spectrum to be identified. S_v is equal to C_v, CP_v, or P_v for standard spectrum.

$$\text{COCAD contrast angle} = \cos^{-1}[(10 \times D_C + D_P + D_{CP}) \div (11 + \omega_{CP})] \qquad (6)$$

where v is the vth virtual ion; V is the total number for one type of virtual ion formed via chain-cut, chain- plus peripheral-cut, or peripheral-cut for a specific LM. ${}_n^{C}B_v$ is equal to 1 if the nth MS/MS peak can be identified as the vth virtual ion formed via chain-cut, or equal to zero if not; ${}_n^{CP}B_v$ or ${}_n^{P}B_v$ has a similar meaning, but for ions formed via chain- plus peripheral-cut or peripheral-cut; N is the total number of peaks in the MS/MS spectrum;

D_C is the dot product between the virtual vectors of U (unknown sample) and S (standard) formed via chain-cut; D_{CP} or D_P is the dot product for chain- plus peripheral-cut or peripheral-cut ions, respectively. D_C, D_{CP}, or D_P in Equation 5 represents the similarity of ions formed via chain-cut, chain- plus peripheral-cut, or peripheral-cut, between an unknown spectrum and a standard spectrum, none of which had a value greater than 1. The vth virtual ion is not used for the calculation of the corresponding D_C, D_{CP}, or D_P if either U_v or S_v is zero. If every C_v, CP_v, or P_v within the vectors is zero, then D_C, D_{CP}, or D_P is assigned the value zero, respectively.

The COCAD contrast angle in Equation 6 represents how well the spectrum of the sample matches the standard: if it is $0°$, the two spectra match exactly; if it is $90°$, the two spectra do not match at all; the smaller the contrast angle between 0 and $90°$, the better the match (Stein and Scott, 1994; Wan et al., 2002). The value is integrated and normalized from dot products D_C, D_{CP}, and D_P (Equation 6). The numeric coefficient 10 in Equation 6 was found to be the best value (2, 20, and 100 were also tested) that emphasizes the fingerprinting feature of chain-cut ions because chain-cut ions are more important for determining the LM structure than are other types of ions. To normalize $[(10 \times D_C + D_{CP} + D_P) \div (11 + \omega_{CP})]$ in Equation 6 to be no more than 1, 11 was used in the denominator of the equation, and ω_{CP} is equal to 1 if at least one MS/MS ion is identified as a chain- plus peripheral-cut virtual ion or equal to 0 if no such ion is identified. No chain- plus peripheral-cut ion is identified in a few LM standard spectra. Therefore, ω_{CP} is introduced in Equation 6 to normalize the COCAD contrast angle to 0 when matching these types of spectra against themselves. Unidentified ions were excluded for matching in Equations 2 to 6.

According to the general rules, for mono-hydroxy-containing LMs, the vth chain-cut ion can be Cc, Cc + 1, Cm − 2, Cm − 1, Cm, Cm + 1, Cm + 2, Mc − 1, Mc, Mm − 2, Mm − 1, Mm, Mm + 1, or Mm + 2. The vth chain- plus peripheral-cut ion is Cc-CO_2, Cc-CO_2 + 1, Cm-H_2O − 2, Cm-H_2O − 1, Cm-H_2O, Cm-H_2O + 1, Cm-H_2O + 2, Mc-H_2O − 1, Mc-H_2O, Mc-CO_2 − 1, Mc-CO_2, Mc-H_2O-CO_2 − 1, or Mc-H_2O-CO_2. The vth peripheral-cut ion is M-H-CO_2, M-H-H_2O, or M-H-H_2O-CO_2. For LMs having multiple functional groups, V is accordingly greater.

5.4. Integrating MS/MS spectra with UV spectra and chromatograms in COCAD to identify LMs

COCAD databases incorporate the LC-UV-MS/MS chromatograms and spectra measured from standard LMs and the segments formed by cleavage illustrated in Fig. 12.3. The features of UV spectra and chromatograms are represented with the UV λ_{max} and CRTs, respectively. The search of COCAD databases was conducted stepwise, as depicted in the logical diagram in Fig. 12.4, from UV λ_{max} to MS/MS spectra, and then to

CRTs. Using COCAD, we identified LXA$_4$ in murine spleen stimulated during exposure to *Toxoplasma gondii* (Fig. 12.6). The search was narrowed down to the sub-database with UV λ_{max} 301 nm (middle inset) and molecular ion m/z 351, which were detected for the unknown peak II in the chromatogram (m/z 251 of MS/MS 351) for the stimulated spleen sample. The CRT (5.2 min) of peak II was then matched against the candidates with the highest MS/MS match score (Fig. 12.6A, left inset). For comparison, searching was also conducted with the system developed via Mass-Frontier (Fig. 12.6C). Both demonstrated that LXA$_4$ was the best match for the unknown peak II. However, COCAD and MassFrontier could not be used alone to identify the structures of unknowns without LC-UV-MS/MS chromatograms and spectra obtained for authentic biologically derived and bioactive standards confirmed with compounds prepared by total organic synthesis (Arita *et al.*, 2005c; Serhan *et al.*, 2006).

6. Theoretical Database for Novel LM and Search Algorithm

Theoretical databases consist of the segments (as illustrated in Fig. 12.5), the UV λ_{max}, and CRTs predicted. Searching against a theoretical database is also performed stepwise, as described in Fig. 12.4, from UV λ_{max}, to MS/MS spectra, and then to CRTs.

Equation 7 is the matching score for an unknown MS/MS spectrum compared with the virtual spectrum based on the segments and fragmentation rule noted above.

$$\text{Matching score} = \left\{ \sum_{f=1}^{F}\left[\sum_{n=1}^{N}\left(^{C}I_n \times {}^{C}_{f}M_n\right) \times \left({}^{f}T_C \div {}^{f}A_C\right)^{0.5}\right] \right.$$
$$+ \sum_{f=1}^{F}\left[\sum_{n=1}^{N}\left(^{CP}I_n \times {}^{CP}_{f}M_n\right) \times \left({}^{f}T_{CP} \div {}^{f}A_{CP}\right)^{0.5}\right]$$
$$\left. + \sum_{n=1}^{N}\left[^{P}I_n \times {}^{P}M_n \times (T_P \div A_P)^{0.5}\right] \right\}$$
$$\div \sum_{n=1}^{N}\left(^{C}I_n + {}^{CP}I_n + {}^{P}I_n\right) \quad (7)$$

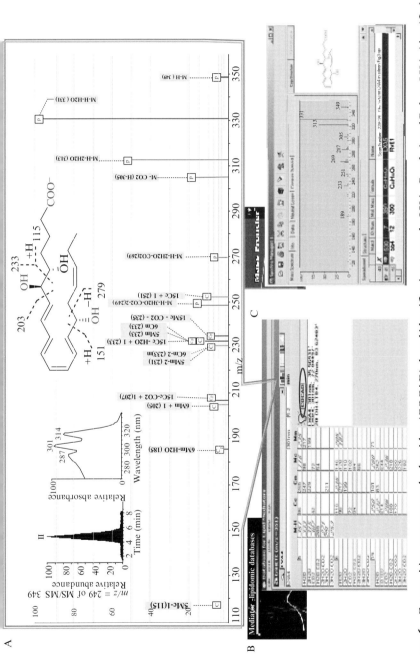

Figure 12.6 Cognoscitive–contrast–angle algorithm (COCAD)–based identification example: LXA$_5$. Panels A and B show LXA$_5$ identified as the best match for the liquid chromatography ultraviolet tandem mass spectrometry (LC-UV-MS/MS) results in the cognoscitive–contrast–angle algorithm system (with highest match-score). LXA$_5$ was also identified in Panel C by searching the database constructed by MassFrontier^{TM}. (See color insert.)

The matching score in Equation 7 summates the weighted intensities of all the identified MS/MS peaks in the spectrum acquired from the sample. The numerator of this formula is composed of three parts:

$$\sum_{f=1}^{F} \left[\sum_{n=1}^{N} \left({}^{C}I_n \times {}^{C}_{f}M_n \right) \times \left({}^{f}T_C \times {}^{f}A_C \right)^{0.5} \right]$$

summating the weighted intensities of MS/MS peaks identified as chain-cut ions;

$$\sum_{f=1}^{F} \left[\sum_{n=1}^{N} \left({}^{CP}I_n \times {}^{CP}_{f}M_n \right) \times \left({}^{f}T_{CP} \div {}^{f}A_{CP} \right)^{0.5} \right]$$

summating the weighted intensities of MS/MS peaks identified as the chain- plus peripheral-cut ions; and

$$\sum_{n=1}^{N} \left[\left({}^{P}I_n \times {}^{P}M_n \times \left(T_P \div A_P \right)^{0.5} \right) \right]$$

summating the weighted intensities of MS/MS peaks identified as the peripheral-cut ions. ${}^{C}_{f}M_n$ is the total number of chain-cut ions via α-cleavage formed from the fth functional group and matched to the nth MS/MS peak. F is the total number of functional groups in one LM. f is counted from the carboxyl terminus of LM. For example, f is 1 for 5-hydroxy, 2 for 6-hydroxy, and 3 for the 15S-hydroxy group present in LXA$_4$. F for LXA$_4$ is 3 (Fig. 12.3). $\sum_{n=1}^{N} ({}^{C}I_n \times {}^{C}_{f} M_n)$ summates the weighted intensities of MS/MS peaks identified as chain-cut ions formed from the fth functional group via α-cleavage. The smallest MS/MS ion detectable in ion-trap MS/MS is generally ∼m/z 95 for LMs with a molecular ion of ∼400 u (Finnigan, 1996). To compensate for the bias caused by the inability to detect an MS/MS ion of m/z less than 95, factors $({}^{f}T_C/{}^{f}A_C)^{0.5}$, $({}^{f}T_{CP}/{}^{f}A_{CP})^{0.5}$, and $(T_P/A_P)^{0.5}$ are used in (Eq. (7)). ${}^{f}T_C$ (or ${}^{f}T_{CP}$) is the total number of chain-cut (or chain- plus peripheral-cut) ions formed from the fth functional group via α-cleavage. T_p is the total number of peripheral-cut ions formed from one LM. ${}^{f}A_C$ (${}^{f}A_{CP}$, or A_P) is the fraction of ${}^{f}T_C$ (${}^{f}T_{CP}$, or T_P) representing the ions within the MS/MS detecting range (m/z from 95 to the m/z of molecular ion). To illustrate this, 20-HETE is a useful example here. The ions formed from the segment 20 Cm and 20 Cm-H$_2$O of 20-HETE are m/z 31 and 13 according to our general fragmentation rules previously noted, which are too small to be detected in ion-trap MS/MS. Consequently, without the use of factors $(^f T_C/^f A_C)^{0.5}$ and $(^f T_{CP}/^f A_{CP})^{0.5}$, 20-HETE could have a lower matching score than other HETEs, even though the MS/MS spectrum is that of 20-HETE. $\sum_{n=1}^{N}(^C I_n + ^{CP} I_n + ^P I_n)$ is used in Equation 7 for normalization to eliminate the impact on the matching scores of the total peak intensities in MS/MS spectra.

The unknown MS/MS spectra were assigned to the LM and/or potentially novel LM that has the highest matching score in addition to having the predicted UV spectral and chromatographic features (e.g., λ_{max} and CRTs). This theoretical system was used successfully to identify 15S-HETE in murine spleen (Lu et al., 2005). This material, displayed at CRT 20.4 min on the chromatogram (at m/z 219 of MS/MS 319, left inset), had a UV λ_{max} 235 nm. Therefore, the search on the theoretical system was narrowed down to the sub-database with molecular ion m/z 319, UV λ_{max} 235 nm, and CRT 21 min. In this case example, 15-HETE gave the highest matching score among all compounds in the sub-database (Lu et al., 2005). The MS/MS peaks identified were annotated with the ion interpretation that also shows a fragmentation mechanism. Segments of 15-HETE that matched the MS/MS peaks according to the fragmentation rules previously noted are italicized (Lu et al., 2005). For comparison, MassFrontier was also used to identify this as 15-HETE.

We used LC-UV-MS/MS chromatograms and spectra acquired from murine tissue and organ extracts spiked with over 34 well-known lipid mediators (i.e., eicosanoids and related compounds) for the evaluation. For every known compound added (i.e., spiked) in all samples, the search of its LC-UV-MS/MS through a database gives a "hit list" showing what compound matches best. The best match has the highest matching score using the MassFrontier or theoretical database, and the smallest contrast angle using COCAD. Hence, the best match is either correct or incorrect. The mean and standard error (S.E.) of percent correct for best match indicated the performance of each database and algorithm. Generally, COCAD offered the greatest percent correct (Lu et al., 2005), which was considered reasonable because COCAD takes into account both MS/MS ion intensities and identities measured from authentic eicosanoid standards and unknowns. The percent correct from the theoretical system is comparable to that of the MassFrontier system (Lu et al., 2005).

Results of these tests demonstrated that it was not necessary to use all the MS/MS ions to identify an LM or other compounds (Lu et al., 2005). Also, in some cases, it is neither obvious nor cost efficient to identify all MS/MS ions. Rigorous identification can be carried out using isotope labeling and derivatization, which may not be feasible or realistic to be applied to routine analyses of biologically relevant samples. Thus, we can identify an LM with precision using one to several chain-cut ions per functional group along with peripheral-cut ions and chain- plus peripheral-cut ions following the

MS/MS fragmentation rules. For example, we use at least four ions for monohydroxy-containing eicosanoids, six ions for dihydroxy-containing eicosanoids, and eight ions for trihydroxy-containing eicosanoids (Fig. 12.6). Of interest, John Beynon et al. (McLafferty et al., 1999) originated the "Eight Peak Index of Mass Spectra" database for manual identification of compounds. For different amounts of lipid mediators added to the samples (1.0, 2.5, and 5.0 ng), the percent identified correctly was not significantly different. Based on the test experiments using the theoretical database and algorithm, the average matching score for correct best match is 2.02, with a standard error of 0.71 ($n = 212$). Therefore, the threshold score for a 95% confidence interval of matching score for correct best match is $2.02 - 1.96 \times 0.71 = 0.62$.

These databases and search algorithms were developed on the basis of LC-UV-ion trap MS/MS data of lipid mediators. The ion intensity patterns of MS/MS spectra generated from an ESI-triple quadrupole mass spectrometer are quite similar to ESI-ion trap MS because the collision energy for both types of instruments is in the low energy region (a few to 100 eV, laboratory kinetic energy of ions) (Aliberti et al., 2002; Murphy et al., 2001; Serhan et al., 2003; Wheelan et al., 1996). Therefore, the constants and algorithms reviewed in this chapter (Lu et al., 2005) may fit the CID spectra from triple-quadrupole MS/MS without the need for substantial modification. For high collision energy (several 10^2 to $\sim 10^3$ eV) CID spectra generated via sector or time-of-flight (TOF)/TOF analyzer, the relative intensity patterns are quite different in comparison with the low energy ones, although many ions occurred for both energy situations (Aliberti et al., 2002; Griffiths et al., 1996; Murphy et al., 2001; Serhan et al., 2003; Wheelan et al., 1996). For example, the peripheral-cut ions are less abundant than the chain-cut ions. For ion-trap and triple quadrupole MS/MS, peripheral-cut ions are more abundant than chain-cut ions. Our constants and algorithms give chain-cut ions more weight than peripheral-cut ions because chain-cut ions are more important to define LM structures. Therefore, they may still fit high collision energy CID spectra. Nevertheless, the constants and algorithms should be thoroughly tested and modified accordingly so that they can be validated for use with other instruments that may give fragmentation patterns of different intensities than the instruments used in the authors' laboratory (Chiang et al., 1999; Devchand et al., 2003; Hong et al., 2003; Serhan et al., 2000, 2002; Takano et al., 1998). The CRTs used in these experiments were obtained with the specified chromatographic conditions (i.e., for a column of 100-mm length and some for 150-mm length) (Lu et al., 2005). The databases and algorithms were programmed so that the new LC-UV-MS/MS data including other chromatographic conditions can easily be entered and used in the databases.

This full cycle of events defines mediator-based lipidomics because it is important to establish both the structure and functional relationships of bioactive molecules in addition to their cataloging and mapping of architectural components of biolipidomics. With this new informatics and lipidomics-based approach that combined LC-PDA-MS/MS, a novel array of endogenous lipid mediators were identified (Serhan et al., 2000, 2002) during the multicellular events that occur during resolution of inflammation. The novel biosynthetic pathways uncovered use omega-3 fatty acids, eicosapentaenoic acid, and docosahexaenoic acid as precursors to new families of protective molecules, termed resolvins and protectins (Figs. 12.1–12.3). These include resolvin E (18R series from EPA) and 17 series resolvin D from DHA (Hong et al., 2003). The vasculature in humans—in particular, endothelial cells during cross-talk with leukocytes—biosynthesizes these mediators via transcellular biosynthesis routes (Serhan et al., 2000). In this novel cell–cell interaction, endothelial cells generate the first biochemical step as illustrated in Fig. 12.2 and then transfer this intermediate 18R-HEPE to leukocytes, which transform this to a potent molecule termed resolvin E1 (RvE1). RvE1 is ~1000 times as potent as native EPA (unoxidized) as a regulator of neutrophils, which stops their migration into inflammatory foci (Serhan et al., 2000, 2002, 2004) (Table 12.1).

Docosahexaenoic acid, which is enriched in neural systems (Salem et al., 2001), is also released and transformed to potent bioactive molecules denoted 10R,17S docosatriene or, more specifically, neuroprotectin D1 protection D1 (NPD1/PD1) (Serhan et al., 2006) and resolvins of the D series (Figs. 12.3 and 12.5). Human brain, synapses, and retina are rich in DHA, a major omega-3 fatty acid. Deficiencies in DHA are associated with altered neural functions, cancer, and even inflammation in experimental animals (Burr and Burr, 1929). Employing our mediator-based lipidomics approach, we found that, on activation, neural systems release DHA to produce neuroprotectin D1 (Table 12.1), which, in addition to stopping leukocyte-mediated tissue damage in stroke, also maintains retinal integrity (Mukherjee et al., 2004). Of interest, fish rich in DHA, such as rainbow trout, generate NPD1 and resolvins of the D series (Hong et al., 2005). These results raise not only the question of their role in fish, but also suggest that the resolvins and protectins are highly conserved structures in the course of evolution.

These structures form the basis for lipidomic databases and informatics, which consist of mass spectra as well as additional physical bioactivity characteristics and search algorithms (Lu et al., 2005). In this section, we review the recent results with resolvin D1 and protectin D1 (Figs. 12.7 and 12.8), as well as other related DHA-derived products without derivatization using low-collision-energy MS/MS acquired on anions generated from ESI of molecules eluted from LC. For structure elucidation and identification, the ion structures and MS/MS fragmentation mechanisms are proposed and confirmed via deuterium-labeled isotopomers of these compounds. RvD1 and PD1, as well as mono-hydroxy-DHA products, were found using this

Table 12.1 Resolvins and protectins: actions in animal disease models

Mediator	Action	Reference
Resolvin E1	Reduces PMN infiltration, murine skin air pouch inflammation, peritonitis	Arita et al., 2005c; Bannenberg et al., 2005; Serhan et al., 2000, 2002
	Gastrointestinal protection in TNBS colitis	Arita et al., 2005b
	Protects in periodontitis, stops inflammation and bone loss	Hasturk et al., 2006
Resolvin D1	Reduces PMN infiltration, murine skin air pouch inflammation	Serhan et al., 2002
	Reduces peritonitis; reduces cytokine expression in microglial cells	Hong et al., 2003
Protectin D1/neuroprotectin D1	Protects in renal ischemic injury	Duffield et al., 2006
	Reduces stroke damage	Marcheselli et al., 2003
	Reduces PMN infiltration	Hong et al., 2003
	Protects from retinal injury	Mukherjee et al., 2004
	Regulates T_H2 cells, apoptosis, and raft formation	Ariel et al., 2005
	Shortens resolution interval in murine peritonitis; regulates cytokines and chemokines	Bannenberg et al., 2005
	Reduces PMN infiltration; reduces peritonitis	Serhan et al., 2006
	Diminished production in human Alzheimer disease and promotes neural cell survival	Lukiw et al., 2005
	Promotes corneal epithelial cell wound healing	Gronert et al., 2005
	Reduces airway inflammation in murine asthma model	Levy et al., 2007

PMN, polymorphonuclear leukocytes; TNBS, trinitrobenzene sulphonic acid.

Lipid Mediator Informatics and Profiling

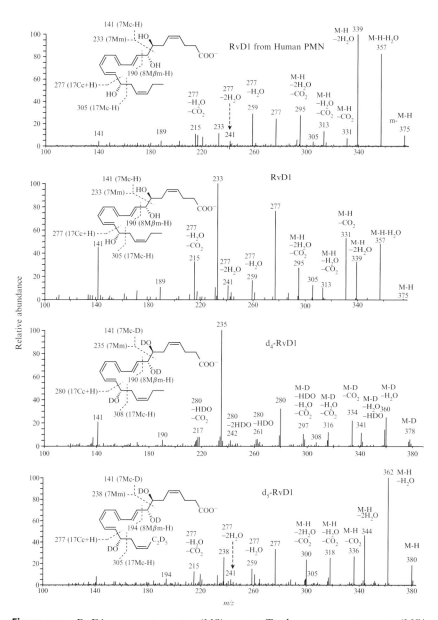

Figure 12.7 RvD1 mass spectrometry (MS) spectra. Tandem mass spectrometry (MS/MS) spectra acquired via collision-induced dissociation (CID) of the electrospray-generated carboxylate anion [M-H or D]⁻ from the liquid chromatographic (LC) peak of resolvin D1 or deuterated RvD1. RvD1, O,O,O,O-d_4-RvD1, and 21,21,22,22,22-d_5-RvD1 were obtained with isolated-enzyme reactions with or without deuterium-exchange, as well as from isolated human polymorphonuclear leukocyte (PMN) (top panel). Insets show the structures and collision-induced dissociation fragmentation patterns (see text).

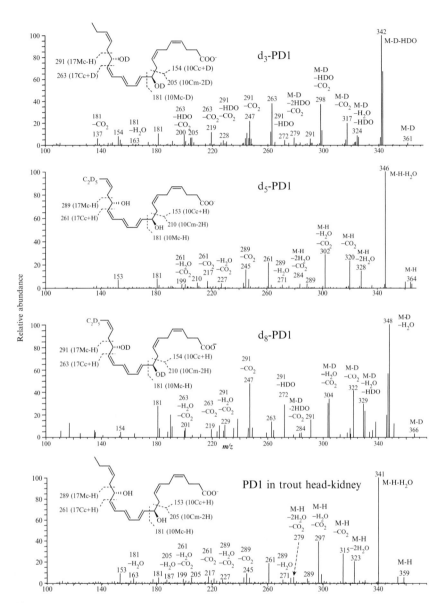

Figure 12.8 Protectin D1 mass spectrometry (MS) spectra. Tandem mass spectrometry (MS/MS) spectra acquired via collision-induced dissociation (CID) of electrospray-generated carboxylate anion [M-H or D]⁻ from the liquid chromatographic (LC) peak of PD1 or deuterated PD1. *Top panel spectra:* PD1 was obtained from trout head-kidney; O,O,O-d_3-PD1, 21,21,22,22,22-d_5-PD1, and O,O,O,21,21,22,22,22-d_8-PD1 were obtained using isolated-enzyme reactions with or without deuterium-exchange, as detailed in Hong *et al.* (2007). Insets show the structures and collision-induced dissociation fragmentation patterns.

LC-UV-MS/MS approach to be generated by human neutrophils, whole blood (Hong et al., 2003), trout head-kidney (Hong et al., 2005), and stroke-injury murine brain tissues (Marcheselli et al., 2003).

To illustrate the interpretation of MS/MS ions generated from RvD1, PD1/NPD1, and other related DHA-derived products, we use the nomenclature recently proposed (Lu et al., 2005) to name the specific "segments," which are fragments formed through hypothetic homolytic-cleavages without hydrogen or group migration. This is depicted in Fig. 12.5 using LXA_4, RvD1, and PD1 structures. A number is used to designate the position of a hydroxy or another functional group on the carbon chain where the cleavage occurs. The letter immediately following the number indicates the side of the functional group on which the cleavage occurs: "c" is for cleavage on the carboxyl side of the functional group and "m" is for cleavage on the methyl side of the functional group. Each cleavage can directly generate two segments. The second letter indicates the side of the cleavage on which the segment forms: "c" is for a segment formed on the carboxyl side of the cleavage and "m" is for a segment formed on the methyl side of the cleavage. Segments formed through β-cleavage or γ-cleavage along the functional group are named by adding β or γ between the two "c"/ "m" letters (Lu et al., 2005).

The ESI in negative mode generated primarily deprotonated molecular ions [M-H or D]$^-$ from resolvin D1 (Figs. 12.7–12.10), Protectin D1, other related DHA-derived compounds, and their deuterium isotopomers eluted from the LC column. In MS/MS experiments, [M-H]$^-$ ions were selected and fragmented via CID. Cleavage on the carbon chain was frequently observed on the α locations toward hydroxy and hydroperoxy. This is similar to the CID fragmentation of derivatives from arachidonic acid and eicosapentaenoic acid (Hong et al., 2003; Murphy et al., 2001; Serhan et al., 2002; Wheelan et al., 1996).

6.1. RvD1, O,O,O,O-d_4-RvD1, and 21,21,22,22,22-d_5-RvD1

RvD1 is 7S,8R,17S-trihydroxy-docosa-4Z,9E,11E,13Z,15E,19Z-hexaenoic-acid, which has three hydroxy groups and six double bonds, four of which are conjugated and bracketed by the 8-hydroxy and 17-hydroxy (see inset of Fig. 12.7). The low collision–energy MS/MS spectrum of the LC peak of the RvD1 shows peripheral-cut ions at m/z 357 [M-H-H_2O] (relative intensity is 47%), 339 [M-H-2H_2O] (32%), 331 [M-H-CO_2] (52%), 313 [M-H-H_2O-CO_2] (10%), and 295 [M-H-2H_2O-CO_2] (26%) (Fig. 12.6). The assignment of these ions is consistent with O,O,O,O-d_4-RvD1 (d_4-RvD1), ions: 360 [M-D-H_2O] (27%), 341 [M-D-H_2O-HDO] (14%), 334 [M-D-CO_2] (22%), 316 [M-D-H_2O-CO_2] (14%), and 297 [M-D-HDO-H_2O-CO_2] (12%); 21,21,22,22,22-d_5-RvD1 (d_5-RvD1) ions: 362 [M-H-H_2O] (100%), 344 [M-H-2H_2O] (46%), 336 [M-H-CO_2] (27%), 318 [M-H-H_2O-CO_2]

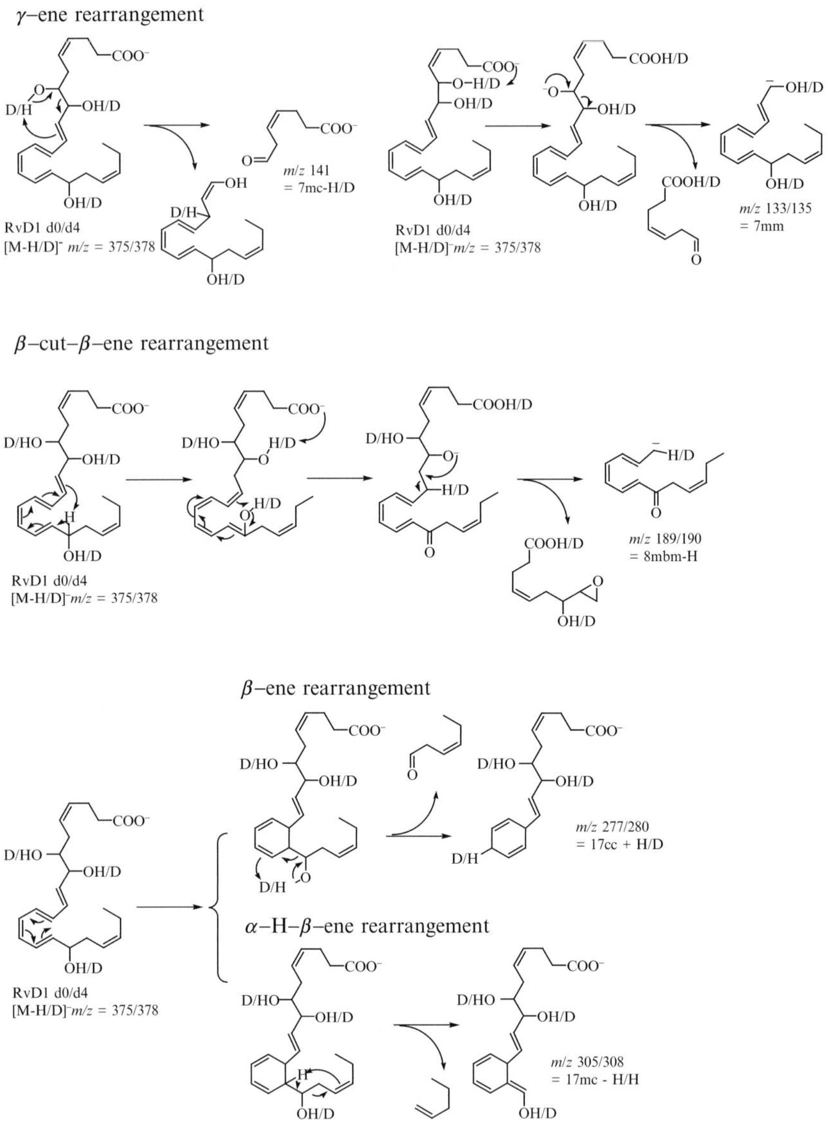

Figure 12.9 Proposed mechanism for RvD1 fragmentation.

(25%), and 300 [M-H-2H$_2$O-CO$_2$] (24%); and d$_4$-RvD1 ions 359 [M-D-HDO] (14%), 342 [M-D-2H$_2$O] (4%), 315 [M-D-HDO-CO$_2$] (8%), and 298 [M-D-2H$_2$O-CO$_2$] (8%). These peripheral-cut ions support the existence of the carboxyl group (for CO$_2$ loss) and hydroxy groups (for H$_2$O loss) in RvD1. However, these ions do not provide the specific information or

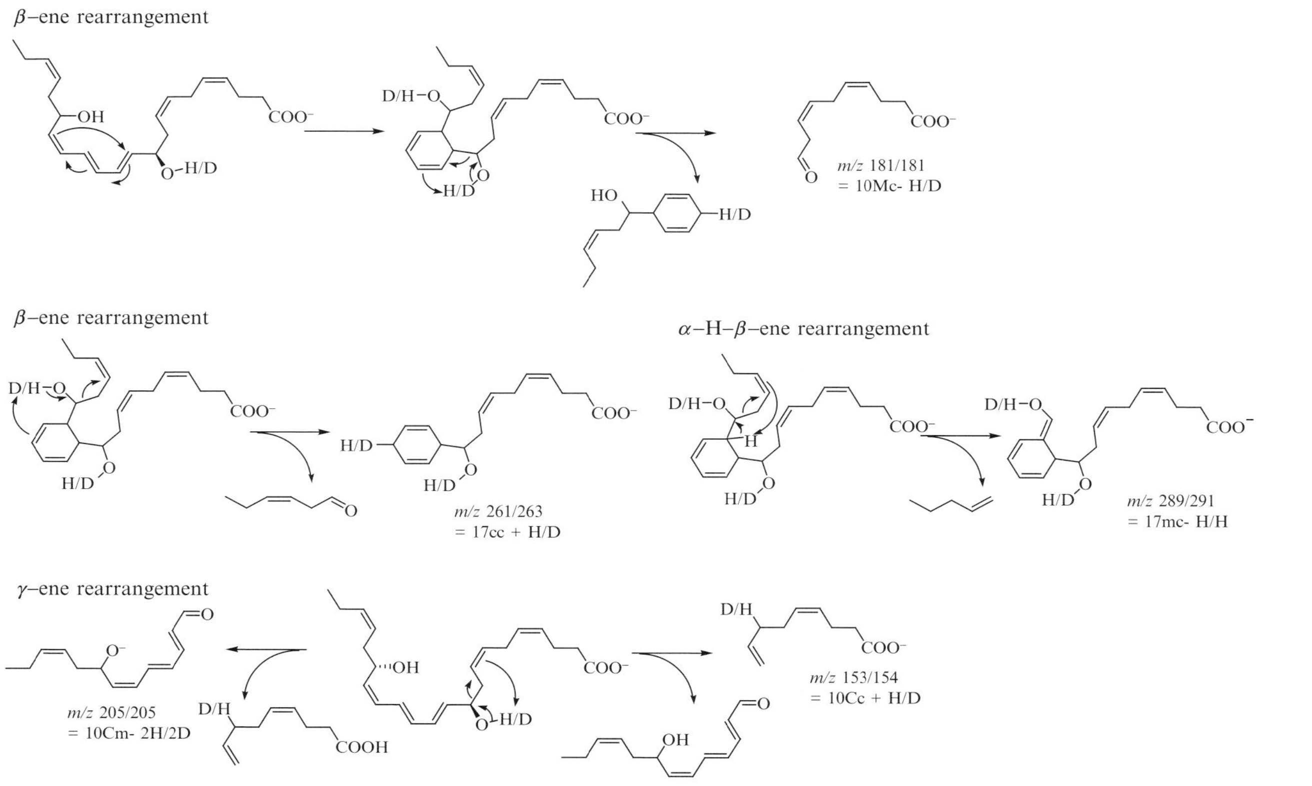

Figure 12.10 Proposed mechanism for PD1 fragmentation. This mechanism illustrates the main mechanism for these eicosapentaenoic acid (EPA)- and docosahexaenoic acid (DHA)-derived mediators (see text for details).

clues for the positions of the hydroxy(s) and double-bonds. The loss of H₂O from d₄-RvD1, whose hydroxy hydrogens were completely deuterated, indicates that the deuteroxy deuterium is exchanged with the carbon chain–bonded hydrogen. The chain-cut MS/MS ions of RvD1 are listed in the next paragraph. Some of them further transfer to chain- plus peripheral-cut ions via loss of water or CO_2. Chain-cut and chain- plus peripheral-cut ions provide the signatures for the positions of functional groups and double bonds.

The ion m/z 141 (44%) of RvD1 is equal to segment 7 mc minus the 7-hydroxy proton (H from 7-OH or 7-OH H), namely [7 mc − H from 7-OH] corresponding to the following mechanism (Fig. 12.9). When the 7-OH H is extracted through a transit six-membered ring to C10 by the double-bond C9,10 (which is at γ position to 7-OH), 7-OH converts to a carbonyl group, the allylic single-bond C7,8 at α position to 7-OH cleaves, and the C8,9 double-bond forms, yielding ion m/z 141 and an enol. We have referred to this process having γ-ene–facilitated α-OH H migration and α-cleavage as a "γ-ene rearrangement." The mechanism was confirmed by ion m/z 141 as [7 mc-deuteron from 7-deuteroxy] (i.e., [7 mc-D from 7-OD]) from d₄-RvD1 and as [7 mc-H from 7-OH] from d₅-RvD1 (Figs. 12.7 and 12.8) (Hong et al., 2007).

In competition with the formation of ion m/z 141, 7-OH H shifts to the carboxyl anion internally, forming 7-alkoxide anion S3a. The negative charge of S3a directs the cleavage of C7,8 allylic single-bond bond at α position, resulting in ion m/z 233 (100%) (corresponding to ions m/z 235 [100%] for d₄-RvD1 and 238 [26%] for d₅-RvD1) (Fig. 12.6), which is equivalent to segment [7 mm]. Also, the deuterium labeling on C21 and C22 of d₅-RvD1 changed the base-peak from ion m/z 233 of RvD1 or m/z 235 (equivalent to [7 mm]) of d₄-RvD1 to ion m/z 362 [M-H-H₂O].

Ion m/z 189 is equal to [8 mm − H] for RvD1, which is equivalent to ion m/z 190 [8 mβm − H] for d₄-RvD1. Their formation corresponds to the following mechanism: initially, H from C17 of the parent ion is extracted to C9 by C9,10 double-bond at β position to 8-OH, and the conjugated tetraene Δ9,11,13,15 converts to Δ10,12,14,16 in S3b; when 17-OH H (or D 17-OD D) moves to C10, this tetraenol S3b changes to tetraenone S3c; meanwhile, the 8-OH H (or 8-OD D for d₄-RvD1) migrates to the carboxylate anion, resulting in the 8-alkoxide anion S3c; the negative charge of the S3c directs the breakage of the C9,10 allylic single-bond at the β position to the 8-alkoxide group, yielding ion m/z 189 (14%) from RvD1, m/z 190 (6%) from d₄-RvD1, and m/z 194 (7%) from d₅-RvD1. We refer to such process with β-cleavage facilitated by α-OH and β-ene as a "β-cut-β-ene rearrangement."

Ion m/z 277 (75%) of RvD1 and ion m/z 280 (32%) of d₄-RvD1 are equivalent to [17cc + H from 17-OH] and [17cc + D from 17-OD], respectively. They correspond to the α-cleavage of the C16,17 bond of their parent ions, of which the β double-bond C15,16 facilitates the cleavage. This is a typical α-hydroxy-β-ene-like rearrangement (abbreviated as β-ene rearrangement) (Murphy et al., 2005). When the conjugated triene Δ11,13,15 of the parent ion forms a six-membered ring in S3d, the β double-bond C15,16 shifts to γ position, forming a C14,15 double-bond, and the vinyl single-bond C16,17 becomes the allylic single-bond C16,17 in S3d. Then, the 17-OH H (or 7-OD D in d_4-RvD1) shifts to C14 of the γ double-bond C14,15, a 17-carbonyl group forms, and the C16,17 allylic single-bond cleaves, producing ion m/z 277 of RvD1 and d_5-RvD1[or 280 of d_4-RvD1]. Thus, this β-ene rearrangement includes conversion of β-ene to γ-ene and α vinyl single-bond to α allylic single-bond and, subsequently, the γ-ene rearrangement, similar to the case of the ion m/z 141 from RvD1 (Fig. 12.7).

When the C16-proton (H) in S3d shifts to C20, instead of the shift of H from 17-OH to C20 (which would have generated ion m/z 307 for d_5-RvD1), the C17,18 allylic single-bond fragments, yielding ion m/z 305 (11%, equivalent to [17 mc − H from C16]) from RvD1. This mechanism is confirmed by ion m/z 308 (5%) from d_4-RvD1 (Fig. 12.9) and ion m/z 305 (6%) from d_5-RvD1. We refer to this process as "α-H-β-ene rearrangement."

The 17-OH in S3d is competitively deprotonated internally by its carboxyl anion to form a C17-alkoxide anion, of which the negative charge directs the breakage of C16,17 allylic single-bond, generating ion m/z 277; similar to that previously discussed for S3d, a charge-remote process for this alkoxide anion, also involving the migration of C16-H to C20 cleaves allylic single-bond C17,18 and generates ion m/z 305 (Fig. 12.9) (Cheng and Gross, 2000). This is also an α-H-β-ene rearrangement. The α-H is less active than H from α-OH because the C-H bond is much less polarized than the O-H bond; thus the abundance of ion m/z 277 is higher than that of ion m/z 305 for RvD1 and d_5-RvD1. The same pattern is observed: ion m/z 280 is more abundant than ion m/z 308 for d_4-RvD1, and ion m/z 277 is more abundant than ion m/z 305 for d_5 RvD1.

The loss of water and/or CO_2 generated ions 259 [277-H_2O] for RvD1 (15%) and d_5-RvD1(16%), 241 [277-$2H_2O$] for RvD1 (11%) and d_5-RvD1 (6%), 215 [277-H_2O-CO_2] for RvD1 (32%) and d_5-RvD1 (15%), and 217 [280-HDO-CO_2] (7%) for d_4-RvD1, which further confirmed the structure assignment of ion m/z 277. Using this LC-UV-MS/MS analysis, RvD1 is biosynthesized and released by human neutrophils. The MS/MS spectrum of a chromatographic peak acquired from the samples of human neutrophils matches the spectrum of synthetic RvD1 (bottom panel, Fig. 12.7), as do the UV spectrum and chromatographic retention time (Hong et al., 2007; Sun et al., 2007).

6.2. PD1, O,O,O-d$_3$-PD1, 21,21,22,22,22-d$_5$-PD1, and O,O,O, 21,21,22,22,22-d$_8$-PD1

Among the six double-bonds of PD1 (protectin D1/neuroprotectin D1:10R,17S-dihydroxy-docosa-4Z,7Z,11E,13E,15Z,19Z-hexaenoic acid), three are conjugated between 10-OH and 17-OH (Serhan et al., 2006). Negative ESI generated ion m/z 359, a deprotonated molecular ion [M-H]$^-$, from PD1 (Fig. 12.8). The MS/MS spectrum at m/z 359 of PD1 from trout head-kidney matches to that acquired from synthetic PD1 (see Fig. 12.1 of Hong et al., 2005). Peripheral-cut ions at m/z 341 [M-H-H$_2$O] (100%), 323 [M-H-2H$_2$O] (20%), 315 [M-H-CO$_2$] (29%), 297 [M-H-H$_2$O-CO$_2$] (39%), and 279 [M-H-2H$_2$O-CO$_2$] (7%) are consistent with the PD1 structure of one carboxylic group and two hydroxy groups. These ions are equivalent to those at m/z 342 [M-D-HDO] (100%), 324 [M-D-H$_2$O-HDO] (7%), 317 [M-D-CO$_2$] (18%), 298 [M-D-HDO-CO$_2$] (37%), and 279 [M-D-2HDO-CO$_2$] (8%) from MS/MS of O,O,O-d$_3$-PD1, respectively. They are further confirmed by ions in MS/MS of d$_5$-PD1 (21,21,22,22,22-d$_5$-PD1) at m/z 346 [M-H-H$_2$O] (100%), 328 [M-H-2H$_2$O] (13%), 320 [M-H-CO$_2$] (23%), 302 [M-H-H$_2$O-CO$_2$] (27%), and 284 [M-H-2H$_2$O-CO$_2$] (6%), respectively; and along with ions from MS/MS of d$_8$-PD1 (O,O,O,21,21,22,22,22-d$_8$-PD1) at m/z 348 [M-D-H$_2$O] (100%), 329 [M-D-H$_2$O-HDO] (30%), 322 [M-D-CO$_2$] (41%), 304 [M-D-H$_2$O-CO$_2$] (35%), and 284 [M-D-2HDO-CO$_2$] (3%), respectively. It is interesting that some of the water loss was as H$_2$O rather than HDO. Loss of HDO is expected for d$_3$-PD1 because D has replaced every hydroxy H. D in the deuteroxy group exchanged with the hydrogen on the carbon chain when the [M-D]$^-$ ion of d$_3$-PD1 was selected and activated for the MS/MS fragmentation in the ion trap, similar to d$_4$-RvD1 (previously shown). The MS/MS chain-cut ions and the formation mechanisms are shown in Figs. 12.7 and 12.9. Those ions formed via loss of H$_2$O from d$_3$-PD1 also manifest the exchange of deuteroxy deuterium with carbon chain–bonded hydrogen.

The MS/MS ion m/z 153 of PD1 is consistent with the fragmentation mechanism of a γ-ene rearrangement: when the 10-OH H shifts to the C7 at γ double-bond C7,8 through a six-membered ring, a carbonyl group forms (Hong et al., 2007) and the C9,10 allylic single-bond at α-position to 10-OH cleaves, yielding the ion at m/z 153 in MS/MS of PD1 (11%) and d$_5$-PD1 (7%). The equivalent ion is at m/z 154 for d$_3$-PD1 (7%) and d$_8$-PD1 (5%). Ions m/z 153 and 154 are equal to [10cc + H from 10-OH] (for PD1 and d$_5$-PD1) and [10cc + D from 10-OD] (for O,O,O-d$_3$-PD1 and d$_8$-PD1), respectively. If, additionally, the 17-OH H or 17-OD D shifts to the carboxyl anion, it yields the neutral molecule S4b and ion m/z 205, equal to [10 cm − 2H from hydroxys] for PD1 (7%)

or [10 cm − 2D from deuteroxys] for d_3-PD1 (7%). Its equivalent ion is at m/z 210 (6%) for d_5-PD1. The assignment of ions m/z 205 and 210 is consistent with ion 187 [205-H_2O] (4%) of PD1.

Ion m/z 181 is equal to [10 mc - H from OH] for PD1 (5%) and d_5-PD1 (10%), or to [10 mc - D from OD] for d_8-PD1 (29%) and d_3-PD1 (10%). The corresponding fragmentation mechanism is a charge-remote β-ene rearrangement with α-OH as 10-OH and β-ene as C11,12 double-bond of the parent ion: the conjugated triene Δ11,13,15 in PD1 forms a six-membered ring in intermediate via the Diels-Alder process, changing the vinyl single-bond C10,11 at position to an allylic single-bond; then the H from 10-OH moves to C13 on the newly formed γ-ene (at C12,13), the allylic single-bond C10,11 breaks, and 10-carbony forms, generating S4d and ion m/z 181. Additional evidence for the composition of ion m/z 181 is ion m/z 163 [181-H_2O] (5%) in the MS/MS spectrum of PD1 (Fig. 12.8).

Ion m/z 261 (20%) in the MS/MS spectrum of PD1 is equivalent to [17cc + H from 17-OH] (Fig. 12.10), generated through a β-ene rearrangement analogous to the formation mechanism for ion m/z 277 from RvD1 (Figs. 12.7 and 12.9). When 17-OH H shifts to C14 in S4c, the allylic single-bond C16,17 breaks, resulting in a carbonyl group in the neutral loss hexen-3-al and ion m/z 261. This is verified by ion m/z 263 from d_3-PD1 (37%) and d_8-PD1 (15%), equal to [17cc + D from 17-OD]; and ion m/z 261 (23%) from d_5-PD1, equal to [17cc + H from 17-OH]. This fragmentation mechanism is further confirmed by ions m/z 217 [261-CO_2] and 199 [261-H_2O-CO_2] from both PD1 (9%, 5%) and d_5-PD1 (7%, 5%), as well as by ions m/z 219 [263-CO_2] (12%) and 200 [263-HDO-CO_2] (9%) from d_3-PD1. Ions m/z 219 [263-CO_2] and 201 [263-H_2O-CO_2] from d_8-PD1 (10%, 11%) are also consistent with this mechanism.

The formation of ion m/z 289 of PD1 corresponds to the α-H-β-ene rearrangement, which is analogous to that for ion m/z 305 from RvD1 (Figs. 12.1 and 12.9), with the shift of C16 H instead of 17-OH H, where α-OH is 17-OH, and γ-ene is C19,20 double-bond. When C16 H shifts to C20 in S4c, the allylic single-bond C17,18 cleaves, yielding a pentaene (neutral loss) and ion m/z 289 (5%) with a 17-enol and a six-membered ring (Figs. 12.7 and 12.9). This fragmentation process was confirmed by ion m/z 289 from d_5-PD1 (4%) and ion m/z 291 from d_3-PD1 (9%) and d_8-PD1 (16%) (Fig. 12.7). Therefore, ions m/z 289 and 291 are equal to [17 mc − H from C16] of PD1 and d_3-PD1, respectively (Figs. 12.8 and 12.10). This process was consistent with the chain- plus peripheral-cut ions formed from m/z 289 and 291 via loss of water and/or CO_2: 271 [289-H_2O], 245 [289-CO_2], and 227 [289-H_2O-CO_2] from PD1 (4%, 10%, 4%) and d_5-PD1 (5%, 18%, 4%); 272 [291-HDO] (5%), 247 [291-CO_2] (23%), and 228 [291-HDO-CO_2] (4%) from d_3-PD1; and 272 [291-HDO] (29%), 247 [291-CO_2] (47%), and 229 [291-H_2O-CO_2] (15%) from d_8-PD1.

The stereoisomers of PD1 were obtained through total organic synthesis (Yang, 2005) and are indistinguishable based only on their MS/MS spectra (Serhan et al., 2006). Most of these can be separated via reverse-phase LC (Serhan et al., 2006). Of interest and most importantly, their bioactivities were found to depend on their stereochemical structures (Serhan et al., 2006).

7. Discussion

7.1. Impact of a functional group and enes on MS/MS fragmentation of various lipid mediators

Using essentially the same MS/MS conditions for RvD1, lipoxin A_4 (5S,6R,15S-trihydroxy-7E,9E,11Z,13E-eicosatetraenoic acid) produces chain-cut ions m/z 115 [5 mc − H], 235 [5 mm], and 251 [15cc + H] [27,29], equivalent to ions m/z 141[7 mc − H], 233 [7 mm], and 277 [17cc + H], respectively, from RvD1 (Fig. 12.7). This corresponds to the same structure of lipoxin A_4 between C5 and C15 as that of RvD1 between C7 and C17, the conjugated tetraene bracketed by two vicinal hydroxys on the carboxyl (c)-side and one hydroxy on the methyl terminus (m)-side. However, the ion m/z 279 [15 mc − H] was not observed for lipoxin A_4. No double bond is located in the γ position to 15-OH on the m side of lipoxin A_4 (i.e., in segment 15 mm) to facilitate generation of this ion through an α-H-β-ene rearrangement that yields ion [17 mc − H] m/z 305 from RvD1. The same phenomenon was observed for 15-hydroxy-eicosatetraenoic acid (15-HETE) versus 17-HDHA and PD1. 15-HETE has no [15 mc − H] ion (m/z 247), but 17-HDHA and PD1 have [17 mc − H] ions as m/z 273 and 289, respectively. Therefore, ion [xmc − H] indicates double bond(s) in xmm segment, with x representing the position of the functional group.

7-HDHA also has the same chain-cut ions as RvD1: [7 mc − H] and [7 mm] around 7-OH (Fig. 12.7). The γ-ene rearrangement for the cleavage of C7,8 allylic single-bond in RvD1 for producing ion m/z 141 is analogous to that for S5k, a transient state of 7-HDHA in MS/MS. However, the shift of C12 H to C8 prior to such cleavage for 7-HDHA is unnecessary for RvD1 because the C7,8 bond in RvD1 is already an allylic single bond. The [7 mm] ions m/z 233 and 201 are generated through the same mechanism from RvD1 and 7-HDHA, respectively. By analogy, 5-HETE has the chain-cut ions at m/z 115 ([5 mc − H]) and 203 ([5 mm]) around 5-OH similar to LXA_4 (Chiang et al., 1998; LIPID MAPS Consortium, 2006; Wheelan et al., 1996), which carries m/z 115 as a signature ion.

Although an 8-OH exists in both RvD1 and 8-HDHA, no MS/MS ions are shared by these. Because of the 7-OH and the conjugated tetraene 9,11,13,15 of RvD1, the 8 cm (equal to 7 mm) segment of RvD1 is involved in formation of ion m/z 233 as 7 mm, not as 8 cm; namely, it follows the fragmentation mechanism to produce ion [7 mm] instead of [8 cm − 2H]. The ion m/z 141, coming from segment 8cc (equivalent to 7 mc) of RvD1, is equivalent to [7 mc − H] instead of [8cc + H]. RvD1 has the [8 mm − H] ion (m/z 189) because the ion can form a conjugated tetraenonyl structure to disperse the carbon anion charge for greater stabilization. In contrast, 8-HDHA does not have the [8 mβm − H] ion (would be m/z 174) due to lack of such stabilization.

Similarities between 17-HDHA and 17-HpDHA are reflected by some features of their respective MS/MS spectra (Table 12.2 and Fig. 12.8). The conversion of the C16,17 vinyl single-bond to an allylic single bond via shift of C12 H to C16 is likely to be the first step of the MS/MS fragmentation (Fig. 12.9). They all have [cc + H] ion m/z 245, although the H is from 17-OH of 17-HDHA and 17-OOH of 17-HpHDHA. They all have [mc − H] ions with the H migrating from C16 to C20. Such similarity was also observed among 15-HETE, 15-hydroperoxy-eicosatetraenoic acid, and 15-oxo-eicosatetraenoic acid (LIPID MAPS Consortium, 2006; MacMillan and Murphy, 1995).

ESI-MS/MS analysis of underivatized ESI-deprotonated molecules of Resolvin D1, Protectin D1, and other DHA-derived products reveals the structures, especially the locations of hydroxy, hydroperoxy, or carbonyl groups, as well as the double-bonds. The definitive fingerprints are the CID MS/MS ions formed via cleavage of the carbon-chain, namely chain-cut ions. These ions are generated via charge-remote and/or charge-directed fragmentation mechanisms (Hong et al., 2007). The CID fragmentation usually consists of several serial and parallel reactions. Each chain-cut ion is equivalent to the corresponding hypothetically homolytic segment (cc, cm, mc, or mm) with addition or extraction of up to two protons. Deuterium labeling facilitated the structural analysis of MS/MS ions and the corresponding fragmentation mechanisms. The molecular ion structures determine the fragmentation mechanisms, which in turn provide the rationale for the assignment of the MS/MS ion structures and, consequently, the molecular ion structures of the DHA-derived compounds. This chapter describes a simple approach for comprehensive structure analysis of DHA-derived products without the need for derivatizing the compounds. The fragmentation rules reported here were used for the development of a database and search algorithm for the theoretic MS/MS spectra of novel bioactive compounds derived from fatty acids (Lu et al., 2005).

Table 12.2 Chain-cut ions and fragmentation mechanisms for mono-HDHA and O,O-d2-mono-HDHA

	-HDHA	Ion m/z	Algorithm	H is from:	Fragmentation rearrangement	Labeled-HDHA	Ion m/z	Algorithm	D/H is from:
	20-	285	20cc+**H**	20-OH	β-ene	d_2-20-	286	20cc+**D**	20-OD
	17-	245	17cc+**H**	17-OH	β-ene	d_2-17-	246	17cc+**D**	17-OD
	16-	233	16cc+**H**	16-OH	γ-ene	d_2-16-	234	16cc+**D**	16-OD
cc[a]	14-	205	14cc+**H**	14-OH	β-ene	d_2-14-	206	14cc+**D**	14-OD
	13-	193	13cc+**H**	13-OH	γ-ene	d_2-13-	194	13cc+**D**	13-OD
	11-	165	11cc+**H**	11-OH	β-ene	d_2-11-	166	11cc+**D**	11-OD
	10-	153	10cc+**H**	10-OH	γ-ene	d_2-10-	154	10cc+**D**	10-OD
	11-	177	11 cm-**2H**	11-OH,12C	α-H-β-ene	d_2-11-	177	11 cm-**D-H**	11-OD, H from 12C
cm	8-	217	8 cm-**2H**	8-OH,9C	α-H-β-ene	d_2-8-	217	8 cm-**D-H**	8-OD, H from 9C
	17-	273	17 mc-**H**	16C	α-H-β-ene	d_2-17-	274	17 mc-**H**	H from 16C
	16-	261	16 mc-**H**	15C	α-H-β-ene	d_2-16-	262	16 mc-**H**	H from 15C
	14-	233	14 mc-**H**	13C	α-H-β-ene	d_2-14-	234	14 mc-**H**	H from 13C
	13-	221	13 mc-**H**	13-OH	β-ene	d_2-13-	221	13 mc-**D**	13-OD
mc	11-	193	11 mc-**H**	11-OH	γ-ene	d_2-11-	193	11 mc-**D**	11-OD
	10-	181	10 mc-**H**	10-OH	β-ene	d_2-10-	181	10 mc-**D**	10-OD
	8-	153	8 mc-**H**	8-OH	γ-ene	d_2-8-	153	8 mc-**D**	8-OD
	7-	141	7 mc-**H**	7-OH	β-ene	d_2-7-	141	7 mc-**D**	7-OD
	4-	101	4 mc-**H**	4-OH	β-ene	d_2-4-	101	4 mc-**D**	4-OD
	11-	149	11 mm	N/A[b]	Charge-direct	d_2-11-	149	11 mm	N/A
mm	8-	189	8 mm	N/A	Charge-direct	d_2-8-	189	8 mm	N/A
	7-	201	7 mm	N/A	Charge-direct	d_2-7-	201	7 mm	N/A

[a] Hypothetical homolytic–segment.
N/A, not applicable.
Note: Low energy MS/MS spectra obtained via collision-induced dissociation (CID) of electrospray-generated carboxylate anion [M-H or D]$^-$ from the LC peak. See Hong et al. (2007).

8. CONCLUSIONS AND NEXT STEPS

We can begin to appreciate at this juncture the temporal differences as well as spatial components within sites of inflammation that are responsible for generating specific local-acting lipid-derived mediators. The immediate future will bring forth without doubt the mapping of local biochemical mediators and the potential impact of drugs, diet, and stress (i.e., hypoxia and ischemia reperfusion within these bionetworks, which are certainly exciting terrain). Also, the results obtained along these lines will no doubt further enable us to appreciate that "the size of the peak doesn't always count!," but rather the bioaction or functional impact within organ systems and its relevance to both health and disease are most important to consider. Hence, transient, seemingly small, quantitatively fleeting members of the lipid mediator pathways discussed herein and their temporal relationship change extensively during the course of a physiologic and/or pathophysiologic response. These changes in magnitude of LM and their interrelationships within a given profile of autacoids/local mediators is a complex network of events that typifies the decoding power of lipid mediator–based informatics-lipidomics.

The main obstacle presently for this area is to devise meaningful ways to visually present the large amounts of data obtained from mapping of biological processes using this and related MS and informatics approaches to the study of lipid-derived mediators. Standardization of data acquisition, instrumentation and instrument settings, synthetic authentic and lipid mediators as well as spectral libraries are critical steps needed toward sharing and unifying the information to achieve an understanding of the vast complexities in the biological processes of interest. Once these challenges are met, we can standardize the approach and methods to obtain results from individuals in health as well as specific disease states (or even during a common cold). It should be possible and likely in the not-too-distant future for a clinician to routinely screen a patient to determine whether their daily dose of aspirin, for example, is effective for them. Moreover, these modest steps taken today shall enable large-scale population studies of lipid mediator profiles, their local levels *in situ*, and their serum PUFA precursor levels. Also, for example, local LM and their relationships to an individual's diet, genomic make-up, disease status, and general host defense status could be analyzed. Given the advances in nanotechnologies and likely increases in sensitivity of mass spectrometers, it should be possible to develop personalized units to continuously monitor an individual's profile of pro-inflammatory LM mediators as well as those involved in resolution of inflammation. Also, it should be possible for a clinician to monitor an individual's metabolic interaction with specific microbes and their own metabolome within (e.g., oral cavity). These raise the question of whether it will it be possible to correct and/or compensate for individual alteration in LM pathways to improve health, modulate disease status and prevent disease via enhancing host defense mechanisms with these and/or related molecules. These represent some of the challenges for LM informatics-lipidomics for the immediate and remote future—certainly fuel for exciting research ahead in this new and emerging field.

8.1. Materials and sample preparation

Biogenic syntheses of RvD1, 17S-HDHA, and 17-HpDHA were carried out (as in Hong *et al.*, 2003; Serhan *et al.*, 2006, 2002) with isolated enzyme(s) (i.e., soybean 15-lipoxygenase and/or potato 5-lipoxygenase purchased from Sigma [St. Louis, MO]). Following these procedures replacing DHA with d_5-DHA (21,21,22,22,22-d_5-DHA) (Cayman Chemical), we prepared d_5-RvD1, d_5-PD1, d_5-17S-HDHA, and d_5-17-HpDHA, and acquired their spectra and chromatograms on LC-UV-MS/MS. The structures of biogenically synthesized RvD1 and PD1 were further confirmed via matching the spectra of MS/MS and UV (coupled to an LC), GC/MS, and chromatographic retention times of RvD1 and PD1, which were prepared by total organic synthesis (Serhan *et al.*, 2006; Sun *et al.*, 2007; Yang, 2005). Some of the deuterium labeled compounds were obtained via deuterium-exchange of labile hydrogens in the hydroxy(s) and carboxylate, which include O,O,O,O-d_4-RvD1, O,O,O,21,21,22, 22,22-d_8-PD1 (from deuterium exchange of labile hydrogens in the two hydroxys and one carboxylate of d_5-PD1), O,O,O-d_3-PD1, and O,O-d_2-HDHAs. The deuterium-exchange was achieved when the samples were analyzed on LC-UV-MS/MS using the same gradient as described above with substitution of MeOD, D_2O, or d-acetic acid for methanol, water, and acetic acid. All other reagents were from available commercial sources at the highest grade.

Human whole blood (venous) and PMN were obtained (as in Hong *et al.*, 2003). Stroke-injury murine brain tissues were provided by N. Bazan's group at Louisiana State University Health Sciences Center. Procedures for trout brain and head-kidney preparation were from Hong *et al.* (2005). Briefly, this procedure is tailored as follows: the samples were centrifuged (3000 rpm, 4°, 15 min) to remove cellular and protein materials; after the supernatants were decanted, the supernatants were diluted with 10 volumes of Milli-Q water. The pH was adjusted to 3.5 with 1 M HCl for C18 solid-phase extraction (SPE). After washing with 15 ml of H_2O and then 8 ml of hexane, the SPE cartridges (C18, 3 ml, Milford, Waters, MA) were eluted with 8 ml of methyl formate, and the effluent was reconstituted into methanol for lipidomic analysis using LC-UV-MS/MS.

8.2. Instrumentation

MS/MS spectra were acquired using HPLC (P4000) coupled to a photodiode-array UV detector and an ion trap (LCQ) MS/MS (Thermoelectron, San Jose, CA). Each lipid mediator (~10 ng) was injected into the column (LUNA C18-2 150 mm × 2 mm × 5 m, Phenomenex, Torrance, CA), which was eluted at 0.2 ml/min and 23°. The mobile phase flowed as C (water:methanol:acetic acid = 65:35:0.01%) from 0 to 50 min; ramped to 100% methanol from 50.1 to 110 min; maintained as 100% methanol for 10 min; then back to C. The UV detector scanned from 200 to 400 nm, providing additional spectral confirmation to ensure the purity of the authentic standards. For RvD1, PD1, mono-hydroxy-DHAs (HDHAs), and HpDHA, the maximum UV absorption wavelengths are 301, 270, and 235 nm, respectively. Conditions for the mass spectrometer are electrospray voltage, 4.3 kV; heating capillary, −39 V; tube lens offset, 60 V; sheath N_2 gas, 1.2 l/min; and auxiliary N_2 gas, 0.045 l/min; collision energy was 35 to 45% (a relative collision energy of 0 to 100% corresponds to a high frequency alternating voltage for resonance excitation from 0 to 5 V maximum); helium gas, 0.1 Pa as a collision gas; and scan range, m/z 95 to 390 (Lu *et al.*, 2005).

ACKNOWLEDGMENTS

We thank M. Halm Small for assistance with manuscript preparation. Work in the author's laboratory is supported in part by National Institutes of Health grants GM38765 and P50-DE016191.

REFERENCES

Aliberti, J., Hieny, S., Reis e Sousa, C., Serhan, C. N., and Sher, A. (2002). Lipoxin-mediated inhibition of IL-12 production by DCs: A mechanism for regulation of microbial immunity. *Nat. Immunol.* **3,** 76–82.

Ariel, A., Li, P.-L., Wang, W., Tang, W.-X., Fredman, G., Hong, S., Gotlinger, K. H., and Serhan, C. N. (2005). The docosatriene protectin D1 is produced by T_H2 skewing and promotes human T cell apoptosis via lipid raft clustering. *J. Biol. Chem.* **280,** 43079–43086.

Arita, M., Clish, C. B., and Serhan, C. N. (2005a). The contributions of aspirin and microbial oxygenase in the biosynthesis of anti-inflammatory resolvins: Novel oxygenase products from omega-3 polyunsaturated fatty acids. *Biochem. Biophy. Res. Commun.* **338,** 149–157.

Arita, M., Yoshida, M., Hong, S., Tjonahen, E., Glickman, J. N., Petasis, N. A., Blumberg, R. S., and Serhan, C. N. (2005b). Resolvin E1, an endogenous lipid mediator derived from omega-3 eicosapentaenoic acid, protects against 2,4,6-trinitrobenzene sulfonic acid-induced colitis. *Proc. Natl. Acad. Sci. USA* **102,** 7671–7676.

Arita, M., Bianchini, F., Aliberti, J., Sher, A., Chiang, N., Hong, S., Yang, R., Petasis, N. A., and Serhan, C. N. (2005c). Stereochemical assignment, anti-inflammatory properties, and receptor for the omega-3 lipid mediator resolvin E1. *J. Exp. Med.* **201,** 713–722.

Ausloos, P., Clifton, C. L., Lias, S. G., Mikaya, A. I., Stein, S. E., Tchekhovskoi, D. V., Sparkman, O. D., Zaikin, V., and Zhu, D. (1999). The critical evaluation of a comprehensive mass spectral library. *J. Am. Soc. Mass Spectrom.* **10,** 287–299.
Bannenberg, G. L., Chiang, N., Ariel, A., Arita, M., Tjonahen, E., Gotlinger, K. H., Hong, S., and Serhan, C. N. (2005). Molecular circuits of resolution: Formation and actions of resolvins and protectins. *J. Immunol.* **174,** 4345–4355.
Bergström, S. (1982). The prostaglandins: From the laboratory to the clinic. *In* "Les Prix Nobel: Nobel Prizes, Presentations, Biographies and Lectures." The Nobel Foundation, pp. 129–148. Almqvist and Wiksell, Stockholm, Sweden.
Burr, G. O., and Burr, M. M. (1929). A new deficiency disease produced by the rigid exclusion of fat from the diet. *J. Biol. Chem.* **82,** 345–367.
Calder, P. C. (2001). Polyunsaturated fatty acids, inflammation, and immunity. *Lipids* **36,** 1007–1024.
Capdevila, J. H., Falck, J. R., Dishman, E., and Karara, A. (1990). Cytochrome P-450 arachidonate oxygenase. *In* "Arachidonate Related Lipid Mediators, Vol. 187" (R. C. Murphy and F. A. Fitzpatrick, eds.), pp. 385–394. Academic Press, San Diego, CA.
Cheng, C., and Gross, M. L. (2000). Applications and mechanisms of charge-remote fragmentation. *Mass Spectrom. Rev.* **19,** 398–420.
Chiang, N., Takano, T., Clish, C. B., Petasis, N. A., Tai, H.-H., and Serhan, C. N. (1998). Aspirin-triggered 15-epi-lipoxin A_4 (ATL) generation by human leukocytes and murine peritonitis exudates: Development of a specific 15-epi-LXA_4 ELISA. *J. Pharmacol. Exp. Ther.* **287,** 779–790.
Chiang, N., Gronert, K., Clish, C. B., O'Brien, J. A., Freeman, M. W., and Serhan, C. N. (1999). Leukotriene B_4 receptor transgenic mice reveal novel protective roles for lipoxins and aspirin-triggered lipoxins in reperfusion. *J. Clin. Invest.* **104,** 309–316.
Devchand, P. R., Arita, M., Hong, S., Bannenberg, G., Moussignac, R.-L., Gronert, K., and Serhan, C. N. (2003). Human ALX receptor regulates neutrophil recruitment in transgenic mice: Roles in inflammation and host-defense. *FASEB J.* **17,** 652–659.
Duffield, J. S., Hong, S., Vaidya, V., Lu, Y., Fredman, G., Serhan, C. N., and Bonventre, J. V. (2006). Resolvin D series and protectin D1 mitigate acute kidney injury. *J. Immunol.* **177,** 5902–5911.
Finnigan, M. (1996). "LCQ™ User Manual." Finnigan Mat, San Jose, CA.
Freire-de-Lima, C. G., Xiao, Y. Q., Gardai, S. J., Bratton, D. L., Schiemann, W. P., and Henson, P. M. (2006). Apoptotic cells, through transforming growth factor-beta, coordinately induce anti-inflammatory and suppress pro-inflammatory eicosanoid and NO synthesis in murine macrophages. *J. Biol. Chem.* **281,** 38376–38384.
Funk, C. D. (2001). Prostaglandins and leukotrienes: Advances in eicosanoid biology. *Science* **294,** 1871–1875.
Griffiths, W. J., Yang, Y., Sjövall, J., and Lindgren, J.Å. (1996). Electrospray/collision-induced dissociation mass spectrometry of mono-, di- and tri-hydroxylated lipoxygenase products, including leukotrienes of the B-series and lipoxins. *Rapid Commun. Mass Spectrom.* **10,** 183–196.
Gronert, K., Maheshwari, N., Khan, N., Hassan, I. R., Dunn, M., and Schwartzman, M. L. (2005). A role for the mouse 12/15-lipoxygenase pathway in promoting epithelial wound healing and host defense. *J. Biol. Chem.* **280,** 15267–15278.
Gronert, K., Kantarci, A., Levy, B. D., Clish, C. B., Odparlik, S., Hasturk, H., Badwey, J. A., Colgan, S. P., Van Dyke, T. E., and Serhan, C. N. (2004). A molecular defect in intracellular lipid signaling in human neutrophils in localized aggressive periodontal tissue damage. *J. Immunol.* **172,** 1856–1861.
Hasturk, H., Kantarci, A., Ohira, T., Arita, M., Ebrahimi, N., Chiang, N., Petasis, N. A., Levy, B. D., Serhan, C. N., and Van Dyke, T. E. (2006). RvE1 protects from local

inflammation and osteoclast mediated bone destruction in periodontitis. *FASEB J.* **20,** 401–403.
Hong, S., Gronert, K., Devchand, P., Moussignac, R.-L., and Serhan, C. N. (2003). Novel docosatrienes and 17S-resolvins generated from docosahexaenoic acid in murine brain, human blood and glial cells: Autacoids in anti-inflammation. *J. Biol. Chem.* **278,** 14677–14687.
Hong, S., Tjonahen, E., Morgan, E. L., Yu, L., Serhan, C. N., and Rowley, A. F. (2005). Rainbow trout (*Oncorhynchus mykiss*) brain cells biosynthesize novel docosahexaenoic acid-derived resolvins and protectins—mediator lipidomic analysis. *Prostaglandins Other Lipid Mediat.* **78,** 107–116.
Hong, S., Lu, Y., Yang, R., Gotlinger, K. H., Petasis, N. A., and Serhan, C. N. (2007). Resolvin D1, protectin D1, and related docosahexaenoic acid-derived products: Analysis via electrospray/low energy tandem mass spectrometry based on spectra and fragmentation mechanisms. *J. Am. Soc. Mass Spectrom.* **18,** 128–144.
Hudert, C. A., Weylandt, K. H., Wang, J., Lu, Y., Hong, S., Dignass, A., Serhan, C. N., and Kang, J. X. (2006). Transgenic mice rich in endogenouse n-3 fatty acids are protected from colitis. *Proc. Natl. Acad. Sci. USA* **103,** 11276–11281.
Jain, A., Batista, E. L., Jr., Serhan, C., Stahl, G. L., and Van Dyke, T. E. (2003). Role for periodontitis in the progression of lipid deposition in an animal model. *Infect. Immun.* **71,** 6012–6018.
Kiss, L., Bieniek, E., Weissmann, N., Schütte, H., Sibelius, U., Günther, A., Bier, J., Mayer, K., Henneking, K., Padberg, W., Grimm, H., Seeger, W., and Grimminger, F. (1998). Simultaneous analysis of 4- and 5-series lipoxgenase and cytochrome P450 products from different biological sources by reversed-phase high-performance liquid chromatographic technique. *Anal. Biochem.* **261,** 16–28.
Levy, B. D., Clish, C. B., Schmidt, B., Gronert, K., and Serhan, C. N. (2001). Lipid mediator class switching during acute inflammation: Signals in resolution. *Nat. Immunol.* **2,** 612–619.
Levy, B. D., Kohli, P., Gotlinger, K., Haworth, O., Horig, S., Kazani, S., Israel, E., Haley, K. J., and Serhan, C. N. (2007). Protectin D1 is generated in asthma and dampens airway inflammation and hyper-responsiveness. *J. Immunol.* **178,** 496–502.
LIPID MAPS Consortium (2006). Lipid classification scheme Accessible at http://www.lipidmaps.org/data/classification/fa.html.
Lu, Y., Hong, S., Tjonahen, E., and Serhan, C. N. (2005). Mediator-lipidomics: Databases and search algorithms for PUFA-derived mediators. *J. Lipid Res.* **46,** 790–802.
Lukiw, W. J., Cui, J. G., Marcheselli, V. L., Bodker, M., Botkjaer, A., Gotlinger, K., Serhan, C. N., and Bazan, N. G. (2005). A role for docosahexaenoic acid-derived neuroprotectin D1 in neural cell survival and Alzheimer disease. *J. Clin. Invest.* **115,** 2774–2783.
MacMillan, D. K., and Murphy, R. C. (1995). Analysis of lipid hydroperoxides and long-chain conjugated keto acids by negative ion electrospray mass spectrometry. *J. Am. Soc. Mass Spectrom.* **6,** 1190–1201.
Majno, G., and Joris, I. (2004). "Cells, Tissues, and Disease: Principles of General Pathology." Oxford University Press, New York.
Mallard, G. W., and Reed, J. (1997). Automated mass spectral deconvolution & identification systems. U.S. Department of Commerce-Technology Admistration, National Institutes of Standards and Technology (NIST) Standard Reference Data Program, Gaithersburg, MD.
Marcheselli, V. L., Hong, S., Lukiw, W. J., Hua Tian, X., Gronert, K., Musto, A., Hardy, M., Gimenez, J. M., Chiang, N., Serhan, C. N., and Bazan, N. G. (2003). Novel docosanoids inhibit brain ischemia-reperfusion-mediated leukocyte infiltration and pro-inflammatory gene expression. *J. Biol. Chem.* **278,** 43807–43817.

McLafferty, F. W., Stauffer, D. A., Loh, Y., and Wesdemiotis, C. (1999). Unknown identification using reference mass spectra: Quality evaluation of databases. *J. Am. Soc. Mass Spectrom.* **10,** 1229–1240.

McLafferty, F. W., and Turecek, F. (1993). Elemental composition. *In* "Interpretation of Mass Spectra" (I. Imfeld and A. Kelly, eds.), pp. 19–34. University Science Books, Mill Valley, CA.

Mukherjee, P. K., Marcheselli, V. L., Serhan, C. N., and Bazan, N. G. (2004). Neuroprotectin D1: A docosahexaenoic acid-derived docosatriene protects human retinal pigment epithelial cells from oxidative stress. *Proc. Natl. Acad. Sci. USA* **101,** 8491–8496.

Murphy, R. C., Fiedler, J., and Hevko, J. (2001). Analysis of nonvolatile lipids by mass spectrometry. *Chem. Rev.* **101,** 479–526.

Murphy, R. C., Barkley, R. M., Zemski Berry, K., Hankin, J., Harrison, K., Johnson, C., Krank, J., McAnoy, A., Uhlson, C., and Zarini, S. (2005). Electrospray ionization and tandem mass spectrometry of eicosanoids. *Anal. Biochem.* **346,** 1–42.

Rodriguez, A. R., and Spur, B. W. (2004). First total synthesis of 7(S),16(R),17(S)-Resolvin D2, a potent anti-inflammatory lipid mediator. *Tetrahedron Lett.* **45,** 8717–8720.

Salem, N., Jr., Litman, B., Kim, H.-Y., and Gawrisch, K. (2001). Mechanisms of action of docosahexaenoic acid in the nervous system. *Lipids* **36,** 945–959.

Samuelsson, B. (1982). From studies of biochemical mechanisms to novel biological mediators: Prostaglandin endoperoxides, thromboxanes and leukotrienes. *In* "Les Prix Nobel: Nobel Prizes, Presentations, Biographies and Lectures" (The Nobel Foundation, ed.), pp. 153–174. Almqvist and Wiksell, Stockholm, Sweden.

Samuelsson, B., Dahlén, S. E., Lindgren, J.Å., Rouzer, C. A., and Serhan, C. N. (1987). Leukotrienes and lipoxins: Structures, biosynthesis, and biological effects. *Science* **237,** 1171–1176.

Serhan, C. N. (1989). On the relationship between leukotriene and lipoxin production by human neutrophils: Evidence for differential metabolism of 15-HETE and 5-HETE. *Biochim. Biophys. Acta* **1004,** 158–168.

Serhan, C. N. (2005). Mediator lipidomics. *Prostaglandins Other Lipid Mediat.* **77,** 4–14.

Serhan, C. N., and Savill, J. (2005). Resolution of inflammation: The beginning programs the end. *Nat. Immunol.* **6,** 1191–1197.

Serhan, C. N., Gotlinger, K., Hong, S., and Arita, M. (2004). Resolvins, docosatrienes, and neuroprotectins, novel omega-3-derived mediators, and their aspirin-triggered endogenous epimers: An overview of their protective roles in catabasis. *Prostaglandins Other Lipid Mediat.* **73,** 155–172.

Serhan, C. N., Clish, C. B., Brannon, J., Colgan, S. P., Chiang, N., and Gronert, K. (2000). Novel functional sets of lipid-derived mediators with antiinflammatory actions generated from omega-3 fatty acids via cyclooxygenase 2-nonsteroidal antiinflammatory drugs and transcellular processing. *J. Exp. Med.* **192,** 1197–1204.

Serhan, C. N., Hong, S., Gronert, K., Colgan, S. P., Devchand, P. R., Mirick, G., and Moussignac, R.-L. (2002). Resolvins: A family of bioactive products of omega-3 fatty acid transformation circuits initiated by aspirin treatment that counter pro-inflammation signals. *J. Exp. Med.* **196,** 1025–1037.

Serhan, C. N., Gotlinger, K., Hong, S., Lu, Y., Siegelman, J., Baer, T., Yang, R., Colgan, S. P., and Petasis, N. A. (2006). Anti-inflammatory actions of neuroprotectin D1/protectin D1 and its natural stereoisomers: Assignments of dihydroxy-containing docosatrienes. *J. Immunol.* **176,** 1848–1859.

Serhan, C. N., Jain, A., Marleau, S., Clish, C., Kantarci, A., Behbehani, B., Colgan, S. P., Stahl, G. L., Merched, A., Petasis, N. A., Chan, L., and Van Dyke, T. E. (2003). Reduced inflammation and tissue damage in transgenic rabbits overexpressing 15-lipoxygenase and endogenous anti-inflammatory lipid mediators. *J. Immunol.* **171,** 6856–6865.

Shen, J., Herderick, E., Cornhill, J. F., Zsigmond, E., Kim, H.-S., Kühn, H., Guevara, N. V., and Chan, L. (1996). Macrophage-mediated 15-lipoxygenase expression protects against atherosclerosis development. *J. Clin. Invest.* **98,** 2201–2208.

Stein, S. E. (1995). Chemical substructure identification by mass spectral library searching. *J. Am. Soc. Mass Spectrom.* **6,** 644–655.

Stein, S. E., and Scott, D. R. (1994). Optimization and testing of mass spectral library search algorithms for compound identification. *J. Am. Soc. Mass Spectrom.* **5,** 859–866.

Sun, Y. -P., Oh, S. F., Uddin, J., Yang, R., Gotlinger, K., Campbell, E., Colgan, S. P., Petasis, N. A., and Serhan, C. N. (2007). Resolvin D1 and its aspirin-triggered 17R epimer: Stereochemical assignments, anti-inflammatory properties and enzymatic inactivation. *J. Biol. Chem.* **282,** 9323–9334.

Takano, T., Clish, C. B., Gronert, K., Petasis, N., and Serhan, C. N. (1998). Neutrophil-mediated changes in vascular permeability are inhibited by topical application of aspirin-triggered 15-epi-lipoxin A_4 and novel lipoxin B_4 stable analogues. *J. Clin. Invest.* **101,** 819–826.

Van Dyke, T. E., and Serhan, C. N. (2006). A novel approach to resolving inflammation. In "Scientific American Presents *Oral and Whole Body Health*," pp. 42–45. Scientific American, New York.

Wan, K. X., Vidavsky, I., and Gross, M. L. (2002). Comparing similar spectra: From similarity index to spectral contrast angle. *J. Am. Soc. Mass Spectrom.* **13,** 85–88.

Wenk, M. R., Lucast, L., Paolo, G. D., Romanelli, A. J., Suchy, S. F., Nussbaum, R. L., Cline, G. W., Shulman, G. I., McMurray, W., and Camilli, P. D. (2003). Phosphoinositide profiling in complex lipid mixtures using electrospray ionization mass spectrometry. *Nat. Biotechnol.* **21,** 813–817.

Wheelan, P. J., Zirrolli, J. A., and Murphy, R. C. (1996). Electrospray ionization and low energy tandem mass spectrometry of polyhydroxy unsaturated fatty acids. *J. Am Soc. Mass Spectrom.* **7,** 140–149.

Xia, S., Lu, Y., Wang, J., He, C., Hong, S., Serhan, C. N., and Kang, J. X. (2006). Melanoma growth is reduced in fat-1 transgenic mice: Impact of omega-6/omega-3 essential fatty acids. *Proc. Natl. Acad. Sci. USA* **103,** 12499–12504.

Yang, R. (2005). Synthetic study of lipid mediators; Doctoral thesis. University of Southern California, Los Angeles.

CHAPTER THIRTEEN

A GUIDE TO BIOCHEMICAL SYSTEMS MODELING OF SPHINGOLIPIDS FOR THE BIOCHEMIST

Kellie J. Sims,[*,†] Fernando Alvarez-Vasquez,[*,†] Eberhard O. Voit,[‡] and Yusuf A. Hannun[†]

Contents

1. Introduction	320
2. System Map: Specifying the Model	322
2.1. Creation of the system's parts list	322
2.2. Determination of dependent and independent variables	322
2.3. Graphical representation of the system's parts list	323
2.4. Specific example: dihydro- and phyto-ceramide dynamics	323
3. Symbolic Equations: From Words and Pictures to Equations	327
4. Numerical Equations: From the Symbolic to a Computational Model	331
4.1. Ceramide synthase	332
4.2. Dihydroceramidase	334
4.3. Inositol phosphorylceramide synthase	336
4.4. Inositol phosphosphingolipid phospholipase C	338
4.5. Hydroxylase	339
4.6. Other parameter calculations	340
5. Model Analysis: Steady State, Stability, and Sensitivity	341
6. Simulation: What Happens If...?	345
7. Conclusion	346
8. Epilogue	346
Acknowledgments	348
References	348

[*] Department of Biostatistics, Bioinformatics, and Epidemiology, Medical University of South Carolina, Charleston, South Carolina
[†] Department of Biochemistry and Molecular Biology, Medical University of South Carolina, Charleston, South Carolina
[‡] The Wallace H. Coulter Department of Biomedical Engineering, Georgia Institute of Technology and Emory University, Atlanta, Georgia

Abstract

The last several years have brought an avalanche of data from various high throughput, genome-wide analyses of yeast and other model organisms. Still, scientists struggle to comprehend the complex behavior of biological systems. One method that has been available for decades but now is more necessary than ever is the mathematical modeling of biological systems. Unfortunately, a chasm of terminology and techniques has separated most biologists from mathematical modelers. This chapter hopes to bridge that gap for metabolic models by delineating the general process used to develop a system of differential equations that describes a biochemical pathway. This modeling process can be generally applied to many biological phenomena. In addition, the specific approach of Biochemical Systems Theory (BST) is demonstrated for the nitty-gritty details of the model equations. These methods are demonstrated using the core section of ceramide metabolism in yeast.

1. INTRODUCTION

The success of contemporary mechanistic approaches in biochemistry and biology has seen the seemingly paradoxical emergence of an increasing need to integrate this information via the discipline known as "systems biology." For example, the achievement in molecularly defining key players in a signal transduction pathway and understanding their biochemical regulation and function leads directly to asking questions on the overall propagation of signals in the entire pathway, such as how the pathway is regulated and how each component affects the system's output. Likewise, biochemical studies have made tremendous strides in dissecting complex metabolic pathways, identifying and characterizing individual enzymes and their regulation, and identifying important metabolites and their individual function. Again, this has led to questions about the overall structure of these pathways and analysis of coordinated regulation and function. Answering such questions does require that we understand each component mechanistically, but also that we develop and employ tools to connect these pieces integratively. This need for integration has inevitably led to mathematical approaches for the analysis of complex biochemical and biological processes. These emerging computational approaches are essential in transforming the descriptive and biochemical phases of "omics" (i.e., Which genes have differential expression for a given condition and how are they regulated? What are the various metabolites in a pathway, and what are their functions?) into a functional phase (i.e., What are the consequences of a certain profile of gene expression or of changes in levels of a cohort of metabolites?).

Our collaborative interdisciplinary research has focused on devising and applying mathematical methodology to the study of sphingolipid metabolism. The interconnected network of sphingolipid metabolism regulates the formation of diverse sphingolipids that function as critical determinants of membrane structure and as a source of bioactive molecules involved in various signaling mechanisms and cell regulatory pathways. Understanding sphingolipid metabolism has thus emerged as a significant area of study; as such, we have devoted considerable effort over the past several years to devise integrative mathematical models that elucidate the functioning of sphingolipid metabolism in *Saccharomyces cerevisiae* (Alvarez-Vasquez et al., 2004, 2005). This chapter will draw on a subset of these results to illustrate the methodology involved, and its utility and limitations.

Mathematical models of biological systems can take many forms and may employ diverse techniques depending on the type of system and purpose of the model. Regardless of the specific details of each model or the mathematical tools used, some generic features are common to most modeling efforts, including the identification of crucial features of the biological system of interest and their mapping into a feasible modeling format; deriving symbolic equations from that map; estimating parameters that allow numerical evaluation of the symbolic equations; diagnosing the model for potential deficiencies; analyzing the repertoire of possible model behaviors; conducting simulations of biologically relevant scenarios; and, finally, interpreting the results in the language of biology. Because of its complexity, the model development process is usually iterative, requiring ongoing discussions of system features and diagnostic results between biologists and mathematical modelers, followed by amendments and refinements to the model, and, ultimately, analysis and interpretation of model behaviors and simulation results.

Numerous distinct techniques are available for mathematical modeling, even within the limited domain of metabolic systems (e.g., Heinrich and Rapoport, 1974; Kacser and Burns, 1973; Savageau, 1969a; Voit, 2000). This chapter will not discuss these various methods in a comparative manner, but instead will illustrate the biochemical modeling process with one set of techniques that has proven very successful—generalized mass action (GMA) models based on BST (Savageau, 1969a,b). This mathematical and computational framework was developed in the late 1960s to study the dynamics of biochemical and gene regulatory systems and has been used to model diverse metabolic pathways, such as the tricarboxylic acid cycle in *Dictyostelium discoideum* (Wright et al., 1992), the Maillard-glyoxylase network (Ferreira et al., 2003), purine metabolism (Curto et al., 1997, 1998) and red blood cell metabolism (Ni and Savageau, 1996a,b), and the sphingolipid pathway in yeast (Alvarez-Vasquez et al., 2004, 2005), to name a few. Some of these models were of considerable size, containing dozens of variables and enzymatic processes.

This chapter is structured to guide the reader step by step through the process of setting up and analyzing a model of a metabolic pathway. We will illustrate the process by focusing on a small portion of yeast sphingolipid metabolism. Along the way, mathematical terminology will be defined, and special considerations of modeling will be discussed briefly. Detailed derivations of formulas are not included here, but can be found in *Computational Analysis of Biochemical Systems* (Voit, 2000) and in a substantial body of literature on BST referenced therein. Some examples will be calculated in detail to assist the reader's understanding. The chapter will conclude with a broader discussion of modeling as a crucial component of the increasing emphasis on systems biology, and will include some resources to assist the reader in pursuing modeling of their particular biological problem.

2. System Map: Specifying the Model

2.1. Creation of the system's parts list

The first and arguably most critical step of metabolic modeling is the creation of an accurate diagram or "map" of the system to be modeled. This step is crucial because the symbolic mathematical equations are subsequently written directly from this map; thus, an incorrect map leads to incorrect equations, which lead to a poor representation of the system and faulty inferences. To begin map construction, it is necessary to specify the system to be modeled and the goal of the model. Often, this requires a compromise between completeness and conciseness; decisions must be made on what will *not* be modeled as much as what will. These specifications will assist in decision making during the development process and help keep the model as simple as possible while retaining a sufficient degree of validity and incorporating all details necessary to answer questions posed at the beginning of the modeling effort. For example, the acyl tails of key sphingolipids can have various lengths. Is this detail critical? Also, yeast sphingolipid metabolism is known to be located primarily in the endoplasmic reticulum and Golgi. Should this information be included in the model? The answer may be *No* or it may be *Yes*, depending on whether the movement between cellular locations or distinct species of ceramide is of particular interest to the modeler (and/or user of the model).

Once the desired scope of the model is specified, the next step is to list all relevant components of the system (e.g., metabolites, enzymes, and co-factors), the flow of materials (fluxes) between components by either reactions or transport, and any known regulatory signals or processes that influence the system. Specific details, such as enzyme activities and metabolite concentrations, are not needed at this point, but become important for the parameterization step described in a later section of this chapter.

2.2. Determination of dependent and independent variables

Once all information on critical parts is compiled, the specific role of each element in the model is to be considered. Notably, the metabolites of primary interest are designated as dependent variables, meaning that their concentrations typically change as a simulation runs its course. The quantification of this change is formulated as one differential equation for each primary metabolite. In contrast, enzymes such as ceramide synthase are usually designated as independent variables since their mass and activity presumably remain unchanged; however, if enzyme production or degradation is an object of study, the designation would change to a dependent variable. Because independent variables do not change in amount, they do not require differential equations of their own; rather, they are included in the equations for dependent variables.

2.3. Graphical representation of the system's parts list

The lists of components and fluxes are initially used to create a network graph (map) using specific meaningful symbols that are well defined. In metabolic pathway analysis, metabolites are designated as nodes (depicted as boxes, ovals, etc.) and the connections between them (fluxes or signals) as arrows. Careful attention should be paid to the types and directions of arrows that connect the metabolites. Solid arrows are reserved for material fluxes and point from one node to another. Additionally, the shape of the arrow indicates whether the reaction is unidirectional (one head and one tail), reversible (double headed), a synthesis or convergence reaction with two substrates (two tails with one head), or requires a cofactor that is altered by the reaction (one straight arrow for the main reaction and a curved arrow that touches the first one for the co-reaction). Dotted arrow tails denoting regulatory signals (i.e., no material is exchanged) begin at a metabolite or pool of metabolites and point to the flux arrow that is modulated by the signal. Additional possibilities for arrow shapes and further discussion on creating proper system maps can be found in Voit (2000).

2.4. Specific example: dihydro- and phyto-ceramide dynamics

The illustration model for this chapter comprises the syntheses of dihydro- and phytoceramide and their immediate subsequent fates during normal exponential growth (Fig. 13.1A). To keep the sample model as small as possible, only the immediate precursors and products are considered, and cellular localization is not relevant. The larger context of these reactions is described in (Alvarez-Vasquez *et al.*, 2004, 2005).

The parts list begins with the metabolites of primary interest, dihydroceramide (DCER) and phytoceramide (PCER); their precursors, dihydrosphingosine (DHS), phytosphingosine (PHS), and C_{26}-CoA; and the main product, inositolphosphorylceramide (IPC). Also involved are six enzymes (ceramide synthase, phytoceramidase, dihydroceramidase, IPC synthase, IPCase, and an α-hydroxylase) and an additional metabolite, phosphatidylinositol (PI), which participates in the IPC synthase reaction (Riezman, 2006; Sims *et al.*, 2004). These components are listed in Table 13.1.

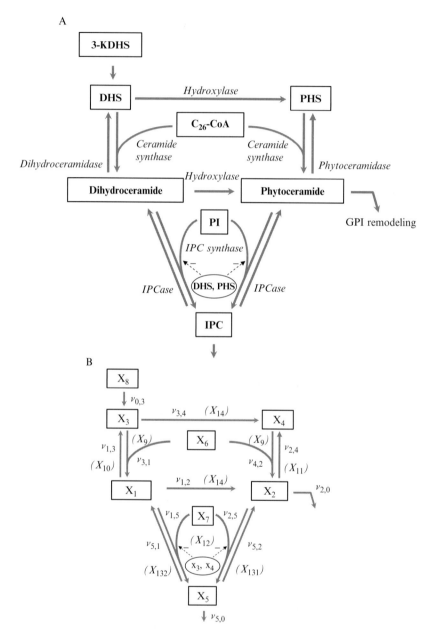

Figure 13.1 (A) Map of the pathway with chemical names. (B) Map of the pathway with indexed variable notation.

Table 13.1 Metabolites and enzymes used in the sample model

	Abbreviation	Notation	Initial value
Metabolites			mol%
Dihydroceramide	DCER	X_1	0.036
Phytoceramide	PCER	X_2	0.052
Dihydrosphingosine	DHS	X_3	0.01
Phytosphingosine	PHS	X_4	0.05
Inositolphosphorylceramide	IPC	X_5	0.102
C_{26}-CoA	C_{26}-CoA	X_6	0.5
Phosphatidylinositol	PI	X_7	4.6
3-Keto-dihydrosphingosine	KDHS	X_8	0.005
Enzymes			U/min
Ceramide synthase	CerSyn	X_9	0.165e-4
Dihydroceramidase	D-Cdase	X_{10}	0.54e-5
Phytoceramidase	P-Cdase	X_{11}	0.198e-4
IPC synthase	IPCsyn	X_{12}	0.33e-3
IPCase		X_{131}, X_{132}	0.15e-3
Hydroxylase		X_{14}	0.17e-3

Note: References for initial values not given in the text can be found in Alvarez-Vasquez et al. (2004) and the supplementary material of Alvarez-Vasquez et al. (2005).

To model the changes in dihydroceramide and phytoceramide, we next identify where materials enter and leave each pool. These influxes and effluxes are diagrammed as solid arrows between origin and destination nodes. Dihydroceramide has two influxes: from DHS + C_{26}-CoA, mediated by ceramide synthase, and from IPC, mediated by IPCase. Three effluxes leave the dihydroceramide pool: toward DHS + free fatty acids (FFA), mediated by dihydroceramidase; toward phytoceramide by hydroxylation; and toward IPC via IPC synthase, which, in the process, utilizes PI and releases diacylglycerol (DAG). Because FFA and DAG are of secondary interest to the purpose of this illustration, they are not included in the model. Phytoceramide has three influxes: from PHS + C_{26}-CoA, mediated by ceramide synthase, from IPC, catalyzed by IPCase, and from DHS via the α-hydroxylase reaction. There are also three effluxes from phytoceramide: toward PHS and FFA via phytoceramidase, toward IPC by IPC synthase (again using PI and producing DAG), and for incorporation as a protein anchor in the process of glycosylphosphatidylinositol (GPI) remodeling (see the first two sections of Table 13.2 and Fig. 13.1A).

Drawing the map of these fluxes immediately shows that DHS, PHS, and IPC each have influxes and effluxes involving the primary metabolites. This indicates that their concentrations will be changing along with those of DCER and PCER. Thus, they are to be declared dependent variables in the

Table 13.2 Fluxes for dihydroceramide and phytoceramide

Influxes for DCER and PCER		
From	To	
DHS (and C_{26}-CoA)	DCER	$v_{3,1}$
IPC	DCER	$v_{5,1}$
PHS (and C_{26}-CoA)	PCER	$v_{4,2}$
IPC	PCER	$v_{5,2}$
DCER	PCER	$v_{1,2}$
Effluxes for DCER and PCER		
From	To	
DCER	PCER	$v_{1,2}$
DCER	DHS	$v_{1,3}$
DCER (and PI)	IPC	$v_{1,5}$
PCER	PHS	$v_{2,4}$
PCER (and PI)	IPC	$v_{2,5}$
PCER	GPI	$v_{2,0}$
Additional fluxes		
From	To	
KDHS	DHS	$v_{0,3}$
DHS	PHS	$v_{3,4}$
IPC		$v_{5,0}$

model system, and each requires its own equation. The model system has thereby grown from two to five equations, illustrating a common issue of modeling—however simple a model is initially envisioned, additional components often become necessary to reach the goals of the modeling effort.

It is important to consider whether any more fluxes or components should be included for these newly declared dependent variables, DHS, PHS, and IPC, such as phosphorylation and dephosphorylation of the long chain bases, which are known from the literature. This potential inclusion requires a nontrivial decision because any fluxes or components added for the sake of completeness require significant amounts of additional data and computational effort, which must be balanced against the purpose of the model and the size and complexity of the system to be developed. As there are always more details that could be included, the risk is imminent that a model expands rampantly and (possibly) with limiting benefit. For the specific illustration here, we use the biochemist's expert opinion and decide to focus on the dynamics of the core ceramides. Thus, we add the following fluxes: hydroxylation of DHS to PHS, one influx into DHS from its precursor 3-ketodihydrosphingosine

(KDHS), and one efflux from IPC to other complex sphingolipids. These are noted in the last section of Table 13.2.

Finally, once the flux structure is set, regulatory signals that influence any of these reactions are added, such as the inhibition of IPC synthase by DHS and PHS. These signals are included in the map as dotted arrows that begin at the ovals denoting the inhibitors and end at the midpoint of the modulated reaction arrow, IPC synthase, as shown in Fig.13.1A.

The system maps thus constructed are shown in Fig. 13.1A and B, along with the parts list of metabolites and fluxes (Tables 13.1 and 13.2; see also next section).

3. Symbolic Equations: From Words and Pictures to Equations

With a proper map in hand and with BST chosen as the modeling framework, the symbolic equations can be written easily. For this step, some mathematical notation must be introduced. In a small system such as this illustration, names of metabolites or alphabetic acronyms could continue to be used as labels; however, even slightly larger systems benefit from the more mathematical terminology of indexed variables. In line with traditional notation, the components of the system will be designated as follows: metabolites are coded as X_i, beginning with dependent variables, and fluxes are written as $v_{i,j}$ where i and j are integers from 1 to n. For instance, if dihydroceramide and phytoceramide are assigned the names X_1 and X_2, respectively, then the flux from dihydroceramide (X_1) to phytoceramide (X_2) mediated by α-hydroxylase is named $v_{1,2}$. Similarly, if dihydrosphingosine and phytosphingosine are named X_3 and X_4, then $v_{1,3}$ represents the flux mediated by dihydroceramidase and $v_{2,4}$ the one by phytoceramidase. The symbolic names for all components and fluxes are given in the third column of the parts and flux lists of Tables 13.1 and 13.2 and are shown in the pathway map (Fig. 13.1B).

With this streamlined notation, the symbolic differential equations for the five dependent metabolites can be written according to simple guidelines. This is accomplished by considering each metabolite, one at a time, and writing down all influxes and effluxes affecting this metabolite. For instance, for dihydroceramide (X_1), we write the following equation:

$$dX_1/dt = (v_{3,1} + v_{5,1}) - (v_{1,2} + v_{1,3} + v_{1,5}). \quad (1)$$

This formulation is a differential equation in which the change (differential) in X_1 on the left-hand side is expressed as a function of fluxes on the right-hand side, even though the system variables that determine the fluxes are not yet explicitly written. Specifically, Equation 1 states that the rate of change in metabolite X_1 (i.e., dX_1/dt) is the algebraic sum of all influxes and effluxes acting on X_1 (in this case two influxes with positive signs and three effluxes with negative signs).

Similarly, the equations for the remaining metabolites read as follows:

$$dX_2/dt = (v_{1,2} + v_{4,2} + v_{5,2}) - (v_{2,4} + v_{2,5} + v_{2,0}) \qquad (2)$$

$$dX_3/dt = (v_{0,3} + v_{1,3}) - (v_{3,1} + v_{3,4}) \qquad (3)$$

$$dX_4/dt = (v_{2,4} + v_{3,4}) - (v_{4,2}) \qquad (4)$$

$$dX_5/dt = (v_{1,5} + v_{2,5}) - (v_{5,1} + v_{5,2} + v_{5,0}). \qquad (5)$$

In total, there seem to be 23 fluxes in the system; however, 10 fluxes appear twice—as an efflux in one equation and as an influx in another equation. For example, since the synthesis of dihydroceramide (X_1) depletes dihydrosphingosine (X_3), the first influx in Equation 1, $v_{3,1}$, is also found as the first efflux in Equation 3. The equality of a paired efflux and its corresponding influx is called a precursor-product constraint. As will be shown in the next section, the existence of these flux pairs, as well as other constraints on the system, simplifies the parameter estimation of the system by reducing the number of parameters.

The steps followed so far more or less apply to most methods for modeling metabolic pathways and can, in principle, be extended to accommodate any complexity of the system under study. At this point, several types of formats could be chosen to proceed with the "mathematical mechanics" of modeling. As mentioned in the introduction, we selected for our illustration the Generalized Mass Action (GMA) representation of BST. This decision is based primarily on BST's biological versatility and intuitive character, which is accompanied by mathematical rigor and multifold biological application and validity.

While it is beyond the scope of this chapter, it should be noted that the appropriateness, utility, and mathematical characteristics of alternative modeling approaches have been formally evaluated and compared numerous times (see Chapter 2 of Voit, 2000). In many cases, little difference exists among such models (e.g., in the quality of data fit), and criteria, such as mathematical convenience, can be used to favor one approach over others. The comparison of nonlinear models in general is a very complicated issue, and the reader is advised to consult the literature (Curto et al., 1998; Shiraishi and Savageau, 1992; Torres and Voit, 2002).

What is clear from these comparative analyses and other considerations is that any mathematical model of a realistic biological phenomenon by necessity must be an approximation, so that the deciding factor for choosing a model becomes the specific type of such an approximation. In BST, the crucial approximation is based on fundamental principles of numerical analysis, with the result that each symbolic process $v_{i,j}$ within the system is substituted with a product of power functions. Thus, rather than trying to determine a kinetic rate law in the tradition of Michaelis and Menten, BST provides a "canonical" formulation that is at once versatile enough to provide a valid representation for essentially all processes encountered in biochemistry, yet computationally tractable enough to yield insights into the complexities of biological systems (for detailed explanation, see Torres and Voit, 2002).

The canonical format of BST provides a template that replaces each flux $v_{i,j}$ with a product, in which each factor represents a single component of the model that affects the particular flux (e.g., metabolites, enzymes, and regulatory signals), and this component is raised to an appropriate power, called a kinetic order. The kinetic order reflects the strength of influence that the given component exerts on the process under study. The kinetic orders are analogous to kinetic orders in elemental chemical kinetics, but can take fractional as well as integer values. Additionally, their signs indicate activation (+) or inhibition (−). Finally, each flux term also includes a rate constant, designated by $\gamma_{i,j}$, that represents the turnover rate of the process. The general form is shown in Equation 6:

$$v_{i,j} = \gamma_{i,j} X_k^{f_{i,j,k}} X_l^{f_{i,j,l}} \ldots X_z^{f_{i,j,z}} \Big|_{\text{op}}. \tag{6}$$

As an illustration, consider the reaction $v_{1,2}$ in the map of Figure 13.1B. This flux is influenced by the concentration of substrate (dihydroceramide, X_1) and mediated by the α-hydroxylase encoded by Sur2p, X_{14}. Thus, the two factors of $v_{1,2}$ are X_1 and X_{14} with exponents $f_{1,2,1}$ and $f_{1,2,14}$, respectively, and the rate constant is $\gamma_{1,2}$:

$$v_{1,2} = \gamma_{1,2} X_1^{f_{1,2,1}} X_{14}^{f_{1,2,14}}. \tag{7}$$

Note that the index numbers for the rate constant $\gamma_{i,j}$ and for the first two indices of the exponent $f_{i,j,k}$ are the same as for the flux $v_{i,j}$. The third index of the exponent equals the index of the effector variable X_k, in this case, 1 and 14.

A more complicated example, $v_{2,5}$, which describes the flux from phytoceramide (X_2) to IPC (X_5), is rewritten in power-law form as shown in Equation 8:

$$v_{2,5} = \gamma_{2,5} X_2^{f_{2,5,2}} X_3^{f_{2,5,3}} X_4^{f_{2,5,4}} X_7^{f_{2,5,7}} X_{12}^{f_{2,5,12}}. \tag{8}$$

This flux is mediated by IPC synthase (X_{12}), utilizes PI (X_7), and is inhibited by DHS (X_3) and PHS (X_4) (Wu et al., 1995). Similar power-law representations are set up for the remaining 11 unique fluxes in the system, as shown in the first two columns of Table 13.3.

Table 13.3 Fluxes for the sample model

Flux	GMA symbolic form	GMA numerical form	Steady-state value[a]
$v_{1,2}$	$\gamma_{1,2} X_1^{f_{1,2,1}} X_{14}^{f_{1,2,14}}$	$736.26 X_1^{0.5} X_{14}$	0.0237
$v_{1,3}$	$\gamma_{1,3} X_1^{f_{1,3,1}} X_{10}^{f_{1,3,10}}$	$3057.42 X_1^{0.5} X_{10}$	0.00313
$v_{1,5}$	$\gamma_{1,5} X_1^{f_{1,5,1}} X_3^{f_{1,5,3}} X_4^{f_{1,5,4}} X_7^{f_{1,5,7}} X_{12}^{f_{1,5,12}}$	$23.55 X_1^{0.973} X_3^{-0.0034} X_4^{-0.0242} X_7^{1.685} X_{12}$	0.00437
$v_{2,0}$	$\gamma_{2,0} X_2^{f_{2,0,2}}$	$3.82 X_2^{0.5}$	0.849
$v_{2,4}$	$\gamma_{2,4} X_2^{f_{2,4,2}} X_{11}^{f_{2,4,11}}$	$2547.85 X_2^{0.5} X_{11}$	0.0115
$v_{2,5}$	$\gamma_{2,5} X_2^{f_{2,5,2}} X_3^{f_{2,5,3}} X_4^{f_{2,5,4}} X_7^{f_{2,5,7}} X_{12}^{f_{2,5,12}}$	$22.73 X_2^{0.962} X_3^{-0.0033} X_4^{-0.023} X_7^{1.685} X_{12}$	0.0251
$v_{3,1}$	$\gamma_{3,1} X_3^{f_{3,1,3}} X_6^{f_{3,1,6}} X_9^{f_{3,1,9}}$	$196097.3 X_3^{0.964} X_6^{0.527} X_9$	0.0264
$v_{3,4}$	$\gamma_{3,4} X_3^{f_{3,4,3}} X_{14}^{f_{3,4,14}}$	$50004 X_3^{0.5} X_{14}$	0.859
$v_{4,2}$	$\gamma_{4,2} X_4^{f_{4,2,4}} X_6^{f_{4,2,6}} X_9^{f_{4,2,9}}$	$827025.2 X_4^{0.8} X_6^{0.527} X_9$	0.8709
$v_{5,0}$	$\gamma_{5,0} X_5^{f_{5,0,5}}$	$0.00065 X_5^{0.5}$	0.0331
$v_{5,1}$	$\gamma_{5,1} X_5^{f_{5,1,5}} X_{132}^{f_{5,1,132}}$	$297 X_5^{-0.972} X_{132}$	0.0048
$v_{5,2}$	$\gamma_{5,2} X_5^{f_{5,2,5}} X_{131}^{f_{5,2,131}}$	$297 X_5^{-0.972} X_{131}$	0.0048
$v_{0,3}$	$\gamma_{0,3} X_8^{f_{0,3,8}}$	$174.67 X_8$	0.875

[a] Steady-state flux values (μmol/min/l) are calculated as explained in the parameter estimation section of the text.

4. NUMERICAL EQUATIONS: FROM THE SYMBOLIC TO A COMPUTATIONAL MODEL

The next step of the modeling process consists of substituting numerical values for the symbols in the model equations. This step is often the most time-consuming and tedious of all parts of the modeling process, as this is where the detailed minutiae of biochemical reactions (e.g., metabolite concentrations, enzyme activities, rate laws, and associated values, such as K_m or K_i) are transformed into numerical values for all parameters (i.e., the $\gamma_{i,j}$s and $f_{i,j,k}$s) in the differential equations for the time-dependent variables; however, first, all the relevant information for each metabolite, enzyme, and reaction must be compiled from the literature and from databases, such as BRENDA (http://www.brenda.uni-koeln.de/) or those listed at Pathguide: the pathway resource list (http://www.pathguide.org/). Often, it will become apparent that custom experiments are required to provide specific data not available in the public domain.

The particularities of estimating parameter values depend critically on the type of data available to the modeler. For the most prevalent data, which characterize reaction steps, the estimation method in BST is quite straightforward and can be calculated by hand or with the support of software packages, such as Maple^{TM} (http://www.maplesoft.com/), Mathcad (www.mathsoft.com/), or Mathematica® (http://www.wolfram.com/). A very different mathematical approach is needed if the data consist of measured time series of metabolites. This situation is still rare and will not be discussed here; however, modern techniques of molecular biology and experimental systems biology are beginning to produce these types of data with increasing prevalence (see Voit and Almeida, 2004; Chou et al., 2006; and Voit et al., 2006, for suitable BST methods applied to time series data).

The five symbolic equations with 23 fluxes of our system seem to suggest that 23 rate constants and 59 kinetic orders must be estimated. Happily, these numbers are substantially reduced by the necessity of fulfilling certain constraints on the system, such as the conservation of precursor-product fluxes mentioned previously. Another constraint is the stoichiometry of fluxes through each node, which needs to be balanced (i.e., the sum of influxes equals the sum of effluxes for each node so the node remains constant) when the system is at steady-state. However, note that relevant situations exist in some models where the system is in a transient phase such that nodes may be growing or shrinking, but the precursor–product relationships still hold true.

In our example, constraints almost halve the number of rate constants and kinetic orders that need to be estimated since 10 fluxes with 28 kinetic orders and 10 rate constants appear twice in the symbolic equations. This leaves an additional three fluxes uniquely present in the equations, each with a rate constant and single kinetic order (Table 13.3).

The experimental data that are available determine which of several methods of parameter estimation to use. The best data, which unfortunately are seldom available, characterize the change in a given flux versus each variable that influences that flux (using the same conditions as in the model specifications). Rate laws such as Michaelis-Menten or Hill are useful if they are known, but again, this is not always the case. Finally, if the system is not adequately characterized and data are scarce, BST has well-proven default values that can be used in the interim and often perform surprisingly well as discussed in Voit (2000). The suitability of the chosen parameters' values is assessed via sensitivity analysis in the next section. To demonstrate some estimation methods, parameters will be determined for the five fluxes in the equation for dihydroceramide. Note that calculations were made using MAPLE, whose internal precision is much more than 10 digits; therefore, numerical values have been truncated to just a few digits for ease of viewing in the following display equations. For instance, the concentration of DCER is given as 0.036 mol% in the section for dihydroceramidase, but the true value 0.036$\overline{1}$, which repeats indefinitely, is used in Maple. Therefore, some calculations may have slightly different answers when performed directly from the display equations, especially for the rate constants.

4.1. Ceramide synthase

The first flux considered is $v_{3,1}$, the synthesis of dihydroceramide (X_1) from dihydrosphingosine (X_3) and C_{26}-CoA (X_6) by ceramide synthase (X_9). The kinetic orders ($f_{3,1,3}, f_{3,1,6},$ and $f_{3,1,9}$) will be estimated first, followed by the rate constant, $\gamma_{3,1}$.

One method to estimate the kinetic orders of a flux uses the rate law for the reaction, if it is known. This is justified since the power-law representation used in BST approximates the rate law, and the two are exactly equal when evaluated at a given operating point (op) that is chosen by the modeler and often coincide with the nominal steady-state point. Assuming that the kinetics of this condensation reaction can be represented by a bi-substrate, Michaelis-Menten rate function gives

$$v_{3,1} = V_{3,1} = V_{\max}\left(\frac{\text{DHS}}{K_{m,\text{DHS}} + \text{DHS}}\right) \times \left(\frac{C_{26}\text{-CoA}}{K_{m,C_{26}\text{-CoA}} + C_{26}\text{-CoA}}\right)$$
$$= V_{\max}\left(\frac{X_3}{K_{m,\text{DHS}} + X_3}\right) \times \left(\frac{X_6}{K_{m,C_{26}\text{-CoA}} + X_6}\right). \quad (9)$$

The kinetic order $f_{i,j,k}$ is given as the relative change in $V_{i,j}$, given a relative change in X_k. Mathematically, this effect is given as the partial derivative of the rate law $V_{i,j}$ with respect to the substrate, enzyme, or modulator X_k of interest at the chosen operating point, multiplied by the fraction of X_k divided by V_{ij}:

$$f_{ijk} = \frac{\partial V_{ij}}{V_{ij}} \bigg/ \frac{\partial X_k}{X_k} = \frac{\partial V_{ij}}{\partial X_k} \times \frac{X_k}{V_{ij}} \bigg|_{\mathrm{op}}. \tag{10}$$

Substituting Equation 9 into Equation 10, differentiating with respect to X_3, and simplifying gives Equation 11. Similarly, differentiation of Equation 10 with respect to X_6 gives Equation 12.

$$f_{3,1,3} = \frac{\partial(V_{3,1})}{\partial X_3} \times \frac{X_3}{V_{3,1}} = \frac{K_{\mathrm{m},X_3}}{K_{\mathrm{m},X_3} + X_3} \tag{11}$$

$$f_{3,1,6} = \frac{\partial(V_{3,1})}{\partial X_6} \times \frac{X_6}{V_{3,1}} = \frac{K_{\mathrm{m},X_6}}{K_{\mathrm{m},X_6} + X_6}. \tag{12}$$

Measured values are then substituted for the variables to obtain an actual number for the kinetic orders. The cellular concentration of DHS was measured for yeast as $X_{3s} = 0.01$ mol% (Cowart, L. A., personal communication) and the cellular C_{26}-CoA concentration as $X_{6s} = 0.5$ mol% (Gaigg et al., 2001). The ceramide synthase $K_{\mathrm{m,DHS}} = 0.27$ mol% (Guillas et al., 2001) and $K_{\mathrm{m,C26-CoA}} = 0.559$ mol% (Shimeno et al., 1998). Substituting these values into Equations 11 and 12 and rounding to three significant digits gives 0.964 and 0.527, respectively, for the kinetic orders of X_3 and X_6 in $v_{3,1}$:

$$f_{3,1,3} = \frac{0.27 \text{ mol\%}}{0.27 \text{ mol\%} + 0.01 \text{ mol\%}} = 0.964 \tag{13}$$

$$f_{3,1,6} = \frac{0.559 \text{ mol\%}}{0.559 \text{ mol\%} + 0.5 \text{ mol\%}} = 0.527. \tag{14}$$

The third kinetic order for this flux is for the enzyme ceramide synthase (X_9). Assuming linear dependence on enzyme activity and concentration (i.e., the concentration is multiplied by a constant C) and solving Equation 10 gives

$$f_{3,1,9} = \frac{\partial(\mathrm{C} \times X_9)}{\partial X_9} \times \frac{X_9}{\mathrm{C} \times X_9} = \mathrm{C} \times \frac{X_9}{\mathrm{C} \times X_9} = 1. \tag{15}$$

Finally, the rate constant $\gamma_{3,1}$ must be estimated. This is accomplished by remembering that the BST term and the rate law have the same value at the chosen operating point. Algebraic manipulation of Equations 6 and 9 shows that a rate constant $\gamma_{i,j}$ can be computed from the flux $V_{i,j}$ divided by the power-law function of the flux and evaluated at the operating point:

$$\gamma_{i,j} = V_{i,j}/(X_k^{f_{i,j,k}} X_l^{f_{i,j,l}} \ldots X_z^{f_{i,j,z}})|_{\text{op}}. \quad (16)$$

The denominator can be calculated from the information previously listed; however, the flux information for this reaction is still needed. Because of the frequent lack of *in vivo* flux information, a flux value for $v_{3,1}$ can be estimated by using the specific enzymatic activity to derive $V_{\max}$ in the Michaelis-Menten representation. This value can then be substituted into the rate law, Equation 9, along with other relevant values, to provide $V_{3,1}$. For that purpose, it is convenient to transform the specific activity units from $\mu M/\min/mg$ to $\mu M/\min/l$ by first estimating the amount of yeast protein present in a liter of solution. Multiplying the biomass present in a liter of medium (3.32 g biomass/l) (Ertugay *et al.*, 1997) by the mg of protein present in 1 g of yeast biomass (350 mg protein/g) gives 1162 mg of protein per liter. Next, multiply this by the specific activity of ceramide synthase, determined as 1.65e-5 μmol/min/mg (Wu *et al.*, 1995), to yield $V_{\max} = 1.5687$ μmol/min/l. Substituting all values into Equation 9 gives $V_{3,1} = 0.0264$ $\mu M/\min/l$. Finally, using Equation 16, $\gamma_{3,1}$ is computed as 196097.3, as shown in Equation 17:

$$\gamma_{3,1} = 0.0264/(0.01^{0.964} \times 0.5^{0.527} \times 1.65\text{e-}5) = 196097.3. \quad (17)$$

The three kinetic orders and the rate constant for the first flux of the model system have thus been estimated to yield

$$v_{3,1} = \gamma_{3,1} X_3^{f_{3,1,3}} X_6^{f_{3,1,6}} X_9^{f_{3,1,9}} = 196097.3 X_3^{0.964} X_6^{0.527} X_9. \quad (18)$$

4.2. Dihydroceramidase

Next, the calculations for the parameters of $v_{1,3}$ (i.e., $f_{1,3,1}, f_{1,3,10}$, and $\gamma_{1,3,1}$) are shown. This reaction produces dihydrosphingosine (X_3) by removing the fatty acyl tail from dihydroceramide (X_1) via dihydroceramidase (X_{10}), which is a complex transmembrane enzyme with reversible kinetics (Mao *et al.*, 2000). However, under the exponential growth conditions for which the model was developed, only the forward reaction is considered. Dihydroceramidase is not yet fully characterized, so the Michaelis-Menten rate law is used to estimate the flux as shown in Equation 19:

$$V_{1,3} = V_{\max}\left(\frac{\text{D-Cdase}}{K_{m,\text{D-Cdase}} + \text{D-Cdase}}\right). \quad (19)$$

The kinetic order $f_{1,3,1}$ is calculated by substitution of Equation 19 into Equation 10 and simplification to produce

$$f_{1,3,1} = \frac{\partial(V_{1,3})}{\partial X_1} \times \frac{X_1}{V_{1,3}} = \frac{K_{m,\text{D-Cdase}}}{K_{m,\text{D-Cdase}} + \text{DCER}}. \quad (20)$$

The concentration for dihydroceramide was estimated at $X_{1s} = 0.036$ mol% for the wild-type yeast strain JK9-3d *in vivo* at 30° (Cungui Mao, personal communication). The specific activity for alkaline dihydroceramidase, X_{10}, is 5.4e-6 μmol/min/mg as measured in microsomes (Mao et al., 2000). Lacking data in yeast, a value similar to the dihydroceramide concentration is used to yield the $K_{m,\text{D-Cdase}} = 0.036$. Inserting these values into Equation 20 and solving gives $f_{1,3,1} = 0.5$, as shown in Equation 21:

$$f_{1,3,1} = \frac{0.036 \text{ mol\%}}{0.036 \text{ mol\%} + 0.036 \text{ mol\%}} = 0.5. \quad (21)$$

In Equation 22, linear dependence on enzyme activity and concentration are again used to give a kinetic order equal to 1 for the dihydroceramidase (X_{10}):

$$f_{1,3,10} = \frac{\partial(C \times X_{10})}{\partial X_{10}} \times \frac{X_{10}}{C \times X_{10}} = 1. \quad (22)$$

As discussed in the previous section, estimating the rate constant $\gamma_{1,3}$ requires using the formula in Equation 16. Recall that this calculation first necessitates estimating the flux $V_{1,3}$ by using the reported kinetics and the relationship between $V_{\max}$ and the specific activity as illustrated in Equations 23 to 25:

$$5.4\text{e-}6 \text{ μmol/min/mg protein} \times 1162 \text{ mg protein/l} \\ = 0.00627 \text{ μmol/min/l} \quad (23)$$

$$V_{1,3} = 0.00627 \times \left(\frac{0.036}{0.036 + 0.036}\right) = 0.00313 \quad (24)$$

$$\gamma_{1,3} = V_{1,3}/(X_1^{f_{1,3,1}} X_{10}^{f_{1,3,10}}) = 0.00313/(0.036^{0.5} \times 5.4\text{e-}6) \quad (25) \\ = 3057.42$$

The fully parameterized term for $v_{1,3}$ is thus given by

$$v_{1,3} = 3057.42 X_1^{0.5} X_{10}. \qquad (26)$$

4.3. Inositol phosphorylceramide synthase

The parameter estimation for $v_{1,5} = \gamma_{1,5} X_1^{f_{1,5,1}} X_3^{f_{1,5,3}} X_4^{f_{1,5,4}} X_7^{f_{1,5,7}} X_{12}^{f_{1,5,12}}$ is complicated by the cooperative activation of inositol phophorylceramide (IPC) synthase by phosphatidylinositol (PI) and ceramide, and by inhibition of the enzyme by the sphingoid bases. Fischl et al. (2000) report a Michaelis-Menten rate law with $K_m = 1.35$ mol% for ceramide, enzymatic activity of 3.3e-4 μmol/min/mg protein, and a Hill coefficient of 3 with $K_m = 5$ mol% for PI.

First, the inhibitory effect for the sphingoid bases was calculated from the equations of (Hayashi and Sakamoto, 1985) for the enzyme-inhibitor allosteric dimeric model of Monod, Wyman, and Changeux (MWC). According to MWC, one may plot the relationship $\sqrt{(V_{max} - v)/v}$ against the inhibitor concentration (I), where V_{max} is the maximum velocity and v represents the velocity or activity. The K_i can be obtained as the negative x-intercept of the linear regression (Hayashi and Sakamoto, 1985). Using the IPC synthase specific activity of 0.5 nmol/min/mg (Wu et al., 1995) as 100%, the $V_{max} = 0.525$ nmol/min/mg (Ko et al., 1995) equals 105%. Plotting $\sqrt{(V_{max} - v)/v}$ against both DHS and PHS, as shown in Fig. 13.2A and B, we obtain $K_{i,DHS} = 0.507/0.1754 = 2.89$ mol% and $K_{i,PHS} = 0.3342/0.1698 = 1.97$ mol%, respectively.

Assuming that the inhibition is competitive (in analogy to the inhibition exerted by DHS and PHS on phosphatidate phosphatase) (Wu et al., 1995), the rate equation is

$$V_{ic} = V_{max} S / (S + K_m (1 + I/K_i)) \qquad (27)$$

and the kinetic order for DHS (X_3) is calculated as

$$f_{1,5,3} = \frac{\partial(V_{ic})}{\partial X_3} \times \frac{X_3}{V_{ic}} = \frac{-K_{m,DCER} \times DHS}{K_{i,DHS} \times DCER + K_{m,DCER}(K_{i,DHS} + DHS)}$$
$$= \frac{-1.35 \times 0.01}{2.89 \times 0.036 + 1.35(2.89 + 0.01)} = -0.0034. \qquad (28)$$

Similarly, we obtain $f_{1,5,4} = -0.0242$ for PHS.

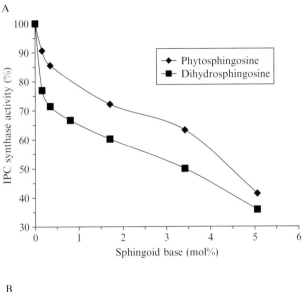

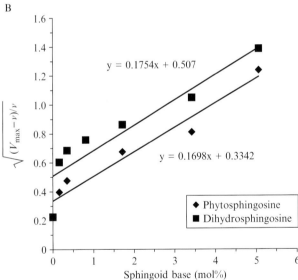

Figure 13.2 (A) Parameter estimation: dependence of inositolphosphorylceramide (IPC) synthase activity on phytosphingosine and dihydrosphingosine concentration (adapted from Wu *et al.*, 1995). (B) Parameter estimation: determination of K_i (see text).

The bi-substrate Hill kinetic equation shown in Equation 29 was used to calculate the remaining kinetic orders in Equations 30 and 31, and the flux for $v_{1,5}$ in Equation 32. This flux used $V_{\max} = $ 3.3e-4 μmol/min/mg protein $\times$ 1162 mg protein/l $=$ 0.383 μmol/min/l. The rate constant is then calculated, as shown in Equation 33. These calculations used PI $=$ 4.6 mol% (Wu et al., 1995) and dihydroceramide as the substrate, but similar calculations must be made when phytoceramide is the substrate:

$$v_{1,5} = V_{1,5} = V_{\max}\left(\frac{\text{DCER}}{K_{m,\text{DCER}} + \text{DCER}}\right) \times \left(\frac{\text{PI}^{\text{Hill}}}{K_{m,\text{PI}}^{\text{Hill}} + \text{PI}^{\text{Hill}}}\right) \quad (29)$$

$$f_{1,5,1} = \frac{\partial(V_{1,5})}{\partial X_1} \times \frac{X_1}{V_{1,5}} = \frac{K_{m,\text{DCER}}}{K_{m,\text{DCER}} + \text{DCER}} = \frac{1.35}{1.35 + 0.036} = 0.973 \quad (30)$$

$$f_{1,5,7} = \frac{\partial(V_{1,5})}{\partial X_7} \times \frac{X_7}{V_{1,5}} = \frac{\text{Hill} \times (K_{m,\text{PI}}^{2 \cdot \text{Hill}} + (K_{m,\text{PI}}^{\text{Hill}} \times \text{PI}^{\text{Hill}}))}{(K_{m,\text{PI}}^{\text{Hill}} + \text{PI}^{\text{Hill}})^2}$$
$$= \frac{3(5^6 + (5^3 \times 4.6^3))}{(5^3 + 4.6^3)^2} = 1.685 \quad (31)$$

$$v_{1,5} = 0.383 \times \frac{0.036}{(1.35 + 0.036)} \times \frac{4.6^3}{(5^3 + 4.6^3)} = 0.00437 \quad (32)$$

$$\gamma_{1,5} = V_{1,5}/(X_1^{f_{1,5,1}} X_3^{f_{1,5,3}} X_4^{f_{1,5,4}} X_7^{f_{1,5,7}} X_{12}^{f_{1,5,12}})$$
$$= 0.00437/(0.036^{0.973} \times 0.01^{-0.0034} \times 0.05^{-0.0242}$$
$$\times 4.6^{1.685} \times 3.3\text{e-4}) \quad (33)$$
$$= 23.55.$$

4.4. Inositol phosphosphingolipid phospholipase C

Calculations to estimate the parameters for the flux $v_{5,1} = \gamma_{5,1} X_5^{f_{5,1,5}} X_{132}^{f_{5,1,132}}$ through the inositol phosphosphingolipid phospholipase C (IPCase) are similar to those for the dihydroceramidase. IPCase is the only known enzyme in yeast for the salvage pathway that recycles complex sphingolipids from the plasma membrane, which has the major concentration of complex sphingolipids (Patton and Lester, 1991). The kinetics of this enzyme is yet to be characterized, so the Michaelis-Menten rate law (Equation 33) is used to estimate the flux and kinetics orders as before with a $K_{m,\text{IPC}} = 3.57$ mol% and specific activity for IPCase equal to 1.5e-4 μmol/min/mg protein (Sawai et al., 2000). The steady-state concentration of IPC, $X_{5s} = 0.102\%$, was estimated assuming that 10% of the non-plasma membrane concentration was located at the synthesis site (Patton and Lester, 1991).

Equations 35 to 38 show the calculations for the kinetic orders, $V_{5,1}$, and the rate constant. The fully parameterized term is shown in Equation 39:

$$V_{5,1} = V_{\max}\left(\frac{\text{IPC}}{K_{m,\text{IPC}} + \text{IPC}}\right) \quad (34)$$

$$\begin{aligned} f_{5,1,5} &= \frac{\partial(V_{5,1})}{\partial X_5} \times \frac{X_5}{V_{5,1}} = \frac{K_{m,X_5}}{K_{m,X_5} + X_5} \\ &= \frac{3.57 \text{ mol\%}}{3.57 \text{ mol\%} + 0.102 \text{ mol\%}} = 0.972 \end{aligned} \quad (35)$$

$$f_{5,1,132} = \frac{\partial(C \times X_{132})}{\partial X_{132}} \times \frac{X_{132}}{C \times X_{132}} = 1 \quad (36)$$

$$V_{5,1} = 1.5\text{e-}4 \times 1162 \times 0.972 = 0.48\text{e-}3 \ \mu\text{mol/min/l} \quad (37)$$

$$\begin{aligned} \gamma_{5,1} &= V_{5,1}/(X_1^{f_{5,1,5}} X_{10}^{f_{5,1,132}}) \\ &= 4.84\text{e-}3/(0.102^{0.972} \times 1.5\text{e-}4) = 297 \end{aligned} \quad (38)$$

$$v_{5,1} = 297 X_5^{0.972} X_{132}. \quad (39)$$

4.5. Hydroxylase

This enzyme catalyzes both the $v_{3,4}$ and $v_{1,2}$ fluxes, but at this time we are interested in only the latter flux. Lacking information about the enzymatic parameters of the yeast hydroxylase, the K_m is assumed to equal the substrate concentration (Cleland, 1970) with classical Michaelis-Menten kinetics:

$$v_{1,2} = V_{1,2} = V_{\max}\left(\frac{\text{DCER}}{K_{m,\text{DCER}} + \text{DCER}}\right). \quad (40)$$

As discussed for previous fluxes, this gives kinetic orders for the substrate, DCER (X_1), and the hydroxylase (X_{14}), as shown in following equations:

$$f_{1,2,1} = \frac{K_{m,X_1}}{K_{m,X_1} + X_1} = \frac{0.036 \text{ mol\%}}{0.036 \text{ mol\%} + 0.036 \text{ mol\%}} = 0.5 \quad (41)$$

$$f_{1,2,14} = \frac{\partial(C \times X_{14})}{\partial X_{14}} \times \frac{X_{14}}{C \times X_{14}} = 1. \quad (42)$$

Since this is the last flux affecting the DCER node in our system, it can be calculated from the stoichiometry of the other fluxes for the node. This calculation is allowed since the model was developed under the assumption of balanced influxes and effluxes at the reference steady-state. Therefore,

$$\begin{aligned} v_{1,2} &= v_{3,1} + v_{5,1} - v_{1,5} - v_{1,3} \\ &= 0.0264 + 0.0048 - 0.00437 - 0.00313 = 0.0237 \end{aligned} \quad (43)$$

and the rate constant is calculated with $X_{14s} = 0.17\text{e-}3$ mol% (Grilley *et al.*, 1998) as

$$\begin{aligned} \gamma_{1,2} &= v_{1,2}/(X_1^{f_{1,2,1}} X_{14}^{f_{1,2,14}}) \\ &= 0.0237/(0.036^{0.5} \times 0.17\text{e-}3) = 736.26. \end{aligned} \quad (44)$$

When fully parameterized, the final term for the DCER equation is given by Equation 45. Combining all the flux terms together yields the numerical equation for DCER, as shown in Equation 46:

$$v_{1,2} = 736.26 X_1^{0.5} X_{14} \quad (45)$$

$$\begin{aligned} dX_1/dt =\ & 196097.3 X_3^{0.964} X_6^{0.527} X_9 + 297 X_5^{0.972} X_{132} \\ & - 3057.42 X_1^{0.5} X_{10} - 736.26 X_1^{0.5} X_{14} \\ & - 23.55 X_1^{0.973} X_3^{-0.0034} X_4^{-0.0242} X_7^{1.685} X_{12}. \end{aligned} \quad (46)$$

4.6. Other parameter calculations

The remaining parameters in the other four system equations are calculated in a similar manner; however, a few deserve special mention. The first flux of the system, from KDHS (X_8) to DHS (X_3), was estimated from knowledge of how palmitoyl-CoA is split between the glycerolipid and sphingolipid pathways, rather than directly from the kinetic equation. This flux was set as 1/7 the palmitoyl-CoA to glycerolipid flux (Alvarez-Vasquez et al., 2004), yielding $v_{0,3} = 0.875 \ \mu M/\min/l$.

The effluxes $v_{2,0}$, and $v_{5,0}$ were calculated from the stoichiometry of their respective nodes. The kinetic orders for the substrates exiting the system through these fluxes were estimated by assuming that the K_m for the reaction was similar to the substrate concentration, as in previous calculations when little is known about the kinetics (Cleland, 1970). Additionally, $v_{3,4}$ and $v_{4,2}$ were also calculated based on the stoichiometry of their corresponding nodes. The remaining fluxes were calculated based on their kinetics. The numerical equations can now be written from the numerical flux terms listed in the third column of Table 13.3, and the GMA model is complete. The calculated $V_{i,j}$ are found in the last column of Table 13.3.

5. Model Analysis: Steady State, Stability, and Sensitivity

Once the system equations are fully parameterized, they are typically entered into a modeling software package that facilitates standard tests for stability and sensitivity. A very intuitive and free program is PLAS© (Ferreira, 2000), which was specifically designed for the exploration of BST models. For expanded analyses of BST models, BSTLab (Schwacke and Voit, 2003) and BSTBox (Goel et al., 2006) were developed as toolboxes to be used with MATLAB. For more generic pathway analyses, other programs, such as COPASI (Hoops et al., 2006), are available (see Alves et al., 2006, and Pettinen et al., 2005, for reviews of such programs). The results discussed in the remainder of the chapter were generated using PLAS.

Although the GMA format is intuitive because each flux has a separate term, some analyses are facilitated by a slight reformulation into so-called S-systems. Specifically, the GMA equations are rewritten in a form that aggregates all incoming fluxes into one power-law term and all effluxes into a second power-law term. The main advantage is that this S-system form becomes linear at steady-state upon logarithmic transformation, thereby permitting the use of tools from linear algebra for the stability and sensitivity analyses (Savageau, 1969b, 1975). This review will not describe the transformation of GMA into S-system equations, and it suffices to say that the methods are straightforward and can be found in Voit (2000); they are also supported by BSTLab. The two systems are equivalent at the chosen operating point. The following analyses were conducted in PLAS with the S-system.

In PLAS, it is easy to perform a variety of analyses. First, we confirm that the model reaches a steady-state, where the net fluxes through all nodes of the pathway are zero (data not shown). We also check that the system is locally stable, meaning that, in response to short-term perturbations, the system returns to the nominal steady state. This type of stability is indicated by the so-called eigenvalues of the system, which are complex numbers with real and imaginary parts. In our example, all the eigenvalues have negative real parts indicating stability, and all imaginary parts are zero, indicating the absence of oscillations in the system dynamics (data not shown). (For a thorough discussion of this topic, see Chapter 6 of Voit, 2000.)

Next, we calculate the sensitivities of the system with respect to parameters or independent variable values. Mathematically, sensitivities are computed via partial differentiation and are easily calculated in PLAS and other programs. (For details on the theory behind sensitivity analysis, please see Chapter 7 of Voit, 2000.) Conceptually, each sensitivity is a number that shows how much an output feature changes if some input feature is altered by a small percentage, usually 1%. As a generic example, consider the sensitivity of the steady-state concentration of the dependent variable X with respect to the rate constant α and suppose the sensitivity is 4. This value means that a 1% increase in α should yield a 4% change in X (i.e., $4 \times 1\% = 4\%$). If the sensitivity value is positive, the changes in α and X are in the same direction, while a negative value indicates changes in opposite direction.

In a robust model, most sensitivities are small in magnitude, and one typically focuses on any sensitivities that are unusually high (maybe >10). A large sensitivity points to one of two features. First, a significant defect may exist in the model. For instance, the model is "precariously sensitive" to small changes in that parameter—perhaps because an important component or process was missing during model design, or because incorrect data were used during model implementation. Either of these indicates the need for revisiting the information available in the literature or from experts in the field. Second, a high sensitivity may indicate a key role of a node in regulating the pathway (or portion of the pathway).

Changes in model parameters (values of rate constants and kinetic orders) correspond to a persistently altered model structure; thus, sensitivities with respect to parameters in some sense characterize robustness against mutational events. By contrast, changes in independent variables (such as enzyme activities) reflect temporary alterations in the environment. Therefore, sensitivities with respect to independent variables are valuable indicators for model robustness against normal fluctuations (and most bench experiments). These latter types of sensitivities are called logarithmic gains (or log gains) and can be calculated for both metabolites and fluxes (Savageau, 1972; Voit, 2000).

Mathematically, a metabolite log gain is defined as the ratio of the percent change in a dependent variable X_i (usually a metabolite) to the percent change in an independent variable X_k (usually an input variable or enzyme activity), while all other independent concentrations and parameters remain unchanged. Therefore, a log gain with a value less than one indicates that the change in the independent variable has little effect on the model output, whereas log gains greater than one indicate amplification. The sign of the log gain again indicates whether changes in input and output are in the same direction (Savageau, 1972; Voit, 2000).

Among the 50 metabolite log gains shown in Table 13.4, 76% are less than one, indicating that perturbations of the independent variables will be attenuated in most of the system. The highest log gains are seen when KDHS (X_8) and the α-hydroxylase (X_{14}) are perturbed. KDHS is the initial entry into the pathway, which may explain its relatively large positive log gains for all dependent variables. Also, in the larger model (Alvarez-Vasquez et al., 2004, 2005), these effects are attenuated as KDHS is more centrally located to the system. The hydroxylase has the second highest log gains overall, but with negative sign for three of the metabolites. As one might expect, these occur for the products of the reaction, PHS and PCER. The highest log gain (−4.65) indicates the notable influence of this enzyme on decreasing DCER levels. The analysis presents a smaller negative effect (−1.94) on IPC, presumably since it is farther away in the pathway.

A flux log gain is defined as the ratio of the percent change in a flux $v_{i,j}$ to the percent change in an independent variable X_k, while all other independent concentrations and parameters remain unaltered. The model has 130 flux log gains (10 independent variables multiplied by 13 fluxes). As with the metabolite log gains, most flux log gains are less than one (81.5%), indicating that the signal produced by a small variation in an enzyme activity will be attenuated in most parts of the model (Table 13.5).

Once again, the larger log gains are associated with the hydroxylase and KDHS. Two fluxes, $v_{1,5}$ through IPC synthase (X_{12}) and $v_{1,3}$ through dihydroceramidase (X_{10}), are most affected by changes in hydroxylase activity, with log gains of −4.52 and −2.32, respectively. For KDHS, the fluxes through IPC synthase ($v_{1,5}$ and $v_{2,5}$) and IPCase ($v_{5,1}$ and $v_{5,2}$) increase the most, with log gains ranging from 1.89 to 3.41. All these fluxes are directly related to the synthesis and degradation of the ceramides and, mostly, to DCER (X_1). This interesting result reflects an asymmetric behavior of the otherwise "parallel" ceramide fluxes in response to perturbations. This asymmetric effect is not exerted by other enzymes of the model. For example, when ceramide synthase (X_9), phytoceramidase (X_{11}), dihydroceramidase (X_{10}), C26-CoA (X_6), or IPCase (X_{131} and X_{132}) are perturbed, the log gains are unremarkable (Table 13.5).

Depending on the results of the model analysis, it may be desirable to modify the original system. For instance, greater detail may be needed for the input to the system or more accurate data may be required to determine the parameters with high sensitivity. In our case, we might want to expand the entry of the model to include serine and palmitoyl-CoA and to

Table 13.4 Metabolite log gains with absolute value greater than 1

		DCER X_1	PCER X_2	DHS X_3	PHS X_4	IPC X_5
C_{26}–CoA	X_6	–	–	–	–	–
PI	X_7	–	–	–	–	1.79
KDHS	X_8	3.54	2.00	1.95	1.22	2.62
CerSyn	X_9	1.61	–	–	−1.28	–
D-Cdase	X_{10}	–	–	–	–	–
P-Cdase	X_{11}	–	–	–	–	–
IPCSyn	X_{12}	–	–	–	–	1.06
IPCase (PCER)	X_{131}	–	–	–	–	–
IPCase (DCER)	X_{132}	–	–	–	–	–
Hydroxylase	X_{14}	−4.65	–	−1.90	–	−1.94

Table 13.5 Flux log gains with absolute value greater than 1

		$v_{3,1}$	$v_{5,1}$	$v_{1,2}$	$v_{1,3}$	$v_{1,5}$	$v_{4,2}$	$v_{5,2}$	$v_{2,4}$	$v_{2,5}$	$v_{2,0}$	$v_{0,3}$	$v_{3,4}$	$v_{5,0}$
C_{26}–CoA	X_6	–	–	–	–	–	–	–	–	–	–	–	–	–
PI	X_7	–	1.74	–	–	1.77	–	1.74	–	1.68	–	–	–	–
KDHS	X_8	1.88	2.54	1.77	1.77	3.41	–	2.54	–	1.89	–	–	–	1.31
CerSyn	X_9	–	–	–	–	1.6	–	–	–	–	–	–	–	–
D-Cdase	X_{10}	–	–	–	–	–	–	–	–	–	–	–	–	–
P-Cdase	X_{11}	–	–	–	–	–	–	–	–	–	–	–	–	–
IPCSyn	X_{12}	–	1.03	–	–	1.05	–	1.03	–	–	–	–	–	–
IPCase(PCER)	X_{131}	–	–	–	–	–	–	–	–	–	–	–	–	–
IPCase(DCER)	X_{132}	–	–	–	–	–	–	–	–	–	–	–	–	–
Hydroxylase	X_{14}	−1.83	−1.88	−1.32	−2.32	−4.52	–	−1.88	–	–	–	–	–	–

acquire better data for the hydroxylase flux. These refinements should be balanced against the goals of the model.

Additionally, during the process of setting up the model, the modeler may realize that an entirely different level of detail is needed in parts of the model. For instance, it may be better to separate ceramide synthase into the known components encoded by LAG1, LAC1, and LIP1 so that "knockouts" of each gene can be simulated. Or, the modeler may decide that it is crucial to include the kinases, phosphatases, and subsequent degradation via the lyase for the sphingoid bases. Again, such decisions are made based on the purpose of the model.

6. SIMULATION: WHAT HAPPENS IF...?

Once a model is at hand and has been mathematically evaluated for its "structural integrity" (as discussed in the "Model Analysis" section), it is time to put it to "use." A model can be beneficial for diverse purposes. For example, diagnosing the model in detail may suggest that a key factor is missing and/or the input biological data are faulty. On the other hand, the sensitivity and log gain analysis may disclose the existence of a key regulatory factor (e.g., an enzyme), such that changes in that variable could have important influences on the pathway (metabolically and functionally). Such a finding may direct the experimentalist to characterize that enzyme in greater detail. Another use of the model is simulations to probe "what if" questions. What would happen if a certain flux is increased, if a precursor is decreased, or if an enzyme activity changes by x-fold? This latter application will be illustrated later. Additionally, the model may be used for targeted manipulations, such as optimizing the yield of some valuable compound. This can be achieved by rerouting metabolism toward the desired metabolite without negatively affecting viability (Torres and Voit, 2002). In a medical or pharmaceutical setting, the model may also be used for improved drug targeting.

As an example of exploring the dynamics of the system, the model was challenged by doubling the KDHS concentration for one minute (Fig. 13.3). The result was a quick increase in the first metabolite of the pathway, DHS. Interestingly, DCER has a slower increase than PCER, which is further down the pathway. This may be explained by the difference in the calculated fluxes of ceramide synthase for the two substrates, DHS ($v_{3,1} = 0.0264$ μM/min/l) and PHS ($v_{4,2} = 0.87$ μM/min/l) (Table 13.3). Also interesting is the slow increase in complex sphingolipids, represented in this model by IPC. This metabolite returns to its initial value at $\sim$100 min (data not shown).

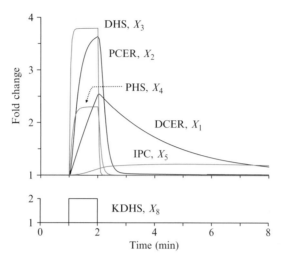

Figure 13.3 Dynamics of the dependent variables after doubling the KDHS (ketodihydrosphingosine, X_8) concentration. Data are normalized against their respective initial values.

7. Conclusion

We have now discussed the entire cycle of model development for a small system of sphingolipid metabolism. Although the estimation of parameter values can be somewhat complicated and tedious, the majority of this modeling process is well within the reach of any competent biologist. However, a few cautionary statements are required.

First, realize that any model is only as good as its design; therefore, do not expect a "VW Bug" to perform like a "BMW Boxster." Second, keep in mind that no matter how good the design, any model is still an approximation of biological reality, and the cell is always right. Third, although this chapter may seem detailed to the reader, it is only an introduction to the topic. Many subtleties and special cases are not included, so the reader is advised to pursue further studies before embarking on any ambitious modeling projects. Finally, in the famous words of George E. P. Box, "Essentially, all models are wrong, but some are useful" (Box and Draper, 1987).

8. Epilogue

Biology is at the dawn of a revolution. In the past, painstakingly focused experiments shed light on ever finer details of selected phenomena and led to an enormous amount of crisp pieces of information. This type of experimentation has produced a wealth of knowledge and will, without any doubt, continue unabated. However, recent years have witnessed the birth of a complementary approach, driven by the development of high-throughput techniques that cover more ground with high efficiency. The enormous volume of these data along with their different features mandated computational approaches to integrate information and ushered in the field of bioinformatics. Superb tools have been produced from bioinformaticians to bring order to the flood of high-density data; however, it is now apparent that bioinformatics alone is insufficient. The missing expertise is now known as systems biology—a methodological framework with the goal of deducing functionality by merging diverse types of quantitative data and qualitative observations into quantitative structures that permit systemic analyses and lead to a deeper understanding of how biological systems work.

We have shown here a small part of this long-term effort by focusing on strategies for formulating mathematical models of metabolic pathways. It has been said many times that modeling is an art, and there is some truth to it. However, our discussion has demonstrated that many steps of the modeling process follow strict guidelines that can be learned and followed by biochemists with a modest amount of training in the computational sciences. (For further reading on BST or modeling in general, see Edelstein-Keshet, 2005; Heinrich and Schuster, 1996; Palsson, 2006; Savageau, 1976; Voit, 1991; or Voit and Schwacke, 2006.) In fact, the lion's share of technical execution is supported by software that is becoming more and more robust and user-friendly. Packages like PLAS, which is freely available for all academic uses, do not require compilation or much more than formulating differential equations in the right syntax. All numerical evaluations are then done in the background, and the only (though crucial) task left is interpretation of results. Thus, the art of modeling is being delegated increasingly to the front and back ends, namely, the definition of model components and processes, and the translation of mathematical results back into the domain of biology. Almost all of the intermediary technical machinery is automated or at least greatly facilitated by simplified model entry and intuitive interfaces.

A good analogy of where the modeling enterprise stands and where it is heading may be biostatistics. Not too long ago, even a lowly t-test was executed by experts on dedicated computers and with complex software. But now it is very easy to use (or abuse) packages on personal computers, and most standard testing is done by subject area scientists who no longer have to worry about the mathematics behind the test but can instead focus on data generation and the interpretation of results. Modeling is a few decades behind, mainly because there are fewer agreed-upon approaches and strategies than in statistics. Nonetheless, such approaches are beginning to crystalize and will enter the mainstream of biology in the not-too-distant future.

ACKNOWLEDGMENTS

KJS thanks Dr. Christopher Clarke for helpful comments on the chapter.
This work was supported in part by a grant from the National Institutes of Health (2 R01 GM063265-06; Y. A. Hannun, PI), and a Molecular and Cellular Biosciences Grant (MCB-0517135; E. O. Voit, PI) from the National Science Foundation. KJS is supported by National Institutes of Health Grants R01 AG16583 and R01 GM62887 (to LMO).

REFERENCES

Alvarez-Vasquez, F., Sims, K. J., Cowart, L. A., Okamoto, Y., Voit, E. O., and Hannun, Y. A. (2005). Simulation and validation of modelled sphingolipid metabolism in *Saccharomyces cerevisiae*. *Nature* **433**, 425–430.

Alvarez-Vasquez, F., Sims, K. J., Hannun, Y. A., and Voit, E. O. (2004). Integration of kinetic information on yeast sphingolipid metabolism in dynamical pathway models. *J. Theor. Biol.* **226**, 265–291.

Alves, R., Antunes, F., and Salvador, A. (2006). Tools for kinetic modeling of biochemical networks. *Nat. Biotechnol.* **24**, (6), 667–672.

Box, G. E. P., and Draper, N. R. (1987). "Empirical Model-Building and Response Surfaces." Wiley, Hoboken, NJ.

Chou, I-C., Martens, H., and Voit, E. O. (2006). Parameter estimation in biochemical systems models with alternating regression. *BMC Theor. Biol. Med. Model.* **3**, 25.

Cleland, W. W. (1970). "Enzymes." Academic Press, New York.

Curto, R., Voit, E. O., Sorribas, A., and Cascante, M. (1998). Mathematical models of purine metabolism in man. *Math. Biosci.* **151**, 1–49.

Curto, R., Voit, E. O., Sorribas, A., and Cascante, M. (1997). Validation and steady-state analysis of a power-law model of purine metabolism in man. *Biochem. J.* **324**, 761–775.

Edelstein-Keshet, L. (2005). "Mathematical Models in Biology." Society for Industrial and Applied Mathematics, Philadelphia, PA.

Ertugay, N., Hamamci, H., and Bayindirli, A. (1997). Fed-batch cultivation of bakers' yeast: Effect of nutrient depletion and heat stress on cell composition. *Folia. Microbiol. (Praha)* **42**, 214–218.

Fell, D. (1997). "Understanding the control of metabolism." Portland Press, London, UK.

Ferreira, A. (2000). Power Law Analysis and Simulation [software]. Accessible at: http://www.dqb.fc.ul.pt/docentes/aferreira/plas.html.

Ferreira, A. E., Ponces, Freire A. M., and Voit, E. O. (2003). A quantitative model of the generation of N(epsilon)-(carboxymethyl)lysine in the Maillard reaction between collagen and glucose. *Biochem. J.* **376**, 109–121.

Fischl, A. S., Liu, Y., Browdy, A., and Cremesti, A. E. (2000). Inositolphosphoryl ceramide synthase from yeast. *Methods Enzymol.* **311**, 123–130.

Gaigg, B., Neergaard, T. B., Schneiter, R., Hansen, J. K., Faergemen, N. J., Jensen, N. A., Anderson, J. R., Friis, J., Sandhoff, R., Schroder, H. D., and Knudsen, J. (2001). Depletion of acyl-coenzyme A-binding protein affects sphingolipid synthesis and causes vesicle accumulation and membrane defects in *Saccharomyces cerevisiae*. *Mol. Biol. Cell.* **12**, 1147–1160.

Goel, G., Chou, I. C., and Voit, E. O. (2006). Biological systems modeling and analysis: A biomolecular technique of the twenty-first century. *J. Biomol. Tech.* **17**, 252–269.

Grilley, M. M., Stock, S. D., Dickson, R. C., Lester, R. L., and Takemoto, Y. J. (1998). Syringomycin action gene *SYR2* is essential for sphingolipid 4-hydroxylation in *Saccharomyces cerevisiae*. *J. Biol. Chem.* **273**, 11062–11068.

Guillas, I., Kirchman, P. A., Chuard, R., Pfefferli, M., Jiang, J. C., Jazwinski, S. M., and Conzelmann, A. (2001). C26-CoA-dependent ceramide synthesis of *Saccharomyces cerevisiae* is operated by Lag1p and Lac1p. *EMBO J.* **20,** 2655–2665.
Hayashi, K., and Sakamoto, N. (1985). "Dynamic Analysis of Enzyme Systems: An Introduction." Springer-Verlag, Tokyo.
Heinrich, R., and Rapoport, T. A. (1974). A linear steady-state treatment of enzymatic chains. General properties, control and effector strength. *Eur. J. Biochem.* **42,** 89–95.
Hoops, S., Sahle, S., Guages, R., Lee, C., Pahle, J., Simus, N., Singhal, M., Xu, L., Mendes, P., and Kummer, U. (2006). COPASI—A COmplex PAthway SImulator. *Bioinformatics* **22,** 3067–3074.
Kacser, H., and Burns, J. A. (1973). The control of flux. *Symp. Soc. Exp. Biol.* **27,** 65–104.
Ko, J., Cheah, S., and Fischl, A. S. (1995). Solubilization and characterization of microsomal-associated phosphatidylinositol: Ceramide phosphoinositol transferase from *Saccharomyces cerevisiae*. *J. Food Biochem.* **19,** 253–267.
Mao, C., Xu, R., Bielawska, A., and Obeid, L. M. (2000). Cloning of an alkaline ceramidase from *Saccharomyces cerevisiae*. An enzyme with reverse (CoA-independent) ceramide synthase activity. *J. Biol. Chem.* **275,** 6876–6884.
Ni, T. C., and Savageau, M. A. (1996a). Application of biochemical systems theory to metabolism in human red blood cells. Signal propagation and accuracy of representation. *J. Biol. Chem.* **271,** 7927–7941.
Ni, T. C., and Savageau, M. A. (1996b). Model assessment and refinement using strategies from biochemical systems theory: Application to metabolism in human red blood cells. *J. Theor. Biol.* **179,** 329–368.
Patton, J. L., and Lester, R. L. (1991). The phosphoinositol sphingolipids of *Saccharomyces cerevisiae* are highly localized in the plasma membrane. *J. Bacteriol.* **173,** 3101–3108.
Pettinen, A., Aho, T., Smolander, O. P., Manninem, T., Saarinen, A., Taattola, K. L., Yli-Harja, O., and Linne, M. L. (2005). Simulation tools for biochemical networks: Evaluation of performance and usability. *Bioinformatics* **21,** 357–363.
Riezman, H. (2006). Organization and functions of sphingolipid biosynthesis in yeast. *Biochem. Soc. Trans.* **34,** 367–369.
Savageau, M. A. (1969a). Biochemical systems analysis. I. Some mathematical properties of the rate law for the component enzymatic reactions. *J. Theor. Biol.* **25,** 365–369.
Savageau, M. A. (1969b). Biochemical systems analysis. II. The steady-state solutions for an n-pool system using a power-law approximation. *J. Theor. Biol.* **25,** 370–379.
Savageau, M. A. (1972). The behavior of in tact biochemical control systems. *Curr. Topics Cell. Reg.* **6,** 63–129.
Savageau, M. A. (1975). Optimal design of feedback control by inhibition: Dynamic considerations. *J. Mol. Evol.* **5,** 199–222.
Sawai, H., Okamoto, Y., Luberto, C., Mao, C., Bielawska, A., Domae, N., and Hannun, Y. A. (2000). Identification of ISC1 (YER019w) as inositol phosphosphingolipid phospholipase C in *Saccharomyces cerevisiae*. *J. Biol. Chem.* **275,** 39793–39798.
Schwacke, J. H., and Voit, E. O. (2003). BSTLab: A Matlab toolbox for biochemical systems theory. Eleventh International Conference on Intelligent Systems for Molecular Biology, Brisbane, Australia.
Segel, L. A. (1984). "Modeling dynamic phenomena in molecular and cellular biology. Cambridge." Cambridge University Press, , New York.
Shimeno, H., Soeda, S., Sakamoto, M., Kouchi, T., Kowakame, T., and Kihara, T. (1998). Partial purification and characterization of sphingosine N-acyltransferase (ceramide synthase) from bovine liver mitochondrion-rich fraction. *Lipids* **33,** 601–605.
Shiraishi, F., and Savageau, M. A. (1992). The tricarboxylic acid cycle in *Dictyostelium discoideum*. *J. Biol. Chem.* **267,** 22934–22943.

Sims, K. J., Spassieva, S. D., Voit, E. O., and Obeid, L. M. (2004). Yeast sphingolipid metabolism: Clues and connections. *Biochem. Cell Biol.* **82,** 45–61.
Torres, N. V., and Voit, E. O. (2002). "Pathway Analysis and Optimization in Metabolic Engineering." Cambridge University Press, Cambridge, UK.
Voit, E. O. (2000). "Computational analysis of biochemical systems: A practical guide for biochemists and molecular biologists." Cambridge University Press, , New York.
Voit, E. O., and Almeida, J. (2004). Decoupling dynamical systems for pathway identification from metabolic profiles. *Bioinformatics* **20,** 1670–1681.
Voit, E. O., Almeida, J., Marino, S., Lall, R., Goel, G., Neves, A. R., and Santos, H. (2006). Regulaltion of glycolysis in *Lactococcus lactis*: An unfinished systems biological case study. *Syst. Biol. (Stevenage).* **153**(4), 286–298.
Wu, W. I., McDonough, V. M., Nickels, J. T., Jr., Ko, J., Fischl, A. S., Vales, T. R., Merrill, A. H., Jr., and Carman, G. M. (1995). Regulation of lipid biosynthesis in *Saccharomyces cerevisiae* by fumonisin B1. *J. Biol. Chem.* **270,** 13171–13178.

CHAPTER FOURTEEN

QUANTITATION AND STANDARDIZATION OF LIPID INTERNAL STANDARDS FOR MASS SPECTROSCOPY

Jeff D. Moore, William V. Caufield, *and* Walter A. Shaw

Contents

1. Introduction	352
2. Lipid Handling Guidelines	352
3. Chemical Characterization of Lipid Stocks	353
4. Preparation of Working Lipid Standards	355
5. Packaging of Lipid Standards	359
6. Quality Control and Stability Testing	361
6.1. Ceramide/sphingoid base internal standard mixture	361
6.2. Triglyceride-d5 internal standard mixture	363
7. Discussion	364
References	366

Abstract

Qualification, preparation, and use of lipid compounds as analytical reference standards are daunting endeavors. The sheer vastness of the number of lipid compounds present in biological samples make it impossible to directly standardize each entity. Available lipid compounds chosen for preparation as standards are difficult to maintain as pure entities of stable concentration due to their physical and chemical interactions. The lipid chemist must understand these constraints for each chosen molecule to construct a standard material, which provides accurate measurement for a practical length of time. We provide methods and guidelines to aid the chemist in these endeavors. These aids include analytical methods for preparation and handling techniques, qualification of candidate materials, packaging, storage, and, finally, stability testing of working standard materials. All information will be provided under the purview of standardization of lipid analysis by mass spectrometry.

Avanti Polar Lipids, Inc., Alabaster, Alabama

1. INTRODUCTION

Lipids function as essential and abundant building blocks of cells and their organelles, neurological second messengers, nutrients, hormones, protein modifiers, and substrates of enzymes, to name a few, and are implicated in many disease processes. They are available from numerous sources and exist in such heterogeneity of structure that they are classified according to molecular functional group similarities, degree of polarity, or biological origin. However, few sources of chemically pure and structurally characterized lipid reference standards exist. None are available for major lipid classes through certified sources, such as the United States Pharmacopoeia, National Formulary, or National Institute of Technology and Standards, nor from international agencies. This is largely due to the difficulty in purifying lipid mixtures from natural sources or synthetically manufacturing sufficient and stable quantities, which are of a single molecular entity. Of those available, they are typically of simple structure, not characterized by modern analytical methods, and generally do not represent lipid structures of current scientific interest. Chemical stability and, therefore, constant concentration is a major concern for any standard material. This is more so for reference standards of lipid molecules due to their reactivity with light, air, moisture, and active surfaces to produce oxidation, hydrolysis, and decomposition. The researcher must locate a source that provides standards for their interests or must qualify available materials as suitable for use as reference standards. The intended final use of these materials as standards for calibration of mass spectrometry techniques dictates a multiple methodology approach to ensure chemical identity and purity. Once chemically characterized, they should be packaged and stored in containers that achieve the best stability for extended use. And lastly, these packaged standards require testing over time to document their continued structural and concentration integrity. Before discussion of analytical procedures, a general overview of handling lipid materials for prevention of avoidable and commonly encountered problems is necessary.

2. LIPID HANDLING GUIDELINES

Lipids, by definition, are soluble in organic solvents, such as chloroform and methanol, but only sparingly soluble, if at all, in water. Ideally, lipid compounds should be stored in glass containers at relatively high concentration (mM to M). Optimal storage conditions will include the absence of light and temperatures below $-15°$. Lipid materials removed from storage should be allowed to reach room temperature before opening for use. This prevents weighing and delivery errors due to changes in density per temperature or evaporation of solvents from solutions. Stock lipid solutions should always be prepared in high quality and fresh solvents of low water content. Organic solvent solutions of lipids should be stored in glass only, as plastics and polymers dissolve and contaminate the lipid solution (Pidgeon *et al.*, 1989). Glass closures and lids that have plastic or rubber cemented into their top should be avoided as well. Closures that have Teflon^{TM} welded into their top are recommended. This contamination is also observed to a lesser extent when lipid solutions are transferred with plastic or polymer pipette tips commonly used in the laboratory. Lipid solutions should be transferred with glass pipettes, glass syringes, or auto-pippettors with Teflon and glass surfaces when the possibility of contamination must be completely avoided. Saturated lipids, which contain no double bonds in their structure, generally produce free-flowing powders amenable to direct handling and weighing. However, extended exposure to air will hydrate many saturated phospholipids. Phosphatidylcholine compounds are a prime example of this through addition of water to the phosphocholine headgroup to a maximum of 4 to 8% water. This hydration contributes to hydrolysis of fatty acids from the glycerol backbone producing "lyso"-phosphatidylcholine and free fatty acid. Atmospheric hydration of monounsaturated and polyunsaturated lipids produce materials that are difficult to accurately weigh. They initially become "sticky," preventing quantitative transfer of weighed amounts, and progressively absorb moisture to a "waxy" appearance when extremely wet. If unsaturated lipids must be handled as solids, use of a dry box should be employed. This atmospheric exposure also introduces oxygen contact with the double bonds of the lipid. Oxidation of unsaturated lipids form peroxides, epoxides, and aldehydes at their double bond positions. The physical appearance of a poor quality or contaminated unsaturated lipid is a waxy yellow substance. It is for these reasons many researchers store lipid stocks with a blanket of inert gas, such as argon or nitrogen, in the head space of containers.

Standards for mass spectrometry are typically very low in final concentrations in the range of μM to nM. Researchers usually obtain excess materials for standard preparation, which are easily handled and weighed by conventional glassware and balances. Therefore, making concentrated stock solutions for chemical characterization and further dilution to working solutions is a practical way to handle and preserve the purity of the lipid for extended future use.

3. Chemical Characterization of Lipid Stocks

The absence of existing reference materials for comparative analysis dictates sufficient chemical analysis to verify theoretical structure and purity. The analytical techniques to evaluate these aspects include well-established and reported methods of lipid analysis, such as thin-layer chromatography, high performance liquid chromatography, gas chromatography, nuclear magnetic resonance, ultraviolet (UV) spectrophotometry, and atomic absorption. Stock solutions should have concentration determination by an empirical method such as phosphorus and nitrogen content or fatty acid content by gas chromatography. These techniques are essential and should not be omitted. Furthermore, focus on the use of mass spectrometry to evaluate lipids as standards should be initiated only after these techniques have provided satisfactory data.

The nonvolatile nature of most lipids requires them to be assayed by mass spectrometers, which have a liquid interface. Simply described, the lipid compound in solution must be liberated from the solvent phase into the gas phase, which is amenable to ion resolution by the mass spectrometer. Our work has been performed on an API 4000 QTRAP^{TM} (Applied Biosystems), a hybrid, triple-quadrupole, linear ion trap mass spectrometer coupled with a Turbospray^{TM} liquid interface. This interface introduces the lipid molecules to the mass spectrometer source and produces positive and negative ions. Initial qualitative identification is performed by infusing a dilute solution on the order of 1 to 10 μg/ml of the compound into the mass spectrometer with a syringe pump at 5 to 20 μl/min and scanning a molecular weight range suitable to detect the exact mass of its positive or negative ion and possible contaminants and degradants. While this is simple in principle, it is complex in practice. Different lipid classes ionize preferentially to positive or negative ions according to individual molecular structure, solvent system, and mass spectrometer instrument settings. The desired scan spectrum, which contains only simple adducts of the lipid's exact mass, is achieved through trial- and-error adjustments of the mass spectrometers ion source settings. Detection of masses that correspond to related degradation products can often be due to in-source fragmentation of the compound and should not be interpreted immediately as contaminants. Intensity comparison of these ions relative to their precursor over a range of instrument settings often helps determine the ions' origin. One must keep in mind that these scanning experiments are solely for qualitative purposes. Only subjective estimates to the degree of contamination should be assigned from mass spectral scan data. Second, the linear ion trap (LIT) is utilized to confirm the exact mass and isotopic ratios as compared to the theoretical molecular formula of the compound. The LIT's mass accuracy is not on the order of some other mass spectrometer designs, such as time of flight, but is reasonably used for this purpose to one decimal place accuracy for singly charged molecules. The LIT should be programmed to achieve full 0.5 to 1.0 u resolution and scanned over a mass range of $M \pm 5$ to 10 u for comparison of isotopic ratios. Finally, the compound should be analyzed to provide fragmentation data that coincides to its theoretical structure. This is performed with triple quadrupole experiments through mass selection of the exact mass molecular ion and collision-activated

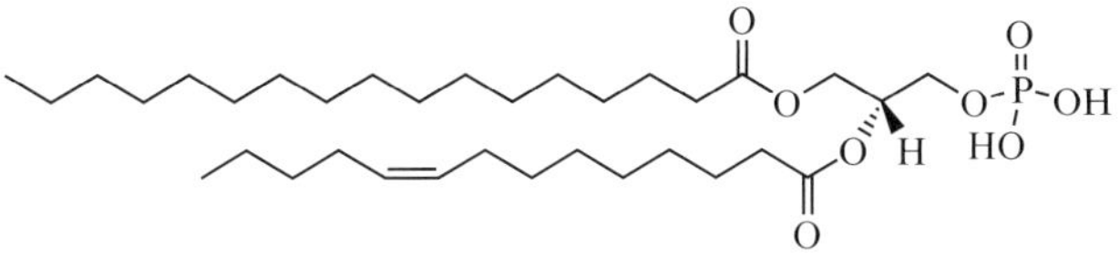

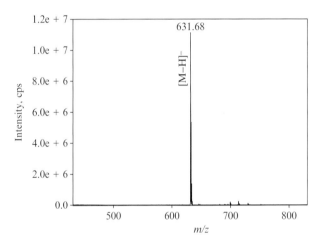

Figure 14.1 Structure of 1-heptadecanoyl-2-(9Z-tetradecenoyl)-sn-glycerol-3-phosphate (17:0–14:1 GPA).

Figure 14.2 Flow infusion of 10 µg/ml of 17:0 to 14:1 GPA at 10 µl/min. Scan used a linear ion trap (LIT) detector [M–H]⁻ = 631.7 u.

dissociation with inert gas and energy to form structurally related fragment ions of the precursor molecule. This collision pattern can be interpreted and is often conformational among lipid classes. (Murphy et al., 2001). Figures 14.1 to 14.4 demonstrate the qualitative mass spectrometry experiments for 1-heptadecanoyl-2-(9Z-tetradecenoyl)-sn-glycero-3-phosphate(17:0–14:1 GPA).

4. Preparation of Working Lipid Standards

Upon satisfactory qualification of molecular structure, purity, and concentration through requisite analytical methods and mass spectrometry, the stock molecule must be diluted to a working concentration practical for direct use in quantitative mass spectrometry methods. Here is where special

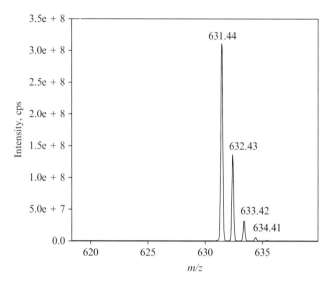

Figure 14.3 Enhanced resolution of 10 μg/ml of 17:0–14:1 GPA using linear ion trap (LIT). Exact mass = 632.44, [M–H]$^-$ = 631.44. Isotopic distribution consistent with [M–H]$^-$ formula of $C_{34}H_{65}O_8P^-$.

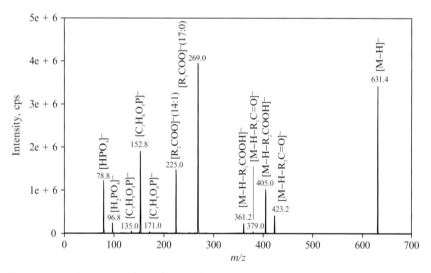

Figure 14.4 Fragmentation of 631.56 (CE = −55 eV) provides molecular structure confirmation.

attention should be given to issues of solvent integrity, solubility, and surface interactions for the working solutions. Solvents used for final preparation should be of the highest quality and recent manufacture. We routinely use high performance liquid chromatography (HPLC)–grade solvents for all lipid standards. Decomposition of aged chloroform to phosgene and hydrochloric acid will immediately destroy lipid compounds at low concentration. We have found that minor contaminants from impure solvents form adducts with polar lipids under certain conditions. An HPLC-grade methanol containing 0.004% formaldehyde produced a corresponding Schiff's base analog in a 10 μg/ml solution of 1-heptadecanoyl-2-(5Z,8Z,11Z,14Z-eicosatetraenoyl)-sn-glycero-3-phosphoethanolamine (17:0–20:4 GPEtn) in a packaged ampoule upon flame sealing at room temperature, as seen in Fig. 14.5. Reformulation with the same source of methanol containing 0.002% formaldehyde along with pre-cooling the solution on dry ice prior to flame sealing reduced the Schiff's base below detection. We routinely filter methanol as a lipid standard diluent through 0.45-μm filters to remove particulates followed by helium sparging to diminish oxygen content and, thus, the extent of lipid oxidation. We discovered that filtering solvents as polar as methanol dissolved nylon filters. Nylon 6.6 was introduced into final lipid solutions, as shown in Fig. 14.6. The use of PTFE filters is recommended.

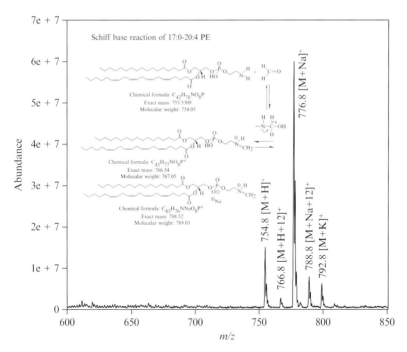

Figure 14.5 Presence of Schiff's base adduct of 17:0 to 20:4 GPEtn at 766.8 and 788.8 u consistent with formation from formaldehyde present in high performance liquid chromatography (HPLC) methanol upon heating.

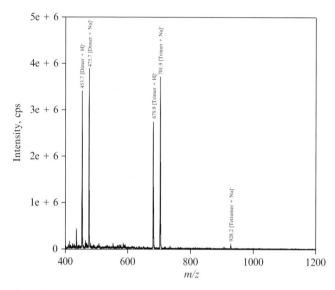

Figure 14.6 Nylon 6.6 mass ions in positive mode from methanol filtered through a 0.45-mm nylon filter.

When possible, we use simple solvents to dissolve lipids at low concentrations. We have used methanol to make working mass spectrometry standards for the phospholipid classes; phosphatidylcholine (GPCho), phosphatidylethanolamine (GPEtn), phosphatidylinositiol (GPIns), phosphatidylserine (GPSer), phosphatidic acid (GPA), lyso-phosphatidylcholine (lyso-GPCho), lyso-phosphatidic acid (lyso-GPA), and cardiolipins (CL), as well as multiple sterols. Mixtures of 1:1 toluene:methanol serve as solvent for stocks and mixtures of triglycerides (TG) and diglycerides (DG) ranging in fatty acid composition from myristic acid (14:0) to eicosapentaenoic acid (20:5). In some instances, a compound will not be stable in its optimum solvent. Solvents for acidic and long chain saturated lipids typically contain water or an ionic buffer that contributes to degradation over time. In these instances, it is best to package these compounds as solids in small quantities by lyophilizing them from t-butanol:water or cyclohexane into their final container. This is true of polyphosphoinositides such as phosphatidylinositol-phosphate (PIP), phosphatidylinositol-bisphosphate (PIP_2), and phosphatidylinositol-trisphosphate (PIP_3), as well as KDO_2 Lipid A and dolichols. For ease of use and assay precision, a standard mixture can be formulated to contain multiple compounds within the same class or according to the focus of the mass spectrometric method. The solubility characteristics of the compounds can often be quite diverse. This is true for ceramides and sphingolipids. Sphingosine-1-phosphate, sphinganine-1-phosphate, and ceramide-1-phosphate required addition of 2.75%/v dimethylamine in 98:2 ethanol:dimethylsulfoxide to dissolve these lipids to respective 20 mg/ml stocks. Sphingomyelin, sphingosine, sphinganine, and ceramide were directly soluble in ethanol at 20 mg/ml. Glucosylceramide and lactosylceramide were dissolved in 65:25:4 (v/v) chloroform:methanol:water. Ten compounds within these classes were added together to form an intermediate stock containing 2.5 mM of each in ethanol. This intermediate stock was further diluted to 25 μM in ethanol as the final working solution. Table 14.1 reports the standard solutions formulated according to their respective solvent and concentration. Primary stocks, intermediate stocks, and working stocks of a compound or mixture should be stored at less than $-15°$ in tightly sealed aliquots of sufficient quantity and concentration whereby their purity can be preserved with constant temperature and minimal exposure to the atmosphere.

5. Packaging of Lipid Standards

The working standard solutions of known identity, purity, and concentration now require packaging for storage and reproducible daily use. The best packaging option for working standards is a quantity sufficient for a single experiment or batch sealed into an inert glass ampoule. A common glass used for pure lipid compounds is pharmaceutical type 1 borosilicate glass. Previous studies performed by us have demonstrated degradation of low concentration solutions stored in glass containers within one hour after packaging. This effect was observed with 18:1, 20:4, and 22:6 lyso-GPCho compounds, when 1 ml was stored as chloroform solutions (0.5 mg/ml) in 10-ml sodium borosilicate glass ampoules. Degradation was not observed when the volume of solution or concentration of solution was increased or when the compound was stored as a lyophilized solid. The cause of degradation was hypothesized to be surface interactions between lipid and glass and/or possible generation of HCl from chloroform vapor upon exposure in flame sealing of the glass ampoule. To rectify these issues, amber ampoules made of highly chemical-resistant glass known as Schott Fiolax^{TM} 8414 were tested as above with flame sealing of ampoules partially suspended in a dry ice/acetone bath ($\sim -70°$). No degradation of the test compounds was detectable by HPLC or mass spectrometry.

An automated repeat pipettor with less than 0.1% variance and more than 99.5% accuracy is used to fill ampoules. This allows rapid filling of 100 to 500 ampoules for sealing and use. Upon preparing these quantities, it is advisable to place the empty ampoules on dry ice to fill. By doing so, the vapor pressure of the volatile solvent is decreased, preventing evaporation as well as thermally protecting the lipid standard awaiting sealing.

Table 14.1 Standard solutions according to concentration and solubility

Methanol (10 μg/ml)	Methanol (50–150 μg/ml)	Toluene:methanol 400 μM
12:0-13:0 GPCho	19:0 Cholesterol ester	20:0-18:0 DG
17:0-20:4 GPCho	Cholesterol(d7)	17:1-18:1(C13) DG
21:0-22:6 GPCho	25-hydroxy-cholesterol(d3)	1,3-20:5 DG-d5
17:0-14:1 GPCho	4-β-hydroxy-cholesterol(d7)	1,3-14:0 DG-d5
12:0-13:0 GPEtn	7α-hydroxy-cholesterol(d7)	1,3-15:0 DG-d5
17:0-20:4 GPEtn	7β-hydroxy-cholesterol(d7)	1,3-16:0 DG-d5
21:0-22:6 GPEtn	5,6α-epoxy-cholesterol(d7)	1,3-17:0 DG-d5
17:0-14:1 GPEtn	6α-hydroxycholestanol(d7)	1,3-19:0 DG-d5
12:0-13:0 GPGro	7-oxocholesterol(d7)	1,3-20:0 DG-d5
17:0-20:4 GPGro	Desmosterol(d6)	1,3-20:2 DG-d5
21:0-22:6 GPGro	24,25-epoxy-cholesterol(d6)	1,3-20:4 DG-d5
17:0-14:1 GPGro	24-hydroxy-cholesterol(d6)	1,3-16:1 DG-d5
12:0-13:0 GPSer	**Lyophilized solids**	1,3-18:0 DG-d5
17:0-20:4 GPSer	**100 μg**	1,3-18:1 DG-d5
21:0-22:6 GPSer	17:0-20:4 PI(3)P	1,3-18:2 DG-d5
17:0-14:1 GPSer	17:0-20:4 PI(4)P	16:0-16:0-18:0 TG
12:0-13:0 GPA	17:0-20:4 PI(5)P	16:0-16:0-18:1 TG
17:0-20:4 GPA	17:0-20:4 PI(3,4)P2	20:5-22:6-20:5 TG-d5
21:0-22:6 GPA	17:0-20:4 PI(3,5)P2	14:0-16:1-14:0 TG-d5
17:0-14:1 GPA	17:0-20:4 PI(4,5)P2	15:0-18:1-15:0 TG-d5
12:0-13:0 GPIns	17:0-20:4 PI(3,4,5)P3	16:0-18:0-16:0 TG-d5
17:0-20:4 GPIns	Nor-dolichol-[13-22]	17:0-17:1-17:0 TG-d5
21:0-22:6 GPIns	**1 mg**	19:0-12:0-19:0 TG-d5
17:0-14:1 GPIns	Kdo2lipid A	20:0-20:1-20:0 TG-d5
13:0 lyso-GPCho	**Ethanol**	20:2-18:3-20:2 TG-d5
17:1 lyso-GPCho	**2.5 mM**	20:4-18:2-20:4 TG-d5
13:0 lyso-GPA	Sphingosine (C17)	**4 μM**
17:1 lyso-GPA	Sphinganine (C17)	18:1c9-18:1c6-18:1c9 TG
	Sphingosine-1-PO4 (C17)	18:1c6-18:1c9-18:1c6 TG
100 μg/ml	Sphinganine-1-PO4 (C17)	d5-TG mixture of 9 cmpds
24:1(3)-14:1 CL	Ceramide (C12)	d5-DG mixture I of 9 cmpds
14:1(3)-15:1 CL	Ceramide (C25)	d5-DG mixture II of 4 cmpds
15:0(3)-16:1 CL	Ceramide-1-PO4 (C12)	
22:1(3)-14:1 CL	Sphingomyelin (C12)	
CL Mixture I of 4 cmpds	Glucosyl(β) C12 ceramide	
	Lactosyl (β) C12 ceramide **25 μM** Cer/Sph mixture of 10 cmpds	

6. Quality Control and Stability Testing

Confidence in use requires the researcher to know that standards do not change concentration upon prolonged storage. Measurement of standard compounds' concentrations packaged at micromolar concentrations require sensitive detection afforded by liquid chromatography tandem mass spectrometry methods. The multireaction monitoring (MRM) technique routinely utilized in quantitative mass spectrometry analysis of small molecules is incorporated upon fast resolution of compounds by liquid chromatography. We provide methods of quantitative measurement for two internal standard mixtures by HPLC-MS/MS.

6.1. Ceramide/sphingoid base internal standard mixture

As described above, a mixture of ceramides and sphingoid bases was prepared to contain 10 lipids each at 25 μM in ethanol and packaged in ampoules for routine use as an internal standard in sphingolipid mass spectrometry methods (Merrill et al., 2005). The lipids were either odd carbon or non-endogenous 12 carbon acids on the d18 backbone of ceramide and of the sphingoid base. The mixture contained (2S,3R,4E)-2-aminoheptadec-4-ene-1,3-diol (C17 So), (2S,3R)-2-aminoheptadecane-1,3-diol (C17 Sa), heptadecasphing-4-enine-1-phosphate (C17 So-1-P), heptadecasphinganine-1-phosphate (C17 Sa-1-P), N-(dodecanoyl)-sphing-4-enine (C12 Cer), N-(pentacosanoyl)-sphing-4-enine (C25 Cer), N-(dodecanoyl)-sphing-4-enine-1-phosphate (C12 Cer-1-P), N-(dodecanoyl)-sphing-4-enine-1-phosphocholine (C12 SM), N-(dodecanoyl)-1-β-glucosyl-sphing-4-eine (C12 GlucCer), and N-(dodecanoyl)1-β-lactosyl-sphing-4-eine (C12 Lac-Cer). The mixed solution from the ampoules and the intermediate stock are volumetrically diluted to an equivalent concentration (25 μM) just prior to analysis by HPLC-MS/MS. For each experiment, three ampoules are assayed with triplicate injections of each. The %RSD for injections is routinely less than 5% for each compound. The HPLCMS/MS conditions are provided in Table 14.2. The MRM transitions for components of this mixture were assigned by experimentation with individual standards and mass spectrometer collisional settings with guidance from the work of Sullards et al. (2003). The [M + H]$^+$ ion of C17 So and C17 Sa was monitored in Q1 and fragmented to selective detection of their respective characteristic fragments of 268.2 and 270.2 u for quantitation. The [M + H]$^+$ ion of C12 SM was selectively monitored by detection of its characteristic 184 u phosphocholine fragment. The d18:1 ceramides of C12 Cer, C12 GlucCer, and C12 LacCer were monitored through the collision of their [M + H]$^+$ ion to the characteristic ceramide base ion of 264.2 u. Unexpected was the optimum MRM for C25 ceramide as the loss of water from the parent [M + H]$^+$ of 664.8 to

Table 14.2 Liquid chromatography tandem mass spectrometry condition for the assay of ceramide/sphingoid base internal standard mixture
HPLC-MS/MS
Column: Mercury C18, 20 × 2 mm, 3 μ, Temp: 30° (Phenomenon, Torrance, CA)
Mobile phase A: 60:40:0.2 (v/v) methanol:water:acetic acid + 10 mM ammonium acetate
Mobile phase B: 60:40:0.2 methanol:chloroform:acetic acid + 10 mM ammonium acetate

Total time (min)	Flow rate (μl/min)	A (%)	B (%)
Equilibrate			
5.0	600	100	0
3.0	600	87	13
4.0	600	60	40
5.5	600	50	50
6.5	600	30	70
10.0	600	30	70

Mass spectrometer settings	
CUR:	10
IS:	5500
TEM:	550°
GS1:	14
GS2:	0
ihe:	ON
CAD:	Medium
EP:	10

Compound	Q1	Q3	DP	CE	CXP
C17 So	286.4	268.2	36	17	16
C17 Sa	288.4	270.2	66	21	14
C17 So-1-P	366.3	250.2	56	25	14
C17 Sa-1-P	368.4	252.3	71	19	6
C12 SM	647.7	184.1	121	35	10
C12 Cer	482.6	264.2	66	33	14
C12 GlucCer	644.7	264.5	66	45	14
C12 LacCer	806.7	264.4	96	57	16
C12 Cer-1-P	562.6	264.5	91	6	6
C25 Cer	664.8	646.7	81	23	18

CAD, collision activation; CE, collision energy; CUR, curtain gas; CXP, Q2 exit potential; DP, declustering potential; Dwell, 150 msec for each compound; EP, exit potential; GS1, nitrogen; GS2, air; ihe, interface heater; IS, ion spray voltage; Q1, quadrupole 1 ion; Q3, quadrupole 3 fragment; TEM, interface temperature.

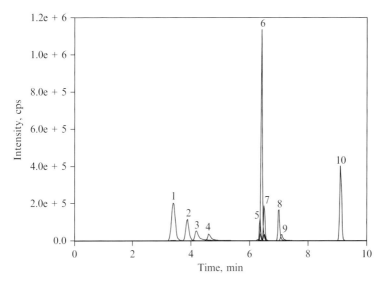

Figure 14.7 Liquid chromatographic tandem mass spectrometry (LC-MS/MS) of Ceramide/Sphingoid base internal standard mixture. 1 = C17 So, 2 = C17 Sa, 3 = C17 So-1-P, 4 = C17 Sa-1-P, 5 = C12 LacCer, 6 = C12 SM, 7 = C12 GlucCer, 8 = C12 Cer, 9 = C12 Cer-1-P, and 10 = C25 Cer. High performance liquid chromatography (HPLC) and mass spectrometer settings are provided in Table 14.2.

646.7 u rather than 264.2 u. The phosphorylated sphingoid bases of C17 So-1-P and C17 Sa-1-P were monitored through the collisional activation of their $[M + H]^+$ ion to their respective 250.2 and 252.2 u fragments. These ampoules were assayed upon packaging and on a schedule of 0.5, 1, 3, 5, 7, 9, and 12 months of storage at -16 to $-24°$. Current analysis of ampoules stored over a 5-month period produced concentration between 22.5 and 27.5 μM for all 10 compounds. The accurate quantitation of the long chain saturated compound C25 Ceramide requires complete re-dissolution into the solvent for both ampoules and intermediate stock when stored at freezer temperatures. This can be accomplished by allowing the solution to reach room temperature equilibrium followed by bath sonication at ambient temperature for 5 to 15 min. A representative LC-MS/MS chromatogram is provided in Fig. 14.7.

6.2. Triglyceride-d5 internal standard mixture

Another mixture containing nine triglycerides deuterated at the position of their glycerol backbone protons with positionally defined acyl groups was formulated to contain 4 μM of each in 1:1 (v/v) toluene:methanol as a readily usable internal standard mixture for triglyceride MS (McAnoy et al., 2005). The mixture contained 1,3-di-(5Z,8Z,11Z,14Z,17Z-eicosapentaenoyl)-2-(4Z,7Z,10Z,13Z,16Z,19Z-docosahexaenoyl)-sn-glycerol-d5

(20:5(2)-22:6 TG-d5), 1,3-ditetradecanoyl-2-(9Z-hexadecenoyl)-sn-glycerol-d5 (14:0(2)-16:1TG-d5), 1,3-dipentadecanoyl-2-(9Z-octadecenoyl)-sn-glycerol-d5 (15:0(2)-18:1 TG-d5), 1,3-dihexadecanoyl-2-octadecanoyl-sn-glycerol-d5 (16:0(2)-18:0 TG-d5), 1,3-diheptadecanoyl-2-(10Z-heptadecenoyl)-sn-glycerol-d5 (17:0(2)-17:1 TG-d5), 1,3-dinonadecanoyl-2-dodecanoyl-sn-glycerol-d5 (19:0(2)-12:0 TG-d5), 1,3-dieicosanoyl-2-(11Z-eicosenoyl)-sn-glycerol-d5 (20:0(2)-20:1 TG-d5), 1,3-di-(11Z,14Z-eicosadienoyl)-2-(6Z,9Z,12Z-octadecatrienoyl)-sn-glycerol-d5 (20:2(2)-18:3 TG-d5), and 1,3-di-(5Z,8Z,11Z,14Z-eicosatetraenoyl)-2-(9Z, 12Z-octadecadienoyl)-sn-glycerol-d5 (20:4(2)-18:2 TG-d5). The mixed solution from the ampoules was assayed against the mixture's intermediate stock at 400 μM, and manually diluted to an equivalent concentration just prior to analysis. For each experiment, three ampoules were assayed with triplicate injections of each. The %RSD for injections is routinely less than 3% for each compound. The HPLC-MS/MS conditions are provided in Table 14.3. The [M + NH$_4$]$^+$ ion of each d5-TG was selected in Q1 and fragmented for detection of its corresponding [M-R$_1$COOH]$^+$ fragment for quantitation. Since each d5-TG was formulated to equimolar concentrations, the dwell time for each was adjusted to achieve near equivalent responses across the compounds' molecular weight distribution in the MRM experiment. A higher dwell time was required to compensate for low ionization/detection of the saturated, long-chain versus polyunsaturated lipids. These ampoules were assayed upon packaging and on a schedule of 0.5, 1, 3, 5, 7, 9, and 12 months of storage at -16 to $-24°$. Current analysis of ampoules stored over a 7-month period produced concentration between 3.6 and 4.4 μM for all nine compounds. A representative HPLC-MS/MS chromatogram is provided in Fig. 14.8.

7. Discussion

The lipid compounds at the specified concentrations listed in Table 14.1 have been prepared as 1-ml solutions packaged in ampoules as described. These compounds have been furnished to the operational cores of the LIPID MAPS glue grant. LIPID MAPS is funded by National Institutes of Health (NIH)/National Institute of Genetic Medical Sciences (NIGMS), Grant 1 U54 GM69338. Each operational core is assigned to measure all lipids within a specified class of the lipidome in the mouse tumor–derived macrophage-like RAW 246.7 cell. In general, measurement of lipids are to be assessed from RAW cell extracts pre- and post-stimulation with KDO$_2$ lipid A, also manufactured, characterized, and packaged by Avanti Polar Lipids (Raetz *et al.*, 2006). Their work is reported in other

Table 14.3 Liquid chromatography tandem mass spectrometry conditions for the assay of d5-TG internal standard mixture
HPLC-MS/MS Conditions
Column: Mercury C8, 20 × 2 mm, 3 μ (Phenomenex, Torrance, CA)
Mobile phase A: 10 m*M*ammonium acetate in 1:9 (v/v) methanol:water
Mobile phase B: 10 m*M* ammonium acetate in methanol

Total time (min)	Flow rate (μl/min)	A (%)	B (%)
0	200	40	60
5	200	40	60

Mass spectrometer settings	
CUR:	10
IS:	5500
TEM:	550
GS1:	50
GS2:	50
ihe:	ON
CAD:	Medium
EP:	10

TG-d5	Q1 mass (amu)	Q3 mass (amu)	Dwell (msec)	Param	
20:5(2)−22:6	993.8	674.4	10	CE	37
14:0(2)−16:1	771.8	526.3	100	CE	33
15:0(2)−18:1	827.8	568.3	300	CE	33
19:0(2)−12:0	857.8	542.4	400	CE	35
17:0(2)−17:1	869.8	582.4	400	CE	41
16:0(2)−18:0	857.8	584.5	450	CE	35
20:0(2)−20:1	996.0	666.5	900	CE	41
20:2(2)−18:3	955.8	630.4	150	CE	43
20:4(2)18:2	949.8	628.4	15	CE	35

CAD, collision activation; CE, collision energy; CUR, curtain gas; CXP, Q2 exit potential; DP, declustering potential; Dwell, 150 msec for each compound; EP, exit potential; GS1, nitrogen; GS2, air; ihe, interface heater; IS, ion spray voltage; Q1, quadrupole 1 ion; Q3, quadrupole 3 fragment; TEM, interface temperature.

chapters of this volume. Each core was supplied with 50 to 100 1-ml ampoules of each lipid internal standard of their respective class assignment. The principle of numerous ampoules is to provide a one-experiment use whereby the ampoule is opened and used directly or further diluted and added to cells prior to extraction. This single-use approach circumvents many, if not all, of the physical and chemical constraints of lipid degradation and interactions. Accompanying each shipment is a lot-specific certificate of

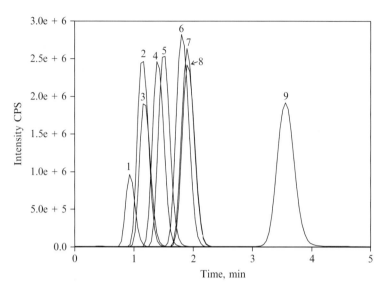

Figure 14.8 Liquid chromatographic tandem mass spectrometry (LC-MS/MS) of triglyceride-d5 internal standard mixture. 1 = 20:5(2)-22:6 TG-d5, 2 = 14:0(2)-16:0 TG-d5, 3 = 20:4(2)-18:2 TG-d5, 4 = 20:2(2)-18:3 TG-d5, 5 = 15:0(2)-18:1 TG-d5, 6 = 17:0(2)-17:1 TG-d5, 7 = 19:0(2)-12:0 TG-d5, 8 = 16:0(2)-18:0 TG-d5, and 9 = 20:0(2)-20:1 TG-d5. High performance liquid chromatography (HPLC) and mass spectrometer settings are provided in Table 14.3.

analysis reporting the methods and results for chemical characterization and concentration determination. Additionally, each lot of ampoules is assayed on a scheduled basis for concentration stability within a specification of 100 ± 5% for individual components and 100 ± 10% for mixtures. Should a lot of internal standard exceed these specifications, the core is notified to discontinue use and a new lot is prepared. Personal communications on the quality, utility, and consistency of the provided materials have been favorable. We have also provided these materials to the general public as a service of the grant. Continued and repeated requests suggest acceptable performance in the hands of researchers.

REFERENCES

McAnoy, A. M., Wu, C. C., and Murphy, R. C. (2005). Direct qualitative analysis of triacylglycerols by electrospray mass spectrometry using a linear ion trap. *J. Am. Soc. Mass Spectrom.* **16,** 1498–1509.

Merrill, A. H., Jr., Sullards, M. C., Allegood, J. C., Kelly, S., and Wang, E. (2005). Sphingolipidomics: High-throughput, structure specific and quantitative analysis of sphingolipids by liquid chromatography tandem mass spectrometry. *Methods* **36,** 207–224.

Murphy, R. C., Fiedler, J., and Hevko, J. (2001). Analysis of nonvolatile lipids by mass spectrometry. *Chem. Rev.* **101,** 479–526.

Pidgeon, C., Apostol, G., and Markovich, R. (1989). Fourier transform infrared assay of liposomal lipids. *Anal. Biochem.* **181,** 28–32.

Raetz, C. R. H., Garrett, T. A., Reynolds, C. M., Shaw, W. A., Moore, J. D., Smith, D. C., Jr., Ribeiro, A. A., Murphy, R. C., Ulevitch, R. J., Fearns, C., Reichart, D., Glass, C. K., et al. (2006). Kdo$_2$-Lipid A of *Escherichia coli*, a defined endotoxin that activates macrophages via TLR-4. *J. Lipid Res.* **47,** 1097–1111.

Sullards, M. C., Wang, E., Peng, Q., and Merrill, A. H., Jr. (2003). Metabolomic profiling of sphingolipids in human glioma cell lines by liquid chromatography tandem mass spectrometry. *Cell Mol. Biol.* **49,** 789–797.

Author Index

A

Abe, A., 87
Abidi, S. L., 147, 148
Adams, J., 90
Aden, Y., 166
Aebi, M., 130, 136
Ahn, S. J., 5
Aho, T., 341
Albanesi, J. P., 41
Albe, K. R., 319
Aliberti, J., 277, 280, 284, 292, 296, 298
Allegood, J. C., 83, 85, 86, 91, 95, 100, 102, 179, 189, 215, 361
Allen, M., 216
Almeida, J., 331
Almeida, R., 216
Alonso, A., 220, 221, 222, 223
Alter, C. A., 40
Alvarez-Vasquez, F., 227, 317, 319, 321, 323, 325, 341, 343
Alves, R., 341
Amado, F. M., 190
Anderson, J. R., 333
Anderson, K. E., 234
Anderson, K. I., 219
Anderson, V. E., 119
Andreyev, A., 61, 72, 74, 180, 268
Angeletti, R. H., 47
Angelova, M. I., 221
Anghel, V., 222
Ann, Q., 90
Annangudi, S. P., 207
Apostol, G., 353
Apweiler, R., 47, 255
Aranda, F. J., 220, 221
Ariel, A., 298
Arita, M., 276, 277, 278, 279, 280, 292, 296, 297, 298
Arkin, A. P., 326, 329
Armstrong, M. D., 31
Arrondo, J. L., 222
Asagami, C., 90
Ascoli, I., 29
Ashida, H., 223
Aubagnac, J. L., 208
Auriola, S., 148, 168
Ausloos, P., 280, 282
Axelrod, D., 221

B

Bacher, A., 118
Badwey, J. A., 288
Baer, T., 282, 292, 297, 298, 306, 308, 312
Bagatolli, L. A., 221, 222
Bahr, U., 110, 187, 192, 200, 216
Baird, B., 23, 190
Bairoch, A., 255
Baker, J. A., 119
Baker, P. R. S., 23
Bannenberg, G., 276, 278, 279, 296
Bannenberg, G. L., 298
Bansal, V. S., 39
Baranowski, R., 60
Barker, W. C., 255
Barkley, R. M., 1, 60, 215, 305
Barylko, B., 41
Basáñez, G., 220, 221
Batista, E. L., Jr., 279
Batoy, S. M., 110
Bayindirli, A., 334
Bazan, N. G., 282, 297, 298, 301
Beale, M. H., 172
Beff, W. E., 2, 8
Behbehani, B., 276, 278, 279, 284, 288, 296
Beletskaya, I., 214
Bendiak, B., 111
Benjamin, T. L., 40
Bergström, S., 283
Berkenkamp, S., 217
Berrie, C. P., 40, 41
Bertilsson, L., 166
Bertini, I., 214
Bertisson, L., 166
Berzat, A. C., 223
Bianchini, F., 277, 280, 292, 298
Bielawska, A., 91, 334, 335, 337
Bielawski, J., 91
Bieniek, E., 283
Bier, J., 283
Binder, M., 216, 224
Bindila, L., 110, 216
Bino, R. J., 172
Björkhem, I., 165
Blackburn, G. M., 214
Bleijerveld, O. B., 208
Bligh, E. G., 29, 122, 149, 151
Bloor, W. R., 29

Bodin, K., 166
Bodker, M., 298
Boeckmann, B., 255
Bogdanov, A., 214
Bogdanov, M., 22
Bohn, T., 148, 168
Bohrer, A., 31, 189
Bolouri, H., 326, 329
Boned, A., 222
Bonner, R., 6
Bonventre, J. V., 298
Borlak, J., 214
Borner, K., 111
Bornstein, B. J., 326, 329
Borst, J., 223
Boschung, M., 223
Boselli, E., 91
Bothorel, P., 221
Botkjaer, A., 298
Botting, R. M., 60
Bowers-Gentry, R., 59, 61
Box, G. E. P., 345
Boxer, S. G., 111
Brady, D. C., 223
Brain, A. P., 221
Brannon, J., 277, 278, 279, 280, 282, 284, 288, 296, 297, 298
Bray, D., 326, 329
Brenna, J. T., 165
Bretillon, L., 166
Breuer, O., 165, 166
Brock, T. G., 60
Brodbeck, U., 223
Broome, U., 166
Browdy, A., 334
Brown, A., 175, 252
Brown, D. A., 222
Brown, H. A., 21, 25, 27, 31, 38, 40, 45, 86, 118, 126, 158, 172, 175, 176, 180, 190, 214, 215, 224, 225, 250
Brugger, B., 186, 191
Brügger, B., 23, 216
Brunelle, A., 111
Brunsveld, L., 119
Buczynski, M. W., 59, 61
Budnik, B. A., 110, 217
Bui, H., 208
Burda, P., 130, 136
Burkard, I., 148, 165, 168
Burns, J. A., 321
Burns, J. L., 220, 221
Burr, G. O., 297
Burr, M. M., 297
Burton, L., 31, 110, 216
Busman, M., 91
Bütikofer, P., 223
Butler, M. H., 319
Byrdwell, W. C., 2, 8

Byrne, M. O., 21, 27
Byrnes, R., 247

C

Caffrey, M., 220, 251
Calaminus, S. D. J., 41, 237
Calder, P. C., 279
Caldwell, K. K., 39
Callaway, S., 223
Callender, H. L., 126
Camilli, P. D., 282
Capdevila, J. H., 277
Carman, G. M., 329, 334, 336, 337, 338
Carpenter, B. K., 165
Carroll, K. K., 130
Carter, K. C., 119
Cascante, M., 321, 328
Casserly, E. W., 165
Catz, I., 217
Caufield, W. V., 351
Ceballos, C., 223
Cerda, B. A., 31, 190
Cerdan, R., 208
Chaminade, P., 111
Chan, L., 276, 278, 279, 284, 288, 296
Chaudhary, N., 130
Cheah, S., 336
Chen, Y., 83
Cheng, C., 8, 305
Cheng, H., 109
Chernushevich, I. V., 31, 110, 187, 191, 192, 200, 216
Chester, M. A., 88, 251
Chestnut, M. H., 220, 221
Cheung, K., 47
Chiang, N., 277, 278, 279, 280, 282, 284, 288, 292, 296, 297, 298, 301, 308
Chiang, T. B., 148, 165
Chiba, S., 61
Chiem, N. H., 147
Chojnacki, T., 119, 130
Chou, I. C., 341
Christie, W. W., 28, 175
Churchill, M. E., 126
Chuuard, R., 331
Ciruela, A., 40
Clark, G. T., 241
Clark, J., 255
Clarke, N. G., 39
Clay, K. L., 126
Cleland, W. W., 339, 341
Clifton, C. L., 280, 282
Cline, G. W., 40, 211, 282
Clish, C. B., 276, 277, 278, 279, 280, 282, 284, 286, 288, 296, 297, 298, 308
Clowers, B. H., 111
Cohen, J. C., 147, 151

Cohen, J. S., 31, 190
Colgan, S. P., 276, 277, 278, 279, 280, 282, 284, 288, 292, 296, 297, 298, 301, 306, 308, 312
Collado, M. I., 220, 221, 222
Collins, E. J., 15
Conchonaud, F., 222
Contreras, F. X., 221
Conzelmann, A., 333
Cooks, R. G., 111, 219
Corda, D., 40, 41
Cornhill, J. F., 279
Cornish-Bowden, A., 326, 329
Costa, J., 5
Costello, C. E., 47, 110, 217
Cotter, D., 175, 225, 247, 252, 254
Covey, D. F., 151
Cowart, L. A., 227, 319, 321, 323, 325, 341, 343
Cox, A. D., 223
Cremesti, A. E., 336
Cribier, S., 221
Cuellar, A. A., 326, 329
Cui, J. G., 298
Cui, Z., 23, 208
Cullis, P. R., 220
Cunin, R., 214
Cunningham, T. W., 39
Curto, R., 321, 328

D

Dahlén, S. E., 277
Dallner, G., 119, 130
Danikiewiczb, W., 119
Darrow, R., 207
David, J. H., 92
Davies, S. S., 207
Dawson, G., 89
Dawson, R. M. C., 39
DeCamilli, P., 40
De Camilli, P., 211
DeCamp, D., 38, 40
Deems, R., 59, 61, 172
Degand, P., 136
DeJong, C. J., 23
de Kruijff, B., 130, 220
Delacour, D., 136
Delannoy, P., 136
Demeester, J., 221
Dennis, E. A., 59, 60, 61, 72, 74, 171, 172, 175, 180, 252, 268
De Smedt, S. C., 221
Deutsch, E. W., 47
Devaux, P. F., 221
Devchand, P. R., 276, 278, 279, 280, 282, 284, 286, 287, 288, 296, 297, 298, 301, 312
DeWald, D. B., 40
DeWitt, D. L., 60
DeWulf, D. W., 6

Dickens, M. J., 221
Dickson, R. C., 340
Diczfalusy, U., 147, 151, 152, 164, 166
Dietschy, J. M., 146
Dignass, A., 278
DiPaolo, G., 40
Di Paolo, G., 211
Dishman, E., 277
Divecha, N., 234
Domae, N., 338
Domingues, M. R., 190
Domingues, P., 190
Dove, S. K., 40, 41, 237
Dowhan, W., 22
Downes, C. P., 40
Doyle, J. C., 326, 329
Dragani, L. K., 41
Draper, J., 172
Draper, N. R., 346
Dreisewerd, K., 217
Drobnik, W., 216
Dronov, S., 326, 329
Duan, D. R., 119
Duchoslav, E., 1, 6, 110, 216
Duffield, J. S., 298
Duffin, K. L., 60, 176, 179
Duncan, L. A., 5
Duncan, M. W., 162
Dunn, M., 298
Dunn, M. J., 249
Dwivedi, P., 111
Dyer, W. J., 5, 29, 122, 149, 151
Dzeletovic, S., 166

E

Eagles, J., 26
Earnest, S., 41
Ebrahimi, N., 298
Edelstein-Keshet, L., 345
Edwards, K., 220
Edwards, P. A., 119
Efremenko, E., 214
Egan, T., 110
Eggens, I., 130
Eichler, J., 214
Einarsson, C., 166
Eisenreich, W., 118
Ejsing, C. S., 6, 31, 110, 187, 192, 200, 216, 219
Ekblad, L., 40
Ekroos, K., 6, 31, 110, 187, 191, 192, 200, 216, 219
Eldho, N. V., 217
Eligh, E. G., 5
Elkin, Y. N., 110, 217
Ellson, C. D., 234
Elmberger, P. G., 130
Elson, E., 221

Emanuel, J. R., 5
Emmons, G. T., 165
Eng, J. K., 47
Enjalbal, C., 208
Entchev, E., 31, 110, 216
Epand, R. M., 222
Epps, D. E., 5
Erben, G., 23, 186, 187, 191, 216
Erdahl, W. L., 2
Ericsson, J., 119
Ericsson, L. H., 31, 187
Ertugay, N., 334
Evans, J. J., 15

F

Faergemen, N. J., 333
Fahy, E., 25, 61, 72, 74, 86, 118, 158, 171, 173, 175, 176, 177, 180, 214, 215, 224, 225, 247, 250, 252, 268
Failla, M. L., 148, 168
Fain, N., 223
Falck, J. R., 207, 277
Faucon, J. F., 221
Fearns, C., 2, 121, 122, 174, 364
Febbraio, M., 207
Fell, D., 345
Feng, S., 41
Fenn, J. B., 26, 148, 187
Fenwich, G. R., 26
Ferguson, G. J., 234
Fernandez, F., 119, 136
Ferreira, A., 321, 341
Ferrer-Correia, A. J., 190
Ferro, S., 255
Feurle, J., 41
Fiedler, J., 26, 31, 262, 280, 282, 284, 287, 288, 296, 301, 355
Fiehn, O., 172
Finnemann, S. C., 207
Finney, A., 326, 329
Finnigan, M., 294
Fiordalisi, J. J., 223
Fischl, A. S., 329, 334, 336, 337, 338
Fitzpatrick, F. A., 80
Fleming, R. C., 110
Florence, C., 119
Folch, J., 29
Forrester, J. S., 27, 31, 45, 126, 180
Fredman, G., 298
Fredman, P., 110
Freed, J. H., 221
Freeman, M. W., 278, 296
Frega, N. G., 91
Fridriksson, E. K., 23, 190
Friis, J., 333
Froesch, M., 110
Fujiwaki, T., 91
Funk, C. D., 60, 277

G

Gaigg, B., 333
Galaev, I., 214
Gale, P. J., 162
Gan, Y. D., 37, 38
Garavito, R. M., 60
Garret, J., 208
Garrett, T. A., 2, 117, 121, 122, 174, 364
Gasteiger, E., 255
Gatta, A., 60
Gaus, K., 221, 222
Gawrisch, K., 217, 297
Gelb, M. H., 119
Gelfand, E., 148, 153, 165
Germain, D. P., 111
German, J. B., 178
Giles, R., 222
Gilmore, R., 121, 140
Gimenez, J. M., 282, 298, 301
Gladyshev, V., 214
Glass, C. K., 2, 25, 86, 118, 158, 172, 174, 175, 176, 214, 215, 224, 225, 250, 252, 364
Goel, G., 341
Goñi, F. M., 213, 220, 221, 222, 223
Gorga, F. R., 40
Gotlinger, K. H., 279, 282, 292, 297, 298, 300, 304, 305, 306, 308, 309, 310, 312
Goto, S., 255
Goto-Inoue, N., 92
Gould, T. A., 126
Gouyer, V., 136
Gray, A., 40, 234
Greene, J. M., 119
Griffiths, W., 214
Griffiths, W. J., 179, 282, 287, 288, 296
Grilley, M. M., 340
Grimm, H., 283
Grimminger, F., 283
Grkovich, A., 61
Gronert, K., 276, 277, 278, 279, 280, 282, 284, 286, 287, 288, 296, 297, 298, 301, 312
Gross, M. L., 8, 291, 305
Gross, R. W., 2, 23, 31, 109, 179, 180, 186, 187, 191, 192, 200, 215
Grundy, S. M., 147
Guages, R., 341
Guan, Y., 190
Guan, Z., 117
Guda, C., 225, 254
Guevara, N. V., 279
Guillas, I., 333
Guillou, H., 234
Gulik-Krzywicki, T., 220
Gunnarsson, T., 40
Günther, A., 283
Guo, X. J., 222
Gutsche, B., 41
Gu Yang, S., 190

H

Hagenhoff, B., 111
Hail, M., 187
Haimi, P., 216, 224
Hall, L. M., 61, 63, 207
Hall, R. D., 172
Hama, H., 40
Hamamci, H., 334
Hamanaka, S., 90
Hammarstrom, S., 89
Han, X., 2, 23, 31, 109, 179, 180, 186, 187, 191, 192, 200, 215
Handa, S., 91
Hankin, J., 60, 215, 305
Hannich, J. T., 31, 110, 216
Hannun, Y. A., 91, 227, 319, 321, 323, 325, 336, 341, 343
Hansen, J. K., 333
Hara, A., 90
Harada, A., 187, 189, 190
Harden, C. S., 31
Harder, T., 221, 222
Hardy, M., 282, 298, 301
Harkewicz, R., 59, 61, 72, 74, 180, 268
Haroldsen, P. E., 126
Harris, M. A., 255
Harrison, K. A., 26, 60, 207, 215, 305
Hassan, I. R., 298
Hasturk, H., 288, 298
Hawkins, P. T., 234
Hayakawa, J., 24, 190
Hayashi, A., 90
Hayashi, K., 336
Haynes, C. A., 83
Hazen, S. L., 207
He, C., 278
He, H. T., 222
Heath, V. L., 40
Hedl, M., 118
Heikinheimo, L., 216
Heinrich, R., 321
Helenius, A., 136, 220
Helms, B., 214, 227
Hendrickson, C. L., 110
Henion, J. D., 176, 179
Henneking, K., 283
Herderich, M., 41, 279
Herman, J., 126
Hermansson, M., 180, 224
Herrmann, A., 221
Hessler, E., 80
Hevko, J., 26, 31, 262, 280, 282, 284, 287, 288, 296, 301, 355
Hieny, S., 284, 296
Hilgemann, D. W., 41
Hill, H. H., Jr., 111
Hillenkamp, F., 217
Hinchliffe, K. A., 40

Hirschmann, H., 25
Hishinuma, T., 61
Hitchingham, L., 80
Hla, T., 60
Ho, P. P., 217
Hobbs, H. H., 147, 151
Hoetzl, S., 218, 219, 222
Hogan, J., 251
Holowka, D., 23, 190
Hong, S., 275, 276, 277, 278, 279, 280, 282, 284, 285, 286, 287, 288, 292, 295, 296, 297, 298, 300, 301, 304, 305, 306, 308, 309, 310, 312, 313
Hoops, S., 341
Hoppe, G., 207
Hoppel, C. L., 6, 24
Horie, K., 60
Horn, D. M., 31, 190
Houjou, T., 187, 189, 190, 191, 192, 200, 201, 237
Houweling, M., 208
Hrycyna, C. A., 119
Hsu, F. F., 2, 31, 33, 179, 189
Hsueh, R. C., 38, 40
Hsueh, Y. W., 222
Huang, H., 255
Hua Tian, X., 282, 298, 301
Hubbard, S. C., 119
Hubley, R., 47
Hucka, M., 326, 329
Hudert, C. A., 278
Huet, G., 136
Hulan, H. W., 91
Hunt, A. N., 241
Hunter, E., 223
Hutcheon, I. D., 111
Hyde, J. S., 222

I

Ieda, J., 92
Ifa, D. R., 111
Ikonen, E., 222
Imagawa, M., 187, 189, 190, 191, 192, 200, 201, 237
Imperiali, B., 136
Inagaki, F., 90
Ingalls, S. T., 6
Inoue, K., 187, 189, 190
Ipsen, J. H., 222
Ireland, A., 255
Irvine, R. F., 40
Isakson, P. C., 60
Ishida, M., 24, 187, 189, 190, 191, 192, 200, 201, 237
Isobe, K. O., 147
Ito, E., 92

IUPAC-IUB Commission on Biochemical
  Nomenclature, 25
Iurisci, C., 40
Ivanova, P. T., 21, 27, 31, 38, 40, 45,
  126, 180, 190
Ivatt, R. J., 119
Ivleva, V. B., 110, 217
Iwabuchi, K., 91
Izgarian, N., 208

J

Jackson, S. N., 110, 111
Jacobsen, P. B., 31
Jacobson, K., 222
Jain, A., 276, 278, 279, 284, 288, 296
Jazwinski, S. M., 333
Jensen, N. A., 333
Jergil, B., 40
Jiang, J. C., 333
Johansson, B., 111, 208, 219
Johnson, C., 60, 215, 305
Jomaa, H., 41
Jones, J. J., 110
Jorde, L. B., 249
Joris, I., 279

K

Kacser, H., 321
Kaga, N., 91
Käkelä, R., 180, 189
Kaluzny, M. A., 5
Kanehisa, M., 255
Kang, J. X., 278
Kanjilal-Kolar, S., 140
Kantarci, A., 276, 278, 279, 284, 288, 296, 298
Kanter, J. L., 217
Karaoglu, D., 121, 140
Karara, A., 277
Karas, M., 110, 187, 192, 200, 216
Karlsson, A., 40
Karlsson, G., 220
Karlstrom, G., 222
Katajamaa, M., 267
Kato, Y., 91
Katsaras, J., 222
Kawasaki, K., 222
Kazuno, S., 91
Kean, E. L., 119
Kelleher, D. J., 121, 140
Keller, S. L., 222
Kellogg, B. A., 119
Kelly, S., 83, 85, 86, 91, 95, 100, 102, 179, 189,
  215, 361
Keppler, D., 187
Kerwin, J. L., 31, 187

Khan, N., 298
Khaselev, N., 207
Kihara, T., 333
Kim, H.-S., 279
Kim, H. Y., 187
Kim, H.-Y., 297
Kinoshita, T., 90, 223
Kirchman, P. A., 333
Kiss, L., 283
Kita, Y., 61
Kitano, H., 326, 329
Kitson, N., 222
Knudsen, J., 333
Ko, J., 329, 334, 336, 337, 338
Koester, M., 187
Kohlwein, S. D., 214
Koivusalo, M., 216
Kolesnick, R. N., 223
Kollmeyer, J., 85, 86
Kondo, T., 126
Kong, H., 190
Kopka, J., 172
Koppel, D. E., 221
Kornfeld, R., 119
Kornfeld, S., 119, 136, 140
Korzeniowski, M., 222
Kostiainen, R., 216
Kouchi, T., 333
Kovtoun, V., 208
Kowakame, T., 333
Kraft, M. L., 111
Krag, S. S., 130
Krank, J., 1, 60, 126, 215, 305
Kriaris, M. S., 6
Kristal, B. S., 172
Kroesen, B. J., 91
Krueger, S., 222
Kühn, H., 279
Kuksis, A., 2
Kummer, U., 341
Kurchalia, T., 110
Kurose, A., 61
Kurzchalia, T., 31, 216
Kushi, Y., 91
Kusumi, A., 222
Kutluay, T., 123, 129
Kuzuyama, T., 118
Kwiatkowska, K., 222

L

Lagerholm, B. C., 222
Lai, C. S., 130
Lange, B. M., 172
Langmann, T., 216, 217, 224

Author Index

Lanzavecchia, S., 223
Laprevote, O., 111
Larsen, Å., 31
Lausmaa, J., 111, 208, 219
Lawrence, P., 165
Lay, J. O., Jr., 110
LeBaron, F. N., 29
Lecureuil, C., 234
Lee, C., 341
Lee, G. H., 91
Lee, M. H., 91
Leeflang, B. R., 213
Lees, M., 29
Lehmann, W. D., 23, 186, 187, 191, 216
Lemmon, M. A., 40
Lenne, P. F., 222
Lesnefsky, E. J., 24
Lester, R. L., 338, 339, 340
Letcher, A. J., 40
Levery, S. B., 85, 215, 216
Levy, B. D., 278, 286, 288, 298
Leyes, A. E., 119
Li, E., 136, 140
Li, J., 247
Li, P.-L., 298
Li, S., 148, 165
Li, X., 15
Lias, S. G., 280, 282
Liebisch, G., 213, 216, 224
Lieser, B., 216
Lindblom, G., 221
Lindgren, J. Å., 277, 282, 287, 288, 296
Linne, M. L., 341
LIPID MAPS Consortium, 308, 309
Lips, D. L., 39, 40
Liskamp, R. M., 219
Litman, B., 297
Little, P. F. R., 249
Liu, Y., 85, 86, 336
Loh, Y., 296
London, E., 222
Longo, M. L., 111
Lopez, R., 255
López-Montero, I., 221
Lu, Y., 275, 278, 280, 282, 284, 285, 286, 292, 295, 296, 297, 298, 300, 301, 304, 305, 306, 308, 309, 310, 312, 313
Lubben, A., 41, 237
Luberto, C., 338
Lucast, L., 40, 211, 282
Lucocq, J. M., 40
Lukiw, W. J., 282, 298, 301
Luna, E. J., 221
Lund, E. G., 147, 151, 152, 164, 166
Lütjohann, D., 218
Luzzati, V., 220
Lytle, C. A., 37, 38

M

Ma, Y. C., 187
Ma, Z., 31, 189
MacDonald, J., 80
MacMillan, D. K., 309
Maeda, Y., 223
Maer, A., 225, 247, 254
Magrane, M., 255
Maheshwari, N., 298
Majerus, P. W., 39, 40
Majno, G., 279
Malherbe, T., 223
Mallard, G. W., 280, 282
Malmberg, P., 111
Mangelsdorf, D. J., 151
Mann, M., 26, 148, 187
Manninem, T., 341
Mansson, J. E., 110, 111
Mao, C., 334, 336, 338
Mao, J., 41
Marai, L., 2
Marathe, G. K., 207
Marcheselli, V. L., 282, 297, 298, 301
Margalit, A., 60
Marguet, D., 222
Markovich, R., 353
Marleau, S., 276, 278, 279, 284, 288, 296
Marnett, L. J., 27
Marsh, D., 220, 222
Marshall, A. G., 110, 190
Martinez, J., 208
Marto, J. A., 190
Matson, W. R., 172
Matsubara, T., 90
Mauerer, R., 217
Mauriala, T., 148, 168
Maxey, K. M., 80
Mayer, K., 283
McAnoy, A. M., 1, 2, 10, 60, 215, 305, 363
McConnell, H. M., 221
McCrum, E. C., 145
McDonald, J. G., 145, 151
McDonough, P. M., 223
McDonough, V. M., 329, 334, 336, 337, 338
McFarland, M. A., 110
McGehee, M. F., 166
McGiff, J. C., 60
McIntyre, T., 207
McLafferty, F. W., 23, 31, 190, 289, 296
McMurray, W., 40, 211, 282
Meath, J. A., 29
Meganathan, R., 121
Meisen, I., 217
Meleard, P., 221
Mendes, P., 172, 341
Meng, C. K., 26, 148, 187
Merched, A., 276, 278, 279, 284, 288, 296
Merrill, A. H., 172

Merrill, A. H., Jr., 25, 83, 86, 91, 95, 100, 102, 118, 158, 175, 176, 179, 189, 214, 215, 224, 225, 250, 252, 269, 329, 334, 336, 337, 338, 361
Merritt, M. V., 5
Meyvis, T. K., 221
Michaelis, S., 119
Michell, R. H., 40
Michelsen, P., 40
Mikaya, A. I., 280, 282
Mikic, I., 223
Milbury, P. E., 172
Mills, T., 222
Milne, S. B., 21, 27, 31, 38, 40, 45, 126, 180
Minh, L. Y., 147
Minkler, P. E., 24
Mirick, G., 278, 279, 280, 282, 284, 288, 297, 298, 301, 312
Miseki, K., 92
Mitchell, C. A., 39
Miyazaki, H., 60
Mizugaki, M., 61
Moehring, T., 110, 216
Moeller, P. D., 91
Momin, A., 85, 86
Monkkonen, J., 148, 168
Moore, J. D., 2, 121, 122, 174, 351, 364
Morgan, E. L., 297, 301, 306, 312
Morita, M., 90
Morris, J. B., 40
Moruisi, K. G., 146
Mouritsen, O. G., 222
Moussignac, R.-L., 276, 278, 279, 280, 282, 284, 286, 287, 288, 296, 297, 298, 301, 312
Moyer, S. C., 110, 217
Mrozinska, K., 222
Mukherjee, P. K., 282, 297, 298
Munter, E., 85, 86
Murayama, K., 91
Murphy, R. C., 1, 2, 10, 25, 26, 31, 60, 61, 63, 86, 118, 121, 122, 126, 148, 153, 158, 165, 172, 174, 175, 176, 179, 186, 207, 214, 215, 224, 225, 250, 252, 262, 280, 282, 284, 287, 288, 296, 301, 305, 308, 309, 355, 363, 364
Musto, A., 282, 298, 301
Muthing, J., 217
Myant, N. B., 146, 151
Myher, J. J., 2

N

Nadeau, D., 247
Nakamura, H., 61
Nakamura, T., 90, 207
Nakanishi, H., 187, 189, 190, 191, 192, 200, 201, 237
Narayana, S., 217

Nasuhoglu, C., 41
Natale, D. A., 255
Neale, J. R., 241
Neergaard, T. B., 333
Ni, T. C., 321, 343
Nickels, J. T., Jr., 329, 334, 336, 337, 338
Nieva, J. L., 223
Nikolau, B. J., 172
Nilsson, C. L., 110
Nilsson, E., 47
Nishijima, M., 185
Nowatzke, W., 31, 189
Numajir, Y., 91
Nussbaum, R. L., 40, 211, 282
Nygren, H., 111

O

Obeid, L. M., 334, 335
O'Brien, J. A., 278, 296
O'Connor, P. B., 110, 217
O'Connor, S. E., 136
Odham, G., 40
Odparlik, S., 288
Oegema, J., 31, 110, 216
O'Hagan, D., 214
Ohira, T., 298
Ohshima, M., 251
Okamoto, Y., 227, 319, 321, 323, 325, 341, 343
Olgun, A., 123, 129
Olson, J. A., 121
Olsson, J., 119
Oosthuizen, W., 146
Opperman, A. M., 146
Oradd, G., 221
Oresic, M., 267
Organisciak, D. T., 207
Ostlund, R. E., Jr., 146
Otvos, J. D., 217
Ouyang, Z., 219

P

Pacetti, D., 91
Pacha, K., 60
Padberg, W., 283
Pahle, J., 341
Palmgren, J. J., 148, 168
Paneda, C., 223
Pang, J., 148, 165
Paolo, G. D., 282
Park, H., 83
Patel, A., 151
Patton, J. L., 338, 339
Pedrioli, P. G., 47
Pencer, J., 222
Peng, Q., 85, 86, 269, 361

Author Index

Petasis, N. A., 276, 278, 279, 282, 284, 288, 292, 296, 297, 298, 300, 304, 305, 306, 308, 309, 310, 312
Peter-Katalinic, J., 110, 216, 217
Peters-Golden, M., 60
Pettinen, A., 341
Pettitt, T. R., 41, 233, 237
Pettus, B. J., 91
Pfefferli, M., 333
Phillips, F. C., 2
Pidgeon, C., 353
Pinckard, R. N., 187
Pittenauer, E., 8
Planey, S., 223
Podrez, E. A., 207
Pohl, A., 221
Polozov, I. V., 217
Postle, A. D., 233, 241
Poulter, C. D., 119
Pratt, B., 47
Preininger, A., 126
Prescott, S., 207
Price, J. H., 223
Prieto-Conaway, M. C., 208
Privett, O. S., 2
Prognon, P., 111
Pulfer, M., 26, 179, 186, 215
Pulfer, M. K., 148, 153, 165

Q
Quinn, P. J., 221

R
Racette, S. B., 146
Raetz, C. R., 2, 86, 118, 140, 174, 175, 214, 215, 224, 225, 252
Raetz, C. R. H., 25, 117, 121, 122, 158, 172, 175, 176, 250, 364
Raggers, R. J., 218, 221
Ramanadham, S., 31, 189
Ramaraju, H., 85, 86
Rapoport, T. A., 321
Rathenberg, J., 216
Raught, B., 47
Reddy, K. M., 207
Reed, J., 280, 282
Reichart, D., 2, 121, 122, 174, 364
Reinhold, V. N., 139, 140
Reinke, M., 60
Reis, A., 190
Reis e Sousa, C., 284, 296
Rentsch, K. M., 148, 165, 168
Reynolds, C. M., 2, 121, 122, 174, 364
Ribeiro, A. A., 2, 121, 122, 174, 364
Richet, C., 136
Riezman, H., 323

Rigneault, H., 222
Rip, J. W., 130
Robbins, P. W., 119
Robinson, W. H., 217
Roditi, I., 223
Rodriguez, A. R., 282
Rodriguez, N., 221
Rodriguez-Concepcion, M., 118
Rodwell, V. W., 118
Roessner-Tunali, U., 172
Roggero, R., 208
Rohdich, F., 118
Rohlfing, A., 217
Rokach, J., 60
Rokukawa, C., 91
Romanelli, A. J., 40, 211, 282
Rosner, M. R., 119
Rotilio, D., 41
Rouzer, C. A., 27, 277
Rowley, A. F., 297, 301, 306, 312
Roy, S., 111
Ruiz-Argüello, M. B., 220
Rupar, C. A., 130
Russell, D. W., 25, 86, 118, 145, 146, 158, 172, 175, 176, 214, 215, 224, 225, 250, 252

S
Saarinen, A., 341
Sacerdoti, D., 60
Saga, T., 61
Sahle, S., 341
Saito, K., 172
Sakamoto, M., 333
Sakamoto, N., 336
Salem, N., Jr., 297
Salomon, R. G., 207
Sampaio, J., 6
Samuel, M., 23
Samuelsson, B., 89, 277, 283
Samuelsson, K., 89
Sandhoff, R., 23, 186, 191, 216, 333
Sandra, P., 2
Saraste, M., 123
Sato, T., 164
Saunders, B., 225, 254
Sauro, H. M., 328, 331
Savageau, M. A., 321, 328, 341, 342, 343
Sawai, H., 338
Sawai, T., 61
Sayre, L. M., 119
Schaloske, R. H., 60
Schenk, B., 119, 136
Schevchenko, A., 31
Schifferer, R., 216
Schlessinger, J., 221
Schmidt, B., 278, 286, 288

Schmidt, J. A., 2
Schmitz, G., 213, 214, 216, 217, 224
Schneiter, R., 333
Schroder, H. D., 333
Schroepfer, G. J., Jr., 148, 165
Schultz, J. A., 110
Schulz, B., 216
Schütte, H., 283
Schutzbach, J. S., 130
Schwacke, J. H., 341
Schwartz, S. J., 148, 168
Schwartzman, M. L., 298
Schwudke, D., 31, 110, 216
Scott, D. R., 280, 282, 291
Seeger, W., 283
Segel, L. A., 347
Seitz, P. K., 166
Sekine, M., 90
Seldomridge, S., 190
Self, R., 26
Sen, A., 221
Seo, S., 164
Serhan, C. N., 227, 275, 276, 277, 278, 279, 280, 282, 283, 284, 285, 286, 287, 288, 292, 295, 296, 297, 298, 300, 301, 304, 305, 306, 308, 309, 310, 312, 313
Serif, A., 123, 129
Seron, T., 223
Seto, H., 118
Seyama, Y., 25, 86, 118, 158, 175, 176, 214, 215, 224, 225, 250
Shan, H., 148, 165
Shan, L., 207
Shaner, R., 83
Shaw, W., 86, 118, 158, 172, 175, 176, 214, 215, 224, 225, 250
Shaw, W. A., 2, 25, 121, 122, 174, 351, 364
Shayman, J. A., 87
Sheeley, D. M., 139, 140
Sheets, E. D., 23, 190
Shell, B. K., 119
Shen, J., 279
Sher, A., 277, 280, 284, 292, 296, 298
Shestopalov, A. I., 172
Shevchenko, A., 6, 110, 187, 191, 192, 200, 216, 219
Shieh, J. J., 176, 179
Shimeno, H., 333
Shimizu, T., 25, 61, 86, 118, 158, 175, 176, 185, 187, 189, 190, 191, 192, 200, 201, 214, 215, 224, 225, 237, 250
Shipkova, P. A., 23, 190
Shiraishi, F., 328, 342
Shono, F., 60
Shulman, G. I., 40, 211, 282
Sibelius, U., 283
Siegel, D. P., 220, 221
Siegelman, J., 282, 292, 297, 298, 306, 308, 312

Sillence, D. J., 218, 221
Silverstein, R. L., 207
Simmons, D. L., 60
Simon, S. A., 223
Simons, K., 6, 31, 110, 187, 191, 192, 200, 216, 220, 222
Sims, K. J., 227, 319, 321, 323, 325, 343
Simus, N., 341
Singhal, M., 341
Six, D. A., 60
Sjövall, J., 282, 287, 288, 296
Sjövall, P., 111, 208, 219
Skotland, T., 31
Smith, A., 175
Smith, D. C., Jr., 2, 121, 122, 174, 364
Smith, L. C., 165
Smith, L. L., 166
Smith, P. B. W., 31
Smith, W. L., 60
Smolander, O. P., 341
Snyder, A. P., 31
Sobel, R. A., 217
Sobota, A., 222
Soeda, S., 333
Soleau, S., 221
Somerharju, P., 180, 189, 216, 224
Song, Q., 111
Sorribas, A., 321, 328
Sot, J., 220, 221
Sparkman, O. D., 280, 282
Spassieva, S. D., 323
Spener, F., 25, 86, 118, 158, 175, 176, 214, 215, 224, 225, 250
Spokas, E. G., 60
Sprong, H., 218, 219, 222
Spur, B. W., 282
St. Pyrek, J., 165
Stafford, G., 208
Stahl, G. L., 276, 278, 279, 284, 288, 296
Stauffacher, C. V., 118
Stauffer, D. A., 296
Stein, S. E., 280, 282, 291
Steiner, W. E., 111
Steinman, L., 217
Stenson, W. F., 146
Stephens, D. L., 61
Stephens, L. R., 234
Sterling, A., 216
Stock, S. D., 340
Stoll, M. S. K., 24
Strzelecka-Kiliszek, A., 222
Stump, M. J., 110
Subczynski, W. K., 222
Subramaniam, S., 25, 86, 118, 158, 171, 172, 173, 175, 176, 177, 214, 215, 224, 225, 247, 249, 250, 252, 254
Suchy, S. F., 40, 211, 282
Sud, M., 175, 247, 252

Suire, S., 234
Sukegawa, K., 91
Sullards, M. C., 83, 85, 86, 91, 95, 100, 102, 179, 189, 215, 269, 361
Sun, M., 207
Suzuki, A., 90, 91, 92
Suzuki, M., 90, 91, 92
Suzuki, N., 61
Suzuki, Y., 92
Sweeley, C. C., 89
Swiezewskaa, E., 119
Symolon, H., 85, 86
Szkopinska, A., 121, 123
Szulc, Z. M., 91

T

Taattola, K. L., 341
Tabas, I., 136, 140
Tabernero, L., 118
Taguchi, R., 24, 185, 187, 189, 190, 191, 192, 200, 201, 237
Tai, H.-H., 282, 288, 308
Taka, H., 91
Takabatake, M., 61
Takada, H., 147
Takahashi, T., 61
Takano, T., 282, 284, 288, 296, 308
Takats, Z., 219
Takeda, K. J., 164
Takemoto, J. Y., 40
Takemoto, Y. J., 340
Taketomi, T., 90, 91
Takeuchi, Y., 24, 190
Tallarek, E., 111
Talmon, Y., 220, 221
Tamanoi, F., 119
Tang, W.-X., 298
Tarao, M., 147
Tardieu, A., 220
Taube, C., 148, 153, 165
Tchekhovskoi, D. V., 280, 282
Teng, J. I., 166
Teshima, K., 126
Tezcan, S., 123, 129
Thewalt, J., 222
Thiele, C., 6, 219
Thomas, M. J., 23, 208
Thompson, B. M., 145
Thompson, N. L., 222
Tian, Q., 148, 168
Tjonahen, E., 280, 282, 284, 285, 286, 295, 296, 297, 298, 301, 306, 309, 312, 313
Tomioka, Y., 61
Torres, N. V., 328, 329, 346
Touboul, D., 111
Toyras, A., 148, 168

Trethewey, R. N., 172
Tserng, K. Y., 6
Tsukamoto, H., 61
Tuininga, A. R., 31, 187
Turecek, F., 289
Turk, J., 2, 31, 33, 179, 189
Turley, S. D., 146
Turunen, M., 119
Tyynela, J., 189

U

Ugarov, M., 110
Uhlson, C., 60, 215, 305
Ulevitch, R. J., 2, 121, 122, 174, 364
Umetani, M., 151
United States Environmental Protection Agency, 162
Uomori, A., 164
Uozumi, N., 61
Uphoff, A., 180, 224
Uran, S., 31
Usuki, S., 110

V

Vaidya, V., 298
Vales, T. R., 329, 334, 336, 337, 338
Valtersson, C., 130
van Blitterswijk, W. J., 223
van der Luit, A. H., 223
van Duyn, G., 130
Van Dyke, T. E., 276, 278, 279, 284, 288, 296, 298
van Meer, G., 213, 214, 217, 218, 219, 221, 222, 227
Van Oostveldt, P., 221
Van Pelt, C. K., 165
Van Voorhis, W. C., 119
Varela-Nieto, I., 223
Varfolomeyev, S., 214
Veatch, S. L., 222
Vega, G. L., 147
Veldman, R. J., 223
Velez, M., 221
Verheij, M., 223
Verkleij, A. J., 130
Via, D. P., 165
Vial, H., 208
Vidavsky, I., 291
Vigneau-Callahan, K. E., 172
Villar, A. V., 223
Vogelzang, M., 47
Voit, E. O., 227, 321, 323, 325, 328, 329, 331, 332, 341, 342, 343, 346, 347
von Eckardstein, A., 148, 165, 168
von Massenbach, B., 223
Vukelic, Z., 110, 216, 217

W

Waechter, C. J., 119, 136
Wakelam, M. J. O., 41, 233, 237
Waldmann, H., 119
Wan, K. X., 291
Wang, C., 24, 190
Wang, E., 83, 85, 86, 91, 95, 100, 102, 179, 189, 215, 269, 361
Wang, H. Y., 110, 111
Wang, J., 220, 278
Wang, T. C., 187
Wang, W., 298
Warren, K. G., 217
Watanabe, K., 60, 251
Watkins, S. M., 178
Watson, A. D., 26
Watson, R., 223
Wawrezinieck, L., 222
Webb, W. W., 221
Weber, P. K., 111
Weerapana, E., 136
Wei, Z., 119
Weinreb, G. E., 222
Weintraub, S. T., 187
Weissmann, N., 283
Wenk, M. R., 26, 40, 211, 282
Wennerstrom, H., 222
Wesdemiotis, C., 296
Westover, E. J., 151
Weylandt, K. H., 278
Wheelan, P. J., 296, 301, 308
White, D. C., 37, 38
White, F. M., 190
Whitehouse, C. M., 26, 148, 187
Wieland, F. T., 23, 186, 191, 216
Wilhelm, M., 41
Wilkins, C. L., 110
Williams, W. P., 221
Wilson, W. K., 148, 165
Wilund, K. R., 147
Wiseman, J. M., 111, 219
Wolf, B. A., 40
Wong, P. Y., 60
Wong, S. F., 26, 148, 187
Woods, A. S., 110, 111
Wright, B. E., 319
Wright, E. M., 223
Wu, C. C., 2, 10, 363
Wu, C. H., 255
Wu, W. I., 329, 334, 336, 337, 338
Wurtz, O., 222
Wynalda, M. A., 80

X

Xia, S., 278
Xiang, Y., 21
Xie, S., 24
Xu, F., 147, 151
Xu, G., 24, 190
Xu, L., 341
Xu, R., 334, 335
Xu, Y., 6

Y

Yamada, M., 92
Yamaguchi, S., 91
Yamakawa, T., 90, 91
Yamamoto, M., 41
Yamamoto, S., 60
Yamashita, K., 60
Yamatani, K., 187, 189, 190, 191, 192, 200, 201
Yamazaki, T., 187, 189, 190, 191, 192, 200, 201, 237
Yamazaki, Y., 92
Yang, C., 151
Yang, J., 24, 109, 190
Yang, Q., 24
Yang, R., 275, 277, 280, 282, 292, 297, 298, 300, 304, 305, 306, 308, 309, 310, 312
Yang, Y., 282, 287, 288, 296
Yasugi, E., 251
Ye, H., 109
Yergey, A. L., 162
Yin, H. L., 41
Yin, J. J., 222
Yli-Harja, O., 341
Yokota, K., 60
Yong Yeng, L., 166
Yoo, J. S., 91
Yost, R. A., 208
Young, A. T., 40
Youshimura, Y., 164
Yu, L., 147, 297, 301, 306, 312
Yu, R. K., 110

Z

Zacharias, D., 223
Zaikin, V., 280, 282
Zakaria, M. P., 147
Zamfir, A., 216
Zamfir, A. D., 110
Zampighi, G. A., 223
Zampighi, L. M., 223
Zanetta, J. P., 136
Zarini, S., 60, 215, 305
Zech, T., 221, 222
Zechner, R., 214
Zemski, B. K., 60
Zemski Berry, K., 215, 305
Zhang, J., 223
Zhang, N., 119
Zhang, Y., 151
Zheng, W., 85, 86
Zhou, Y., 247
Zhu, D., 280, 282
Zirrolli, J. A., 296, 301, 308
Zsigmond, E., 279
Zuckermann, M. J., 222

Subject Index

A

AA, *see* Arachidonic acid
Acetonitrile, effects on mass spectrometry signal intensity, 165
Arachidonic acid, metabolism, 60
Auto-oxidation, sphingolipids, 165–166

B

Biochemical systems theory, metabolic modeling, 320, 327, 329, 339

C

Ceramides, *see* Signaling pathway lipidomics; Sphingolipids
Ceramide synthase, metabolic modeling, 332–334
Chip-based mass spectrometry, sphingolipids, 110
Chiral chromatography, eicosanoid liquid chromatography-mass spectrometry, 63
Cholesterol, *see* Sterols
COCAD, *see* Cognoscitive-contrast-angle algorithm and database
Coenzyme Q, *see* Prenols
Cognoscitive-contrast-angle algorithm and database, lipid mediator identification
    contrast angle calculation, 290–291
    liquid chromatography-ultraviolet spectra-tandem mass spectrometry data integration, 291–292
    mass spectral ion identification, 286–288
    novel lipid mediator theoretical database and search algorithm, 292, 294–297, 301
    tandem mass spectrometry ion intensity modification according to identities, 289–290
Cyberlipid, features, 225, 250

D

DAC, *see* Diacylglycerol
Diacylglycerol
    mass spectrometry analysis
        cell culture, 4
        cell stimulation and harvesting, 4–5
        DNA quantification, 5
        instrumentation and data acquisition, 6
        Kdo_2-lipid A preparation, 4
        lipid extraction
            solid phase extraction of lipid classes, 5–6
            total lipids, 5
        materials, 3
        overview of approaches, 2
        qualitative analysis with tandem and triple mass spectrometry, 7–14
        quantitative analysis, 11, 15, 17–18
        spectrophotometric enzyme assay, 2
Differential scanning calorimetry, lipid phases, 220
Dihydroceramidase, metabolic modeling, 334–336
Dihydroceramide, dynamics modeling, 323–327
Dolichols, *see* Prenols
DSC, *see* Differential scanning calorimetry

E

Eicosanoids
Eicosanoids prospects, 319
    arachidonic acid metabolism, 60
    enzyme-linked immunosorbent assay, 60–61
    identification
        cognoscitive-contrast-angle algorithm contrast angle calculation, 290–291
        liquid chromatography-ultraviolet spectra-tandem mass spectrometry data integration, 291–292
        mass spectral ion identification, 286–288
        tandem mass spectrometry ion intensity modification according to identities, 289–290
    databases, 286
    fragmentation mechanisms, 302, 304, 308–310
    instrumentation, 313
    liquid chromatography-ultraviolet spectra-tandem mass spectrometry, 283–286
    logic diagram, 284–286
    MassFrontier search algorithm, 282
    novel lipid mediator theoretical database and search algorithm, 292, 294–297, 301
    prospects, 311–319
    sample preparation, 312
inflammation mediation, 279
mass spectrometry

Eicosanoids prospects (*cont.*)
  extraction, 62, 80
  liquid chromatography-mass spectrometry
    chiral chromatography, 63
    internal standards, 64–72
    lower limit of detection, 78–79
    multi-reaction monitoring transition selection, 74–77
    quantitative analysis, 63–64, 74
    recoveries, 79
    reverse phase liquid chromatography, 62
    running conditions, 63
    stereoisomer detection, 77
  overview, 61
  sample collection, 62, 79–80
  metabolomics and mediator lipidomics informatics, 276–278
  transgenic animal engineering, 278–279
Electron ionization mass spectrometry, sphingolipids, 89
Electron microscopy, ultrastructural localization of lipids, 218
Electrospray ionization mass spectrometry
  advantages in lipidomics, 186–187
  inside-out sphingolipidomics and liquid chromatography-tandem mass spectrometry
    ceramide 1-phosphate analysis, 106
    (dihydro)ceramide, (dihydro) sphingomyelin, and (dihydro) monohexosyl-ceramide analysis in positive ion mode, 104–109
    electrospray ionization, 90, 109
    extraction, 98–99
    galactosylceramide analysis, 105–106
    ganglioside analysis, 107–109
    glucosylceramide analysis, 105–106
    materials, 100
    matrix-assisted laser desorption, 91–92
    multiple reaction monitoring for quantitative analysis, 101–103
    overview, 92–93, 95
    sample collection and preparation, 98–99
    sphingoid base analysis in positive ion mode, 102–103
    subspecies characterization prior to quantitative analysis, 100–101
    triple mass spectrometry of complex glycosphingolipids, 108–109
    work flow, 97
  prenol liquid chromatography-electrospray ionization mass spectrometry
    coenzyme Q, 123, 126, 129–130
    dolichol, 130–131, 133, 136
    dolichol diphosphate-linked oligosaccharides, 136, 140
    extraction, 121–123
    instrumentation, 122
    materials, 121
  signaling pathway lipidomics with liquid chromatography-electrospray ionization tandem mass spectrometry
    diagnostic ions, 238–239
    instrumentation, 237
    phospholipid synthesis analysis with stable isotopes
      data analysis, 245
      labeling in cells and *in vivo*, 242
      lipid extraction, 243
      multiple reaction monitoring, 244
      native and newly synthesized phospholipids, 244
      overview, 241–242
      running conditions, 243–244
    reverse phase high-performance liquid chromatography
      ceramide, diradylglycerol, and monoradylglycerol, 239–240
      phosphoinositides, 240
      phospholipids, 237–239
  sphingolipid liquid chromatography-electrospray ionization mass spectrometry
    acetonitrile effects on signal intensity, 165
    extraction
      auto-oxidation concerns, 165–166
      bulk lipids, 149–151
      clean-up, 166
      residual insoluble material, 167–168
      solid-phase extraction, 152–153
    instrumentation, 168
    internal standards, 150, 158–159, 163–164
    materials, 148–149
    overview, 147–148
    peak analysis, 157, 159, 161–162, 168
    quantitative analysis, 156–157, 162–163
    reverse phase high-performance liquid chromatography, 153, 164
    running conditions, 153–154
    saponification of extracts, 151
    selected reaction monitoring, 159, 161–162
  sterol liquid chromatography-electrospray ionization mass spectrometry
    acetonitrile effects on signal intensity, 165
    extraction
      auto-oxidation concerns, 165–166
      bulk lipids, 149–151
      clean-up, 166
      residual insoluble material, 167–168
      solid-phase extraction, 152–153
    instrumentation, 168
    internal standards, 150, 158–159, 163–164
    materials, 148–149
    overview, 147–148
    peak analysis, 157, 159, 161–162, 168

Subject Index

quantitative analysis, 156–157, 162–163
reverse phase high-performance liquid chromatography, 153, 164
running conditions, 153–154
saponification of extracts, 151
selected reaction monitoring, 159, 161–162
ELISA, *see* Enzyme-linked immunosorbent assay
ELSD, *see* Evaporative light scattering detection
Enzyme-linked immunosorbent assay, eicosanoids, 60–61
ESI-MS, *see* Electrospray ionization mass spectrometry
European Lipidomics Initiative
    goals, 214
    imaging lipids, 218–219
    lipidomics
        databases, 224–226
        expertise platform, 226
        overview, 215
        software requirements, 224
    physical property analysis
        amphiphiles, 220
        bilayer and membrane structure, 221
        fusion, 222–223
        membrane fluidity, 221
        mesomorphism, 220
        rafts and domains, 222
    prospects, 226–227
    signaling lipid studies, 223
    technology development, 215–217
    workshops and meetings, 214
Evaporative light scattering detection, sterols, 147

F

FAB-MS, *see* Fast atom bombardment mass spectrometry
Fast atom bombardment mass spectrometry, sphingolipids, 90
Flux analysis, stable isotope labeling, 208
Fourier transform ion cyclotron mass spectrometry
    lipid identification with direct injection, 199
    mass accuracy, 188–189
    sphingolipids, 110
FT-MS, *see* Fourier transform ion cyclotron mass spectrometry

G

Gangliosides, *see* Sphingolipids
Gas chromatography-mass spectrometry
    sphingolipids, 147
    sterols, 147
GC-MS, *see* Gas chromatography-mass spectrometry
Generalized mass action model, metabolic modeling, 321, 328, 341

Glycerophospholipids, *see* Phospholipids
Glycosphingolipids, *see* Sphingolipids

H

High-performance liquid chromatography
    chiral chromatography for eicosanoids, 63
    lipid separation and reproducibility, 189–190
    liquid chromatography-mass spectrometry, *see* Eicosanoids; Prenols; Protectins; Resolvins; Signaling pathway lipidomics; Sphingolipids; Sterols
    phospholipids, 34
HPLC, *see* High-performance liquid chromatography

I

Inositol phosphorylceramide synthase, 336–338
Inositol phosphosphingolipid phospholipase C, 338–339
Inside-out sphingolipidomics, *see* Sphingolipids
Internal standards, *see* Mass spectrometry
Ion mobility mass spectrometry, sphingolipids, 110–111

K

KEGG, *see* Kyoto Encyclopedia of Genes and Genomes
Kyoto Encyclopedia of Genes and Genomes
    features, 225
    lipid pathways, 268–269

L

Linear ion trap, internal standard characterization, 354–355
LIPIDAT, features, 250–251
Lipid Bank, database features, 187–188, 225, 249
Lipid Data Bank, features, 225
Lipid Library, features, 225, 249–250
LIPID MAPS, *see* Lipid Metabolites and Pathways Strategy
Lipid Metabolites and Pathways Strategy
    consortium members and functions, 172–173
    data collection, display, and analysis tools, 264–267
    drawing and naming tools, 260–262
    eicosanoid tandem mass spectrometry data, 74
    funding, 171
    infrastructure building in lipidomics, 173–174, 248–249
    inside-out sphingolipidomics, *see* Sphingolipids
    lipid classification scheme and nomenclature, 174–177, 224, 249–251

Lipid Metabolites and Pathways Strategy (cont.)
  lipid pathways
    bioinformatics resources, 268–269
    drawing tools, 269–270
    lipid profiling tools, 267–268
    lipid structure
      databases, 251–253
      representation, 253–254
    mass spectra prediction tools, 262–263
    mass spectrometry initiatives, 178–180
    metadata management tools, 263
    overview, 225, 248–249
    phospholipid tandem mass spectrometry data, 33
    prospects, 181–182, 270, 272
  Proteome Database
    contents, 254
    searching, 255–256
    sterol standards, 164
    Web site and user interface, 256–260
Lipid microarray, principles and applications, 217
Lipidomics, see Eicosanoids; European Lipidomics Initiative; Lipid Bank; Lipid Metabolites and Pathways Strategy; Lipid Search; Mass spectrometry; Protectins; Resolvins; Signaling pathway lipidomics; Sphingolipids
Lipid Search
  database selection
    adduct ions, 196
    mass surveys, 197
    mass tolerance, 196–197
    overview, 196
    positive versus negative ion modes, 196
  lipid quantitative analysis, see also specific lipids
    compensation techniques, 198–199
    overview, 197–198
    origins, 192
    overview, 192–193
    sample searches, 193–196
Liquid secondary ionization mass spectrometry, sphingolipids, 90
LIT, see Linear ion trap
LMBP, lipid pathway drawing, 269

M

MALDI-MS, see Matrix-assisted laser desorption mass spectrometry
MassBank, development, 208
MassFrontier, lipid mediator search algorithm, 282
Mass imaging
  phospholipids, 208–209
  resolution, 219
  sterols, 111
Mass spectrometry, see also specific techniques
  eicosanoids, see Eicosanoids

glycerophospholipids, see Phospholipids
glyceryl lipids, see Diacylglycerol; Triacylglycerol
historical perspective, 26
internal standards for lipids
  availability, 352, 364–366
  chemical characterization, 353–355
  eicosanoids, 64–72
  handling guidelines, 352–353
  Lipid Metabolites and Pathways Strategy initiatives, 164, 364–366
  packaging, 359
  preparation, 355–359
  quality control and stability testing
    ceramide/sphingoid base mixture, 361, 363
    triglyceride-d5, 363–364
  solution concentration and solubility, 360
  sterols, 150, 158–159, 163–164
Lipid Metabolites and Pathways Strategy initiatives, 178–180
lipid identification requirement
  fragment ion characterization, 189
  liquid chromatography separation and reproducibility, 189–190
  mass accuracy, 188–189
lipidomics approaches
  databases, see Lipid Metabolites and Pathways Strategy; Lipid Search
  focused approach, 191–192, 200–201, 204–207
  global and untargeted approach, 190–191, 200
  targeted approach, 192, 207–208
prenols, see Prenols
principles, 26–27
sphingolipids, see Sphingolipids
sterols, see Sterols
Matrix-assisted laser desorption mass spectrometry, inside-out sphingolipidomics
  ceramide 1-phosphate analysis, 106
  (dihydro)ceramide, (dihydro)sphingomyelin, and (dihydro)monohexosyl-ceramide analysis in positive ion mode, 104–109
  extraction, 96, 98–99
  galactosylceramide analysis, 105–106
  ganglioside analysis, 107–109
  glucosylceramide analysis, 105–106
  materials, 100
  matrix-assisted laser desorption, 91–92
  multiple reaction monitoring for quantitative analysis, 101–103
  overview, 92–93, 95
  sample collection and preparation, 98–99
  sphingoid base analysis in positive ion mode, 102–103
  subspecies characterization prior to quantitative analysis, 100–101

Subject Index

triple mass spectrometry of complex glycosphingolipids, 108–109
work flow, 97
Mediator lipidomics, *see* Eicosanoids; Protectins; Resolvins
Mesomorphism, lipids, 220
Metabolic modeling, *see* Sphingolipids
MRM, *see* Multiple reaction monitoring
MS, *see* Mass spectrometry
MS3, *see* Triple mass spectrometry
Multiple reaction monitoring, *see* Eicosanoids; Signaling pathway lipidomics; Sphingolipids

N

Nanoelectrospray ionization, sphingolipids, 110
NMR, *see* Nuclear magnetic resonance
Nuclear magnetic resonance, lipoproteins, 217

P

Phosphoinositides, *see* Signaling pathway lipidomics
Phospholipids
  chromatographic separation, 23
  classes, 23–24
  functional overview, 22, 24–25
  mass spectrometry
    computational analysis of complex mixtures
      direct injection data, 45–47
      liquid chromatography-mass spectrometry data, 47–51, 53–54
      overview, 43, 45
    direct injection analysis, 30–33
    extraction
      cells, 29–30
      overview, 27–28
      solvents, 29
      tissues, 30
    ionization techniques, 23, 27
    liquid chromatography-mass spectrometry, 33, 35
  nomenclature, 25
  polyphosphoinositides
    extraction
      cultured cells, 37
      tissues, 37–38
    mass spectrometry
      deacylation of GPIns*Pn* lipids, 39
      direct injection analysis, 38–39
      liquid chromatography-mass spectrometry of deacylated compounds, 39–41
  signaling, *see* Signaling pathway lipidomics
Phytoceramide, dynamics modeling, 323–327
PLAS software, metabolic modeling, 341–345, 347
Polyphosphoinositides, *see* Phospholipids

Prenols
  biosynthesis, 118
  conjugates, 121
  liquid chromatography-electrospray ionization mass spectrometry
    coenzyme Q, 123, 126, 129–130
    dolichol, 130–131, 133, 136
    dolichol diphosphate-linked oligosaccharides, 136, 140
    extraction, 121–123
    instrumentation, 122
    materials, 121
  structures, 118–120
Prostaglandins, *see* Eicosanoids
Protectins
  biosynthesis, 281
  identification
    cognoscitive-contrast-angle algorithm
      contrast angle calculation, 290–291
      liquid chromatography-ultraviolet spectra-tandem mass spectrometry data integration, 291–292
      mass spectral ion identification, 286–288
      tandem mass spectrometry ion intensity modification according to identities, 289–290
    databases, 286
    fragmentation mechanisms, 302, 304, 308–310
    instrumentation, 313
    liquid chromatography-ultraviolet spectra-tandem mass spectrometry, 283–286
    logic diagram, 284–286
    MassFrontier search algorithm, 282
    novel lipid mediator theoretical database and search algorithm, 292, 294–297, 301
    prospects, 311–319
    sample preparation, 312
  inflammation mediation, 279
  metabolomics and mediator lipidomics informatics, 276–278
  protectin D1
    animal disease models and actions, 298
    fragmentation mechanism, 304
    mass spectra, 300
    novel lipid mediator identification, 304–308
  transgenic animal engineering, 278–279

R

Relative response factor, sterol quantification, 156–157
Resolvins
  biosynthesis, 280–281

Resolvins (cont.)
  identification
    cognoscitive-contrast-angle algorithm
      contrast angle calculation, 290–291
      liquid chromatography-ultraviolet
        spectra-tandem mass spectrometry
        data integration, 291–292
      mass spectral ion identification,
        286–288
      tandem mass spectrometry ion intensity
        modification according to
        identities, 289–290
    databases, 286
    fragmentation mechanisms, 302, 304,
      308–310
    instrumentation, 313
    liquid chromatography-ultraviolet spectra-
      tandem mass spectrometry, 283–286
    logic diagram, 284–286
    MassFrontier search algorithm, 282
    novel lipid mediator theoretical database and
      search algorithm, 292, 294–297, 301
    prospects, 311–313
    sample preparation, 312
  inflammation mediation, 279
  metabolomics and mediator lipidomics
    informatics, 276–278
  resolvin D1
    animal disease models and actions, 298
    fragmentation mechanism, 302
    mass spectra, 299
    novel lipid mediator identification, 301,
      304–305
  resolvin E1 animal disease models and actions,
    298
  transgenic animal engineering, 278–279
RRF, see Relative response factor

S

Secondary ion mass spectrometry, sphingolipids,
  111
Signaling pathway lipidomics
  bioinformatics resources, 268–269
  lipid extraction
    acidified phosphoinositide/lysolipid
      extraction, 236–237
    general lipid extraction, 235–236
  liquid chromatography-electrospray ionization
    tandem mass spectrometry
    diagnostic ions, 239
    instrumentation, 237
    phospholipid synthesis analysis with stable
      isotopes
      data analysis, 245
      labeling in cells and in vivo, 242
      lipid extraction, 243
      multiple reaction monitoring, 244
    native and newly synthesized
      phospholipids, 244
    overview, 241–242
    running conditions, 243–244
  reverse phase high-performance liquid
    chromatography
    ceramide, diradylglycerol, and
      monoradylglycerol, 239–240
    phosphoinositides, 240
    phospholipids, 237–239
  overview, 235
  pathway drawing tools, 269–270
SOFA Database, features, 226
Sphingolipids
  biosynthesis, 93–94
  classification, 87
  mass spectrometry
    chip-based nanospray techniques, 110
    electron ionization mass spectrometry, 89
    electrospray ionization mass spectrometry,
      90
    fast atom bombardment mass spectrometry,
      90
    Fourier transform mass spectrometry, 110
    imaging mass spectrometry, 111
    ion mobility mass spectrometry, 110–111
    liquid chromatography-tandem mass
      spectrometry and inside-out
      sphingolipidomics
      ceramide 1-phosphate analysis, 106
      (dihydro)ceramide, (dihydro)
        sphingomyelin, and (dihydro)
        monohexosyl-ceramide analysis in
        positive ion mode, 104–109
      electrospray ionization, 90, 109
      extraction, 90, 98–99
      galactosylceramide analysis, 105–106
      ganglioside analysis, 107–109
      glucosylceramide analysis, 105–106
      materials, 100
      matrix-assisted laser desorption, 91–92
      multiple reaction monitoring for
        quantitative analysis, 101–103
      overview, 92–93, 95
      sample collection and preparation,
        98–99
      sphingoid base analysis in positive ion
        mode, 102–103
      subspecies characterization prior to
        quantitative analysis, 100–101
      triple mass spectrometry of complex
        glycosphingolipids, 108–109
      work flow, 97
    liquid secondary ionization mass
      spectrometry, 90
    nanoelectrospray ionization, 110
    secondary ion mass spectrometry, 111
  metabolic modeling

biochemical systems theory, 328, 329, 331, 341
determination of variables, 322–323
dihydroceramide dynamics, 323–327
generalized mass action model, 321, 328, 341
numerical equations and computation
ceramide synthase, 332–334
dihydroceramidase, 334–336
hydroxylase, 339–340
inositol phosphorylceramide synthase, 336–338
inositol phosphosphingolipid phospholipase C, 338–339
overview, 329–330
performance analysis of model, 341–345
phytoceramide dynamics, 323–327
PLAS software, 341–345, 347
prospects, 345–346
simulations, 344
symbolic equations and fluxes, 327–330
system parts list
creation, 322
graphical representation, 323
nomenclature, 86–89
structures, 86–88
Sterols
Spingolipids metabolic modeling overview, 329–330
Spingolipids metabolic modeling prospects, 346–347
evaporative light scattering detection, 147
functions, 146, 148
gas chromatography-mass spectrometry, 147
liquid chromatography-electrospray ionization mass spectrometry
acetonitrile effects on signal intensity, 165
extraction
auto-oxidation concerns, 165–166
bulk lipids, 149–151
clean-up, 166
residual insoluble material, 167–168
solid-phase extraction, 152–153
instrumentation, 168

internal standards, 150, 158–159, 163–164
materials, 148–149
overview, 147–148
peak analysis, 157, 159, 161–162, 168
quantitative analysis, 156–157, 162–163
reverse phase high-performance liquid chromatography, 153, 164
running conditions, 153–154
saponification of extracts, 151
selected reaction monitoring, 159, 161–162

T

TAC, *see* Triacylglycerol
Tissue imaging mass spectrometry, *see* Mass imaging
Triacylglycerol
mass spectrometry analysis
cell culture, 4
cell stimulation and harvesting, 4–5
DNA quantification, 5
instrumentation and data acquisition, 6
Kdo_2-lipid A preparation, 4
lipid extraction
solid phase extraction of lipid classes, 5–6
total lipids, 5
materials, 3
overview of approaches, 2
qualitative analysis with tandem and triple mass spectrometry, 7–14
quantitative analysis, 11, 15, 17–18
spectrophotometric enzyme assay, 2
Triple mass spectrometry
complex glycosphingolipids, 108–109
diacylglycerol, 7–14
triacylglycerol, 7–14

U

Ubiquinones, *see* Prenols

V

VANTED, lipid pathway drawing, 269

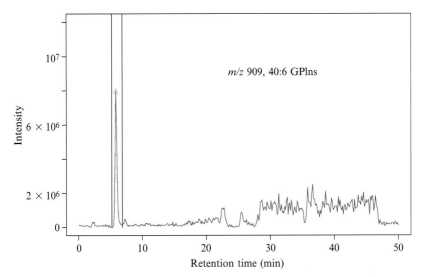

Pavlina T. Ivanova et al., Figure 2.11 Generation of extracted ion chromatograms (XIC; m/z 909.06–909.96) from Fortran code as described in the text. The horizontal axis is retention time, and the vertical axis ion intensity. The vertical black lines indicate the region of integration dynamically selected by the code. The red triangle indicates the peak(s) automatically identified by the algorithms described in the text.

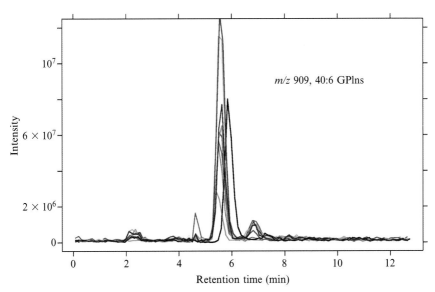

Pavlina T. Ivanova et al., Figure 2.12 A zoomed-in section including the extracted ion chromatogram (XIC) from Fig. 2.11 (m/z 909.06 to 909.96), which also includes extracted ion chromatograms generated from nine other full-scan spectra files, indicative of the variation in amplitude and peak location across these files. Peak areas from the 10 files are automatically aligned as described and output in ASCII format for further analysis (normalization and quantification). Many mass-to-charge (m/z) values in glycerophospholipid (GPL) full-scan spectra contain extracted ion chromatograms with multiple peaks at different retention times.

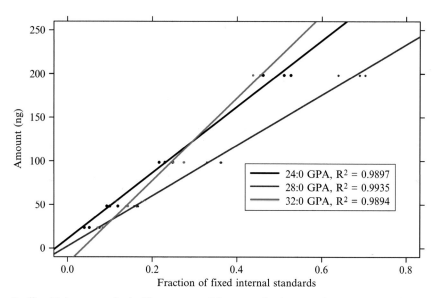

Pavlina T. Ivanova et al., Figure 2.13 Three standard curves for 24:0, 28:0, and 32:0 glycerophosphatidic acid (GPA) (black, green, and red, respectively) generated as described in the text using the automated code. The horizontal axis refers to peak areas normalized to the mean of four fixed odd-carbon internal standards, and the vertical axis is the amount added in nanograms (ng).

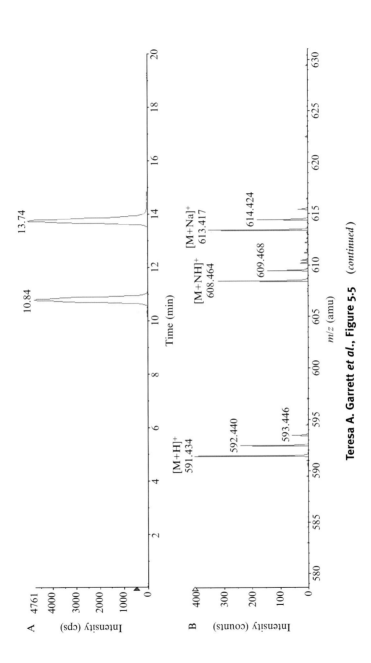

Teresa A. Garrett *et al.*, Figure 5.5 (*continued*)

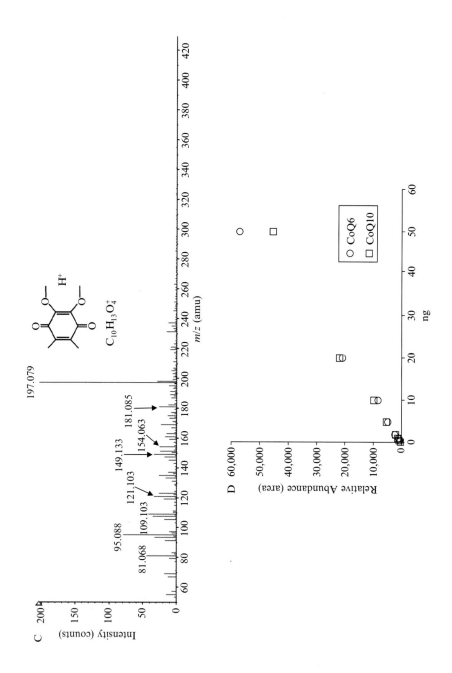

Teresa A. Garrett et al., Figure 5.5 Positive-ion liquid chromatography-mass spectrometry (LC-MS) analysis of CoQ6 and CoQ10 standards. (A) The extracted ion current (EIC) for the $[M+NH_4]^+$ ions of CoQ6 (608.464 m/z) and CoQ10 (880.734 m/z) standards. Under these liquid chromatography conditions, the CoQ6 elutes at about 10 minutes while the CoQ10 elutes at about 14 minutes. (B) The mass spectrum of the material eluting from minutes 10.6 to 11.1, representing the CoQ6 standard. The CoQ6 form $[M+H]^+$, $[M+NH_4]^+$, and $[M+Na]^+$ adduct ions. (C) The collision-induced mass spectrometry (CID-MS) of the $[M+H]^+$ ion of the CoQ6 standard. The major fragment ion (m/z 197.079) corresponds to the quinone ring and is formed by the elimination of the isoprenoid chain (Teshima and Kondo, 2005). (D) The plot of the peak area of the extracted ion current of the $[M+H]^+$ ion of CoQ6 and CoQ10 versus the nanogram (ng) of standard injected onto the Zorbax C-8 reverse phase column. The peak areas of the CoQ6 and the CoQ10 were similar when equivalent amounts of standard were analyzed.

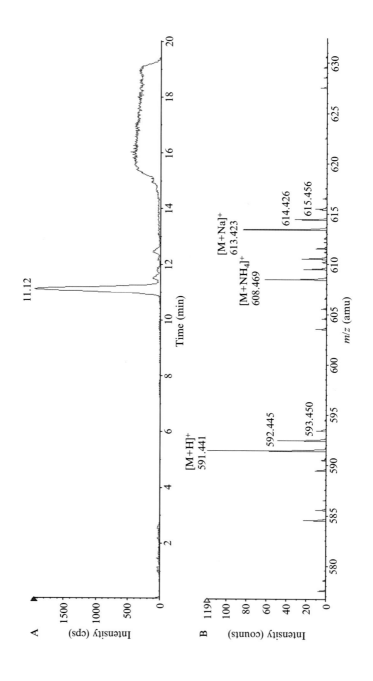

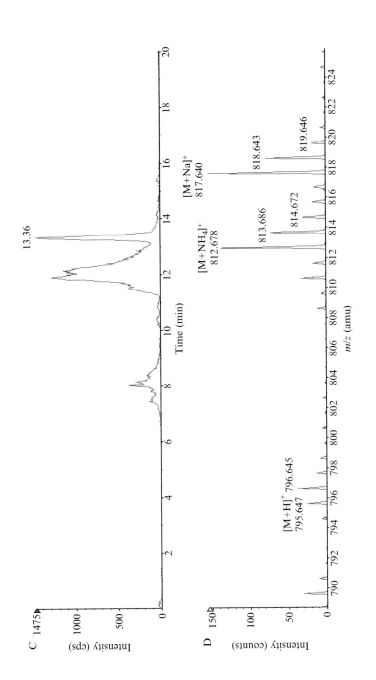

Teresa A. Garrett et al., Figure 5.6 Positive ion liquid chromatography-mass spectrometry (LC-MS) analysis of CoQs in RAW lipid extracts. (A) The extracted ion current (EIC) for the [M+NH$_4$]$^+$ ion of CoQ6 standard after 1 μg of standard is co-extracted in the presence of RAW cells from one 150-mm plate. The retention time (11.12 min) is consistent with the retention time seen when the standard is injected alone (Fig. 5.5A). (B) The mass spectrum of the material eluting between 10.9 and 11.3 min. The [M+H]$^+$, [M+NH$_4$]$^+$, and [M+Na]$^+$ ions of the CoQ6 standard are easily resolved. (C) The extracted ion current of the [M+NH$_4$]$^+$ ion of the major CoQ of RAW cells, CoQ9. The CoQ9 elutes at approximately 13.3 min under these liquid chromatography conditions. The peaks at approximately 8 and 12 min are due to isobaric lipids found in the lipid extract and are not due to CoQ9. (D) The mass spectrum of the material eluting between 13.3 and 14.3 min. The [M+H]$^+$, [M+NH$_4$]$^+$, and [M+Na]$^+$ ions of the CoQ9 standard are detected.

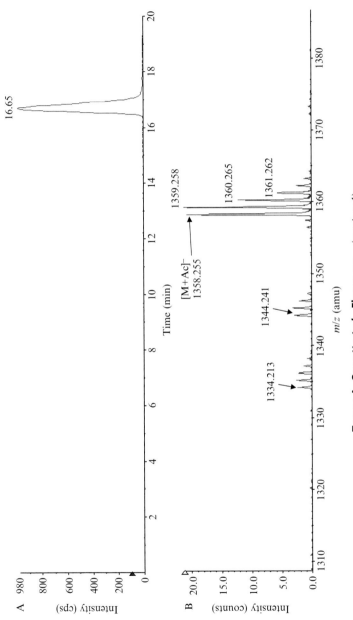

Teresa A. Garrett et al., Figure 5.9 (continued)

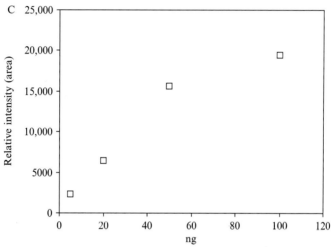

Teresa A. Garrett et al., Figure 5.9 Negative-ion liquid chromatography-mass spectrometry (LC-MS) analysis of nor-dolichol standard. (A) The extracted ion current (EIC) for the [M+Ac]⁻ ions of the nor-dolichol-19 (1358.255 m/z) standard. Under these liquid chromatography conditions, the nor-dolichol-19 elutes at approximately 16.7 min. The shorter nor-dolichols, nor-dolichol-17 and 18, elute slightly earlier, while the nor-dolichol-20 elutes slightly later. (B) The mass spectrum of material eluting between 16.3 to 17.3 min, representing mainly nor-dolichol-19. The small peaks at m/z 1334.213 and 1344.241 are likely due to loss of additional carbons from the dolichol during the preparation of the nor-dolichol standard. (C) The plot of the peak area of the extracted ion current of the [M+Ac]⁻ ion of nor-dolichol-19 versus the nanogram (ng) of standard injected onto the Zorbax C-8 reverse-phase column.

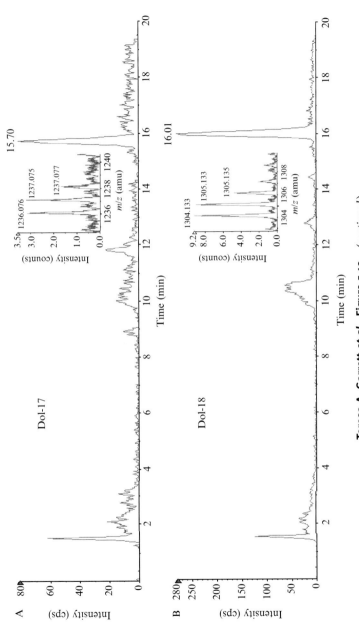

Teresa A. Garrett *et al.*, Figure 5.10 (*continued*)

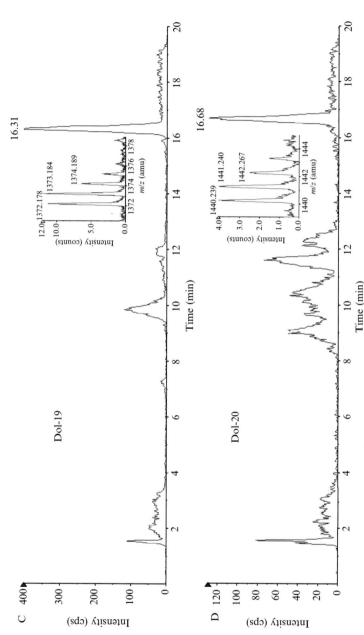

Teresa A. Garrett et al., Figure 5.10 Negative ion liquid chromatography–mass spectrometry (LC–MS) analysis of dolichols in RAW lipid extracts. (A–D) The extracted ion current (EIC) of the acetate adducts of dolichol-17, -18, -19, and -20, respectively. For each, the peak labeled with a retention time corresponds to the relevant extracted ion current peak for a given dolichol. The inset on each panel shows the mass spectra of each of the dolichols detected in the RAW lipid extract.

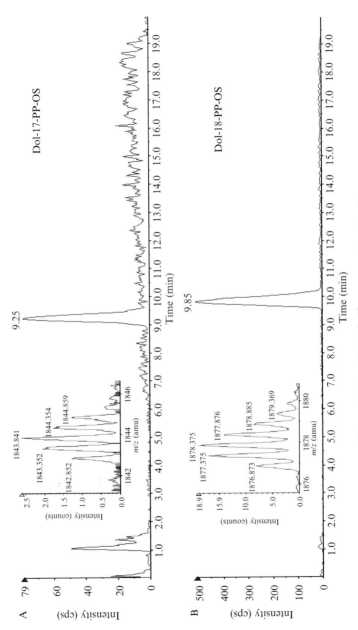

Teresa A. Garrett *et al.*, Figure 5.11 (*continued*)

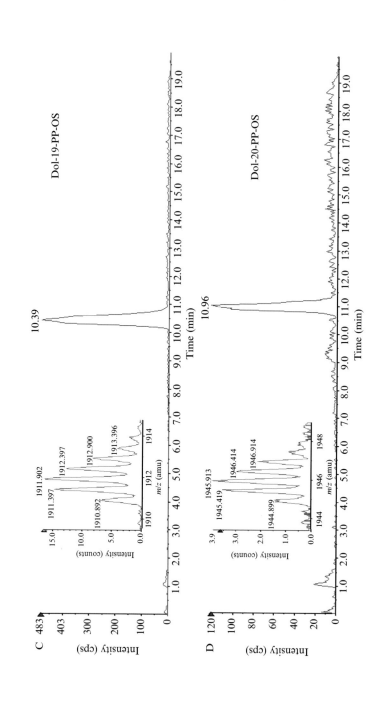

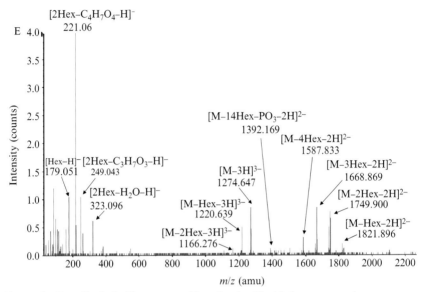

Teresa A. Garrett et al., Figure 5.11 Negative-ion liquid chromatography-mass spectrometry (LC-MS) analysis of purified Dol-PP-(GlcNAc)$_2$(Man)$_9$(Glc)$_3$. (A–D) The extracted ion current (EIC) of Dol-PP-(GlcNAc)$_2$(Man)$_9$(Glc)$_3$ with 17, 18, 19, and 20 isoprenoid units on the dolichol, respectively. The inset shows the mass spectrum of the [M-2H]$^{2-}$ ion for the Dol-PP-(GlcNAc)$_2$(Man)$_9$(Glc)$_3$ detected in the purified sample. (E) The collision-induced mass spectrometry (CID-MS) of the most abundant isotope of the [M-3H]$^{3-}$ ion of the Dol-19-PP-(GlcNAc)$_2$(Man)$_9$(Glc)$_3$ (m/z 1274.647). Sequential loss of 162 u, which corresponds to the loss of one, two, three, or four hexoses, is observed. A fragment ion corresponding to Dol-19-P is also detected as a singly charged ion at m/z 1392.169. Fragment ions corresponding to a single hexose (m/z 179.051) and two hexoses (m/z 323.096) are detected in the low mass region. The remaining low mass fragment ions are most likely due to cross-ring cleavage as described by Sheeley and Reinhold (1998).

Kara Schmelzer et al., Figure 7.1 The LIPID MAPS chemistry-based approach defines lipids as molecules that may originate entirely or in part by carbanion-based condensations of (A) thioesters and/or by (B) carbocation-based condensations of isoprene units.

Several MS analytical methods in metabolomics

Untargeted methods (global methods) without preliminary expectation (LC-MS/MS)

For high-content molecules (with low sensitivity)

Possible to find unexpected molecules; need structural identify

Focused method (neutral loss scan, precursor ion scan)

For medium-content molecules (with medium sensitivity)

Possible to identify small amount of molecules (within focused molecules)

Targeted method (targeted to specified individual molecules. theoretically expanded MRM)

For low-content molecules (with high sensitivity)

Possible to find new structural isomers (not able to find other than targeted molecules)

Ryo Taguchi et al., Figure 8.1 Several mass spectrometric analytical methods in metabolomics. High-content metabolites exist in relatively few molecular species, while low content metabolites exist in a large number of species. These three methods (i.e., untargeted, focused, and targeted) can be available in a more or less comprehensive manner.

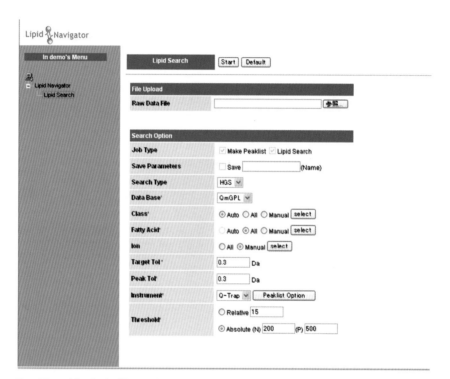

Ryo Taguchi et al., Figure 8.2 Lipid Search with Lipid Navigator, a search engine for lipid identification (http://lipidsearch.jp). The user selects the proper description in each box of the indicated search condition.

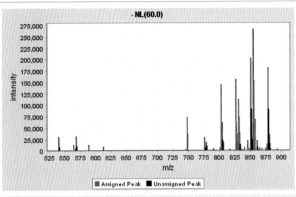

Ryo Taguchi et al., Figure 8.3 Results of Lipid Search with mass spectrometry law data of headgroup survey. Molecular species with summed number of carbon chain and double bond both on sn-1 and sn-2 are indicated as search results for headgroup survey law data.

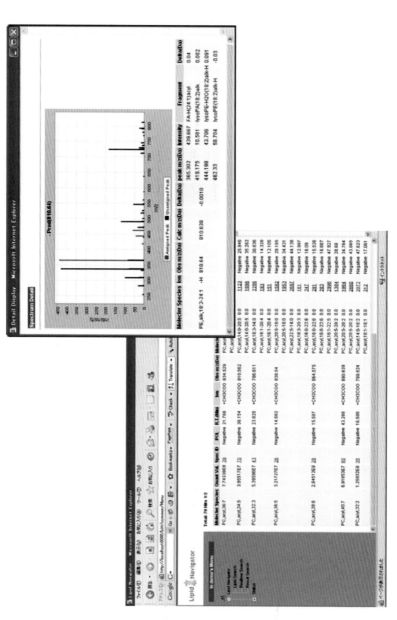

Ryo Taguchi et al., Figure 8.4 Results of Lipid Search for liquid chromatography mass spectrometry law data of molecular weight-related ion survey. Search results of molecular weight-related ion survey law data from liquid chromatography mass spectrometry are also indicated in the same manner as in Figure 8.3. The user can also obtain results of molecular species with individual pairs of fatty acyl chains on sn-1 and

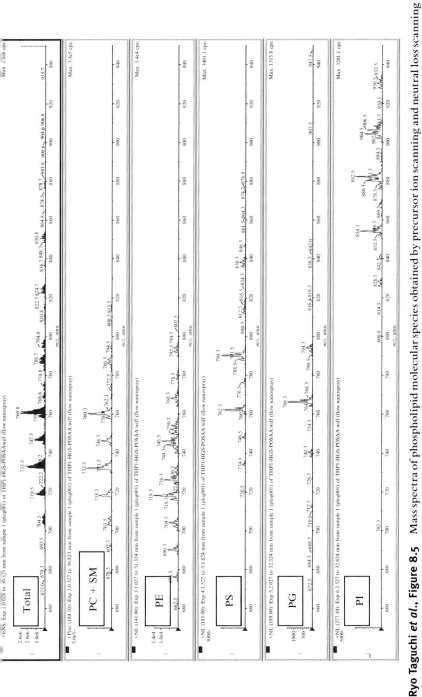

Ryo Taguchi et al., Figure 8.5 Mass spectra of phospholipid molecular species obtained by precursor ion scanning and neutral loss scanning of each headgroup of phospholipids in the positive ion mode (Taguchi et al., 2005). These are application results of a headgroup-specific survey (HGS) for phospholipids extracted from THP-1 cells. Choline-containing lipids such as PC and SM, and PE, PS, PG, and PI were separately detected without liquid chromatographic separation by specific precursor or neutral loss caused by class-specific structure.

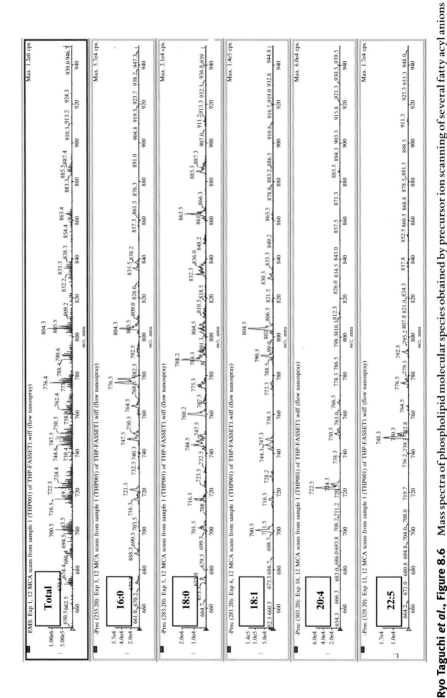

Ryo Taguchi et al., Figure 8.6 Mass spectra of phospholipid molecular species obtained by precursor ion scanning of several fatty acyl anions in the negative ion mode (Taguchi et al., 2005). These results were obtained by fatty acid–specific survey (FAS) for same samples. These data were automatically obtained by applying mass spectrometry law data to our search engine, as described in the text.

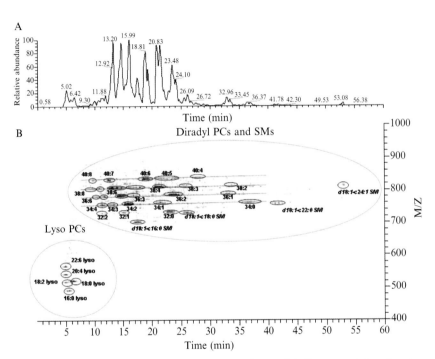

Ryo Taguchi et al., Figure 8.7 Two-dimensional map of choline-containing phospholipids detected by reverse-phase liquid chromatography electrospray ionization mass spectrometry with headgroup survey in the positive ion mode. (A) Total ion chromatogram of precursor for m/z 184. (B) Two-dimensional map. By reverse-phase liquid chromatography/electrospray ionization tandem mass spectrometry analysis, approximately 80 peaks were detected.

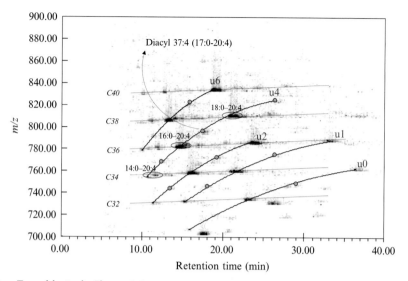

Ryo Taguchi et al., Figure 8.8 Elution characteristics depending on the number of fatty acyl chains and unsaturated bonds. u0, diacyl 14:0–16:0/16:0–16:0/16:0–18:0; u1, diacyl 16:0–16:1/16:0–18:1/18:0–18:1; u2, diacyl 14:0–18:2/ 16:0–18:2/18:0–18:2; u4, diacyl 14:0–20:4/ 16:0–20:4/18:0–20:4; u6, diacyl 14:0–22:6/16:0–22:6/18:0–22:6. The molecules, which are marked with a closed circle, were supposed to be diacyl molecular species with odd number acyl chains, not even number alkyl- or alkenyl-acyl species, from their retention time.

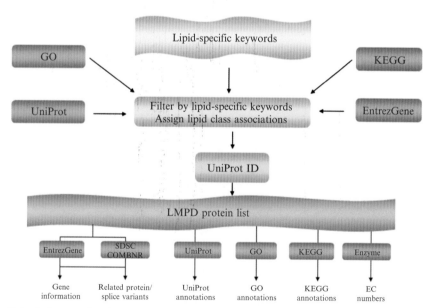

Eoin Fahy et al., Figure 11.2 Overview of the LIPID MAPS Proteome Database (LMPD).

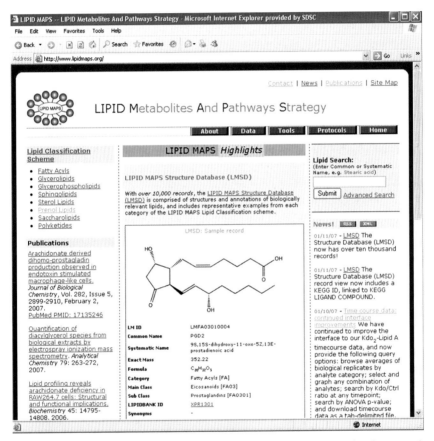

Eoin Fahy et al., Figure 11.3 LIPID MAPS public website (http://www.lipidmaps.org).

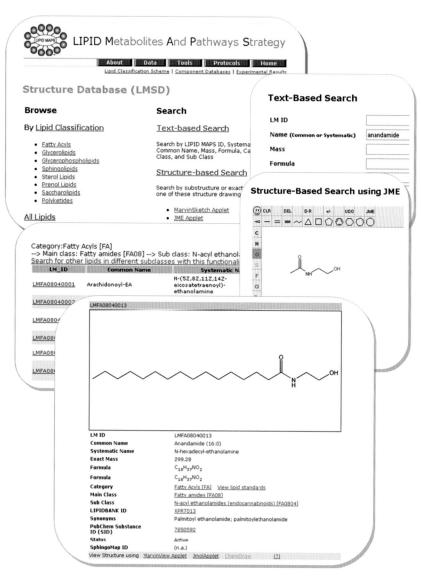

Eoin Fahy *et al.*, **Figure 11.4** Online interfaces for searching the LIPID MAPS Structure Database (LMSD).

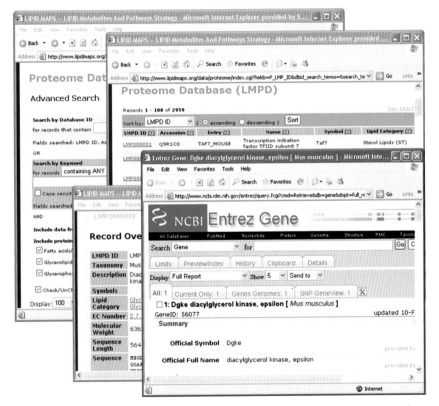

Eoin Fahy et al., Figure 11.5   Online interface to LIPID MAPS Proteome Database (LMPD) at the LIPID MAPS website.

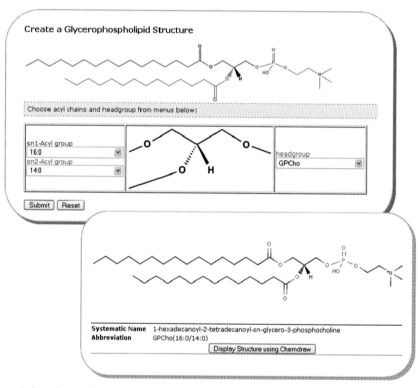

Eoin Fahy et al., Figure 11.6 Online structure-drawing tool for sphingolipids.

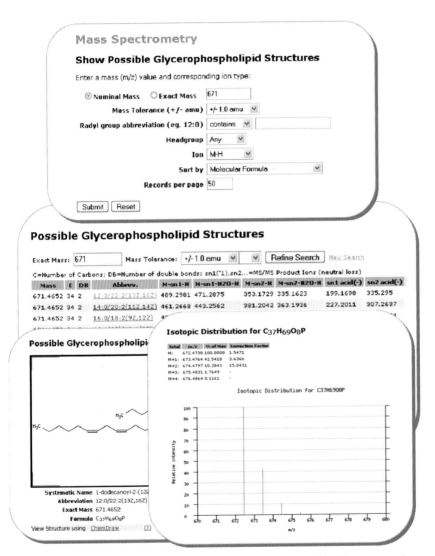

Eoin Fahy et al., Figure 11.8 Online utility to predict possible lipid structures from mass spectrometry (MS) data.

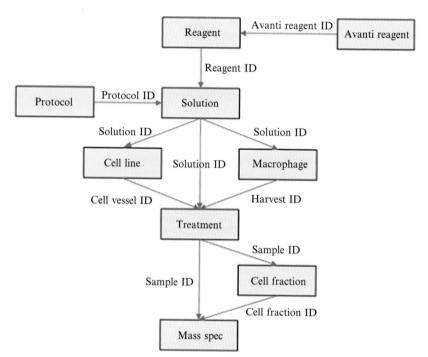

Eoin Fahy *et al.*, **Figure 11.9** Simplified sequential flow of Laboratory Information Management System (LIMS) usage. Rectangles represent modules of functionality as described in the text. Arrow labels indicate sets of one or more identifiers that are generated by the module at the base of the arrow and entered into the immediately following target module.

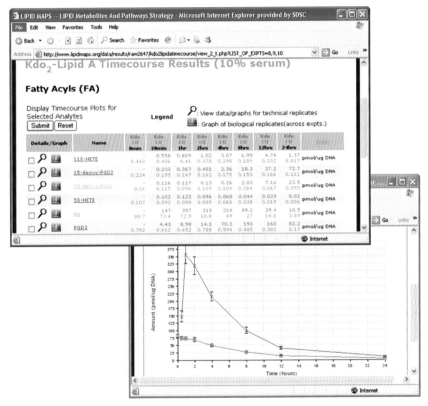

Eoin Fahy *et al.*, **Figure 11.10** Online interface for viewing lipidomics mass spectrometry (MS) data.

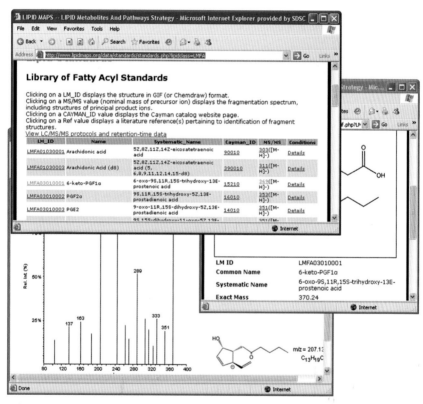

Eoin Fahy *et al.*, **Figure 11.11** Online interface to Lipid Standards Library and associated data.

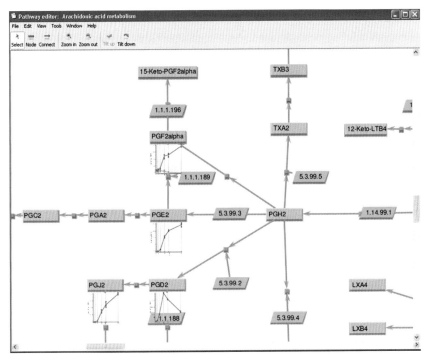

Eoin Fahy et al., Figure 11.12 Screenshot of the Biopathways editor showing a portion of the arachidonate pathway with overlaid timecourse data.

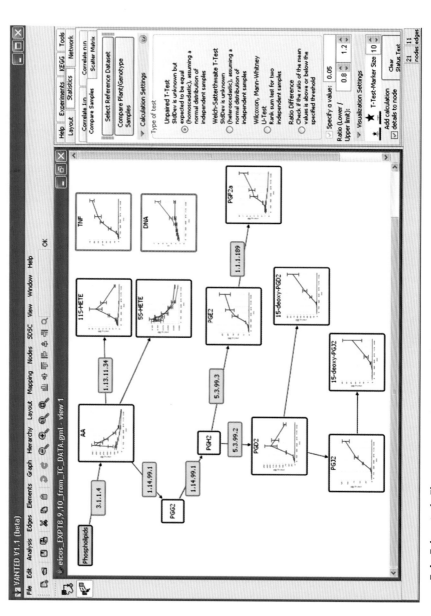

Eoin Fahy et al., Figure 11.13 VANTED application showing arachidonate pathway and time-course data.

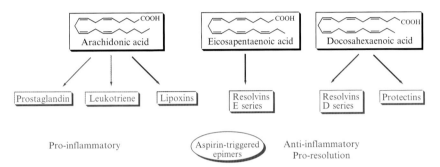

Charles N. Serhan et al., Figure 12.1 Lipid-derived mediators in programmed resolution of acute inflammation. *Precursors of Lipid Mediators*: Arachidonic acid is the precursor to eicosanoids that have distinct roles as proinflammatory mediators. The prostaglandins and leukotrienes each play specific actions pivotal to the progression of inflammation. Arachidonic acid–derived epoxyeicosatetraenoic acids (EETs) produced via P450 (Funk, 2001; Samuelsson et al., 1987) and ω-3 PUFA P450 epoxides may also play roles (Arita et al., 2005a; Capdevila et al., 1990). Cell–cell interactions, exemplified by platelets-leukocytes within blood vessels and/or PMN–mucosal interactions, enhance generation of lipoxins that serve as endogenous anti-inflammatory mediators self-limiting the course of inflammation (Serhan and Savill, 2005). The essential omega-3 fatty acids eicosapentaenoic acid and docosahexaenoic acid (C20:5 and C22:6) are converted to two novel families of lipid mediators, resolvins and protectins, that play pivotal roles in promoting resolution. Resolvins of the E series are generated from eicosapentaenoic acid (e.g., RvE1), and resolvins of the D series (e.g., resolvin D1) are generated from DHA as well as the protectins, such as neuroprotectin D1. *Aspirin-triggering epimers of lipid mediators*: Aspirin impacts the formation of lipoxins and resolvin, by acetylating COX-2 (e.g., in human vascular endothelial cells that stereoselectively can generate, in the case of RvE1 biosynthesis, 18R-HPEPE, which is picked up via transcellular cell–cell interactions by leukocytes and converted in a lipoxygenase-like mechanism to RvE1). The complete stereochemistry of RvE1 and at least one of its receptors were established (Arita et al., 2005c; see Fig. 12.2). The biosynthesis of RvE1 can also be initiated by P450-like enzymes in microbes (Serhan et al., 2000). Aspirin also influences the biosynthesis of D-series resolvins. Aspirin catalytically switches COX-2 to a 17R-lipoxygenase-like mechanism that generates 17R-containing series of resolvin D and protectins (e.g., neuroprotectin D1/protectin D1; see text).

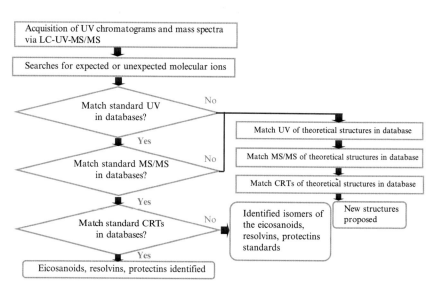

Charles N. Serhan et al., Figure 12.4 Logic diagram for developing LC-UV-MS/MS-based mediator lipidomic databases and search algorithms for polyunsaturated fatty acids (PUFA)-derived lipid mediators (LMs): eicosanoids, resolvins, and protectins (Lu et al., 2005). CRT is the chromatographic retention time.

LM name	Molecularion	Peripheral-cut segments				Chain-cut segments						
LXA$_4$ RT: 5.2min 300nm	[M-H=351.3]	5Cc=86	5Cm=265	5Mc=116	5Mm=235	6Cc=116	6Cm=235	6Mc=146	6Mm=205	15Cc=250	15Cm=101	15Mc=280	15Mm=71
-H$_2$O	333		247	98	217	98	217	128	187	232	83	262	
-2H$_2$O	315		229		199		199	110		214		244	
-3H$_2$O	297		211									226	
-CO$_2$	307	42		72		72		102		206		236	
-H$_2$O-CO$_2$	289			54		54		84		188		218	
-2H$_2$O-CO$_2$	271							66		170		200	
-3H$_2$O-CO$_2$	253											182	

Charles N. Serhan et al., Figure 12.5 Liquid chromatographic ultraviolet tandem mass spectrometry (LC-UV-MS/MS) database layout: naming lipidmediated segments for PD1, RvD1, and LXA$_4$. These examples depict RvD1, LXA$_4$, and PD1, formed via chain-cut, peripheral-cut, and chain- plus peripheral-cut for interpretation of MS/MS fragmentation (see text for further details).

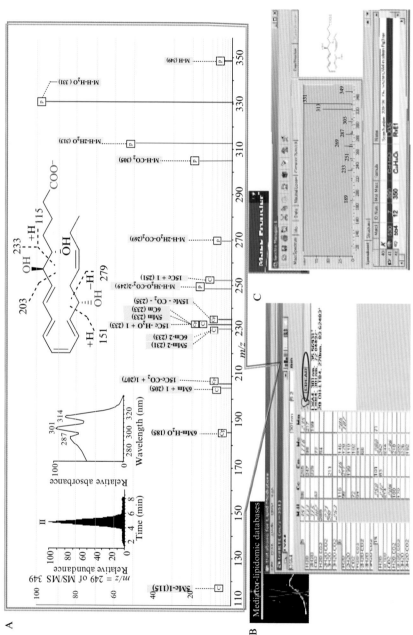

Charles N. Serhan et al., Figure 12.6 Cognoscitive-contrast-angle algorithm (COCAD)-based identification example: LXA_5. Panels A and B show LXA_5 identified as the best match for the liquid chromatography ultraviolet tandem mass spectrometry (LC-UV-MS/MS) results in the cognoscitive-contrast-angle algorithm system (with highest match-score). LXA_5 was also identified in Panel C by searching the database constructed by MassFrontier[TM].